VOCATIONAL MATHEMATICS

James L. Southam
Professor
Business Information and Computing Systems
San Francisco State University

Cynthia L. Nahrgang
The Blake School
Hopkins, Minnesota

FOURTH EDITION

MB18DB/MB18DBU
PUBLISHED BY
SOUTH-WESTERN PUBLISHING CO.
CINCINNATI, OH   WEST CHICAGO, IL   DALLAS, TX   LIVERMORE, CA

Copyright © 1990

by SOUTH-WESTERN PUBLISHING CO.
Cincinnati, Ohio

ALL RIGHTS RESERVED

The text of this publication, or any part thereof,
may not be reproduced or transmitted in any form
or by any means, electronic or mechanical, including
photocopying, recording, storage in an information
retrieval system, or otherwise, without the prior
written permission of the publisher.

ISBN: 0-538-60218-X (MB18DB)
       0-538-60219-8 (MB18DBU)

Library of Congress Catalog Card Number: 88-64141

1 2 3 4 5 6 7 8 K 6 5 4 3 2 1 0 9

Printed in the United States of America

PREFACE

VOCATIONAL MATHEMATICS, Fourth Edition, is a text-workbook in fundamental business mathematics. The text assumes only background knowledge in addition, subtraction, multiplication, and division of whole numbers. The book has two basic components: the rules of and practice with the fundamental mathematical principles, and the use of these principles for common applications. The mathematical fundamentals are explained in Unit 1 (Chapters 1 through 5); the applications are discussed in Unit 2 (Chapters 6 through 12).

Some of the topics in the text have been rearranged to provide a more natural development. The discussion of charts and graphs has been moved to Unit 1 because of its importance to a fundamental background in mathematics. The chapter on time-payment purchases and short-term loans has been relocated; it now follows Chapter 7, on calculating interest. Section 12.2, on depreciation, has been completely changed to introduce the Modified Accelerated Cost Recovery System (MACRS), which is the depreciation method that is required on tax returns by the Tax Reform Act of 1986. A discussion of calculators and examples using calculators have been integrated into both Units 1 and 2. Finally, more work space has been provided in the assignments for the student's convenience.

TO THE STUDENT

This book has been written to help you learn and/or practice the fundamental mathematical concepts that are absolutely essential in modern careers. Some suggestions and comments to guide you as you proceed through the text-workbook are given below. We recommend that you periodically turn back to this page and review these suggestions.

Text: Chapters 1 through 5 represent the fundamental mathematics that will be invaluable to you in your career. Each chapter is divided into Sections, and each Section is further subdivided into Parts. After you have completely studied a Part, go back and rework each example. Ask your instructor for help when necessary. Chapters 1 through 5 should be studied in the order that they are presented.

Chapters 6 through 12 will help you apply your newly learned mathematics in typical situations. This experience will be doubly beneficial. You will get practice for your new skills, and you will become familiar with much of the terminology and procedures used by a wide variety of organizations.

Assignments: To assist you in your work, a set of assignments has been developed to parallel the organization of the text-workbook itself. There is one assignment for each Section. The assignment is subdivided into the same Parts as the Section. You should ask your instructor whether you should do the assignment one Part at a time or whether you should first study all the Parts and then do the entire assignment for the Section.

Answers to odd-numbered problems are provided on pages 385–398 in this text-workbook. Check your work after each Part. Everyone can learn from his or her mistakes; when you do find that you made a mistake, study the problem carefully to determine why you made the mistake. Go back to the examples and text material if you have any difficulties.

Do all of your work in the space provided. Many problems will require more than one step in the solution, so show all of your work. You and your instructor will be unable to retrace your work if you skip any steps.

Calculators: Assignments requiring calculators have been added to the Fourth Edition. Other assignments may or may not require the use of a calculator. Your instructor will have a policy regarding the use of calculators on assignments and tests. If you are using this book on your own, the following are two suggestions that you might consider:

1. Do all problems without a calculator. Then, *before* looking at the answers in the back of the book, check your work with the calculator.
2. Do the odd-numbered problems without a calculator. Then, use the calculator to check your work. Next, do the even-numbered problems with a calculator, and then do them without a calculator to check your work.

Calculators can differ widely from manufacturer to manufacturer. One manufacturer may even make various calculators that have different functions and operating techniques. Therefore, you should read and understand the operating manual provided with your calculator. Always keep the manual handy for quick reference.

It is important to realize that a calculator does not replace the problem-solving skills needed to solve a mathematics problem. An understanding of mathematics principles is still necessary to minimize errors and maximize the efficiency of the calculator. Therefore, the authors have included discussions of the calculator to enhance the mastery of the mathematics skills presented in this textbook.

Tests: Several tests have been prepared to further assist you in your work. A test is meant to be a learning experience. The value of a test is that it helps you identify the topics that you have mastered as well as those you have not.

One Achievement Test is provided for each of the 12 chapters in the text-workbook. Additionally, because of the importance of the material, one Check Test has been written for every Section of Chapters 1 through 5. If you score below 80 on a Check Text, review the text, examples, and problems to increase your understanding. Your instructor may have you take a retest on the Section. After taking all of the Check Tests for a chapter, you should be ready to take the Achievement Test for that chapter.

Progress Record: The inside cover of the text-workbook contains your own personal Progress Record. You should use it to maintain a record of your scores on the various assignments and tests.

Vocational Careers: For your interest, we have categorized some of the current careers for which VOCATIONAL MATHEMATICS might serve as a valuable preparation. For some areas of career interests, further mathematics may be required. The word *business* is a broad one; almost all of the careers listed below, which might be termed *business*, may be found in government and non-profit organizations as well as in profit-making corporations.

Business/Office/Marketing

Accounting
Banking
Data/Information Processing
Merchandising
Office Management
Production/Inventory Management
Sales and Marketing

Technical

Commercial Art
Color Separation
Dental Laboratory Technician
Electro-mechanical Technology
Electronic Publishing
Electrical/Electronics/Laser Technology
Environment Chemical Technology
Metallurgy Technology
Plastics Technology
Quality Control

Trade/Industrial

Architectural Drafting
Carpentry
Construction
Heating, Cooling, and Refrigeration
Industrial Drafting
Landscaping
Production Machinist
Surveying

James L. Southam
Cynthia L. Nahrgang

CONTENTS

UNIT 1 FUNDAMENTALS

CHAPTER 1 WORKING WITH DECIMALS

Section 1.1	**The Decimal Number System**	**2**
Part A	Reading and Writing Whole Numbers	2
Part B	Writing Numbers from Words	3
Part C	Reading and Writing Decimals	4
Part D	Writing Decimals from Words	5
Part E	Comparing Decimals	6
Part F	The Decimal: The Number Used by the Calculator	7
Section 1.2	**Addition of Decimals**	**13**
Part A	Adding Decimals Vertically	13
Part B	Adding Long Columns of Decimals	14
Part C	Adding Decimals Horizontally	14
Part D	Adding Decimals with the Calculator	14
Section 1.3	**Subtraction of Decimals**	**21**
Part A	Subtracting One Decimal from Another	21
Part B	Subtracting a Decimal from a Whole Number	21
Part C	Subtracting a Whole Number from a Decimal	22
Part D	Subtracting Decimals with the Calculator	22
Part E	Checking Calculator Tapes and Adding Machine Tapes	23
Part F	Applications	25
Section 1.4	**Multiplication of Decimals**	**33**
Part A	Multiplying a Decimal by a Decimal or a Whole Number	34
Part B	Zeros to the Left of a Product	34
Part C	Rounding Off to a Given Number of Decimal Places	35
Part D	Multiplying by Ten and Its Multiples	35
Part E	Multiplying Decimals with the Calculator	37
Part F	Applications	38
Section 1.5	**Division of Decimals**	**45**
Part A	Review of Division with Whole Numbers	45
Part B	Dividing a Decimal by a Whole Number	46
Part C	Dividing a Whole Number by a Decimal	47
Part D	Dividing a Decimal by a Decimal	47
Part E	Dividing One Whole Number by Another to Several Places	48
Part F	Dividing by Ten and Its Multiples	49
Part G	Dividing Decimals with the Calculator	49
Part H	Applications	50

CHAPTER 2 WORKING WITH FRACTIONS

Section 2.1	**Types of Fractions**	**55**
Part A	Changing a Fraction to Higher Terms	55
Part B	Reducing a Fraction to Lowest Terms	56
Part C	Changing Mixed Numbers into Improper Fractions and Changing Improper Fractions into Mixed Numbers	58
Section 2.2	**Addition of Fractions**	**63**
Part A	Adding a Fraction or a Mixed Number to a Whole Number	63
Part B	Adding Fractions or Mixed Numbers with Common Denominators	63
Part C	Finding the Least Common Denominator	63
Part D	Adding Fractions with Unlike Denominators	65
Part E	Adding Mixed Numbers with Unlike Denominators	65
Part F	Applications	66
Section 2.3	**Subtraction of Fractions**	**71**
Part A	Subtracting Fractions Without Borrowing	71
Part B	Subtracting Fractions by Borrowing	72
Part C	Applications	73
Section 2.4	**Multiplication of Fractions**	**79**
Part A	Multiplying Two or More Fractions	79
Part B	Multiplying a Fraction by a Whole Number	80
Part C	Multiplying a Mixed Number by a Fraction, a Whole Number, or a Mixed Number	80
Part D	Multiplying Mixed Numbers—An Alternate Method	81
Part E	Applications	81
Section 2.5	**Division of Fractions**	**89**
Part A	Dividing by a Fraction	89
Part B	Dividing by a Whole Number	90
Part C	Dividing by a Mixed Number	90
Part D	Simplifying Complex Fractions	91
Part E	Applications	92
Section 2.6	**Equivalent Forms**	**101**
Part A	Changing a Fraction to a Decimal	101
Part B	Changing a Decimal to a Fraction in Its Lowest Terms	102
Part C	Adding and Subtracting Fractions and Decimals	103
Part D	Multiplying Fractions and Decimals	103
Part E	Dividing Fractions and Decimals	105
Part F	Applications	108

CHAPTER 3 WORKING WITH PERCENTS

- **Section 3.1** **Percent Conversions**..................115
  - Part A Converting a Percent to a Decimal or a Whole Number..................115
  - Part B Converting a Percent to a Fraction or a Mixed Number..................116
  - Part C Converting a Decimal to Its Percent Equivalent..................119
  - Part D Converting a Fraction or a Mixed Number to Its Percent Equivalent.....119
- **Section 3.2** **Finding the Percentage**..............125
  - Part A Multiplying by Percents Between 1% and 100%..................125
  - Part B Using Fractional Equivalents of Percents..................126
  - Part C Multiplying by Percents Less than 1%..................127
  - Part D Multiplying by Percents Greater than 100%..................128
  - Part E Applications..................128
- **Section 3.3** **Finding the Rate**..................137
  - Part A Finding Percents Between 1% and 100%..................137
  - Part B Finding Percents Less Than 1%......139
  - Part C Finding Percents Greater Than 100%..................139
  - Part D Applications..................140
- **Section 4.4** **Finding the Base of a Percentage**...147
  - Part A Finding the Base When the Given Percent Is Between 1% and 100%....147
  - Part B Finding the Base When the Given Percent Is Less Than 1%..................149
  - Part C Finding the Base When the Given Percent Is Greater Than 100%.......149
  - Part D Applications..................150

CHAPTER 4 WORKING WITH WEIGHTS AND MEASURES

- **Section 4.1** **Working with the Customary and the Metric Systems**..................155
  - Part A Changing to the Next Smaller or the Next Larger Unit..................158
  - Part B Converting from One System to the Other..................163
  - Part C Adding Measures..................165
  - Part D Subtracting Measures..................166
  - Part E Multiplying Measures..................166
  - Part F Dividing Measures..................168
- **Section 4.2** **Applications Using Weights and Measures**..................175
  - Part A Perimeter, Area, and Volume..........175
  - Part B Ounces, Pounds, and Tons...........178
  - Part C Units, Dozens, and Gross............178
  - Part D C, Cwt, and M..................179

CHAPTER 5 ESTIMATIONS, GRAPHS, AND SHORTCUTS

- **Section 5.1** **Estimations**..................185
  - Part A Rounding Off to the Nearest Unit....186
  - Part B Rounding Off to the Nearest Ten, Hundred, Thousand, etc...........186
  - Part C Estimating Sums and Differences.....187
  - Part D Estimating Products..................188
  - Part E Estimating Quotients..................189
  - Part F Checking Calculator Answers by Estimating..................190
- **Section 5.2** **Graphs**..................199
  - Part A Line Graphs..................199
  - Part B Bar Graphs..................200
  - Part C Circle Graphs..................204
- **Section 5.3** **Shortcuts**..................213
  - Part A Adding Digits from Left to Right.....213
  - Part B Subtracting Digits from Left to Right..................214
  - Part C Adding Two Fractions Whose Numerators Are 1..................215
  - Part D Subtracting Two Fractions Whose Numerators Are 1..................215
  - Part E Multiplying and Dividing by 0.1, 0.01, 0.001, etc..................215
  - Part F Multiplying and Dividing by 50 and 25..................216
  - Part G Multiplying by 11..................217
  - Part H Some Miscellaneous Shortcuts.......217

UNIT 2 APPLICATIONS

CHAPTER 6 KEEPING A CHECKING ACCOUNT

- **Section 6.1** **Checks and Deposits**..................228
  - Part A Keeping Records of a Checking Account..................228
  - Part B Writing Checks..................229
  - Part C Endorsing Checks..................230
  - Part D Clearing of Checks..................231
  - Part E Making Deposits..................231
  - Part F Transferring Funds Electronically.....231
- **Section 6.2** **Bank Statements and Reconciliation Statements**..................237
  - Part A Understanding the Monthly Bank Statement..................237
  - Part B Preparing the Reconciliation Statement..................239

CHAPTER 7 CALCULATING INTEREST

- **Section 7.1** **The Interest Period**..................247
  - Part A Finding the Exact Time Between Two Dates..................248
  - Part B Finding the Due Date and Exact Time When the Interest Period Is Stated in Days..................249
  - Part C Finding the Due Date and Exact Time When the Interest Period Is Stated in Months..................250
- **Section 7.2** **Calculating Simple Interest**........255
  - Part A Ordinary Interest..................255
  - Part B Exact Interest..................258
  - Part C Monthly Interest Rates..................259
- **Section 7.3** **Calculating Compound Interest**.....267
  - Part A Compounding Interest at Different Periods..................267
  - Part B Using Compound-Interest Tables.....268
  - Part C Calculating Compound Amount Factors..................272

CHAPTER 8 CALCULATING TIME-PAYMENT PLANS AND SHORT-TERM LOANS

- **Section 8.1** **Time-Payment Plans** **279**
  - Part A Finding the Amount of the Finance Charge 280
  - Part B Finding the Amount of the Monthly Installment 281
  - Part C Finding the Annual Percentage Rate (APR) 282
- **Section 8.2** **Short-Term Commercial Loans** **287**
  - Part A Finding the Interest Charge, Due Date, and Maturity Value 287
  - Part B Discounting Non-Interest-Bearing Notes 288
  - Part C Discounting Interest-Bearing Notes ... 290

CHAPTER 9 PURCHASE ORDERS AND INVOICES, CASH DISCOUNTS, AND TRADE DISCOUNTS

- **Section 9.1** **Purchase Orders and Invoices** **297**
  - Part A Preparing and Checking Purchase Orders 297
  - Part B Preparing and Checking Invoices 298
- **Section 9.2** **Cash Discounts** **303**
  - Part A The Discount Periods for Cash Discounts 303
  - Part B Calculating Cash Discounts 305
- **Section 9.3** **Trade Discounts** **311**
  - Part A Finding the Net Price with a Single Trade Discount 311
  - Part B Finding the Net Price with a Chain of Discounts 313
  - Part C Using Net Price Factors for Chain Discounts 314

CHAPTER 10 SELLING GOODS

- **Section 10.1** **Selling Price, Cost, and Markup Rate** **321**
  - Part A Cost as the Base for Markup 322
  - Part B Selling Price as the Base for Markup 324
  - Part C Converting Markup Rates from One Base to Another 326
- **Section 10.2** **Profit and Loss** **333**
  - Part A Net Profit 333
  - Part B Net Loss 335
  - Part C Adjusting the Selling Price 336
  - Part D Using Price Tag Codes 337

CHAPTER 11 CALCULATING GROSS PAY FOR PAYROLLS

- **Section 11.1** **Time-Basis Payment and Payroll Deductions** **343**
  - Part A Straight-Time (or Day-Rate) Pay 343
  - Part B Overtime Pay on a Weekly Basis 344
  - Part C Overtime Pay on a Daily Basis 344
  - Part D Overtime Pay for Salaried Employees 345
  - Part E Payroll Deductions 346
- **Section 11.2** **Commissions and Piecework Methods of Payment** **351**
  - Part A Various Plans of Commission Payment 351
  - Part B Various Plans of Piecework Payment 352

CHAPTER 12 INVENTORY VALUATION, COST OF GOODS SOLD, AND DEPRECIATION

- **Section 12.1** **Inventory Valuation and Cost of Goods Sold** **359**
  - Part A Specific Identification 360
  - Part B First-In, First-Out (FIFO) 360
  - Part C Last-In, First-Out (LIFO) 360
  - Part D Average Cost 361
  - Part E Finding the Cost of Goods Sold 362
  - Part F Estimating Inventory Value at Cost ... 363
- **Section 12.2** **Calculating Depreciation** **373**
  - Part A Straight-Line Method (SL) 374
  - Part B Sum-of-the-Years-Digits Method (SOYD) 374
  - Part C Declining-Balance Method (DB) 375
  - Part D Comparison of Depreciation Methods 377
  - Part E Modified Accelerated Cost Recovery System (MACRS) 378

Answers to Odd-Numbered Problems **385**

Glossary **399**

Index **404**

UNIT 1
FUNDAMENTALS

1   Working with Decimals

2   Working with Fractions

3   Working with Percents

4   Working with Weights and Measures

5   Estimations, Graphs, and Shortcuts

CHAPTER 1

Working with Decimals

All business transactions involve the basic operations of arithmetic: addition, subtraction, multiplication, and division. In the business world, these operations are performed on both decimal numbers and whole numbers. Today, many businesses use computers and calculators for most of their calculations. These machines accept decimal numbers as input and give the answers in decimal numbers. More than ever before, it is important to be certain of yourself when working with decimals.

Section 1.1 of this chapter will review reading, writing, and comparing decimal numbers. Sections 1.2 through 1.5 will show you how to add, subtract, multiply, and divide decimal numbers.

SECTION 1.1 The Decimal Number System

The word **decimal** usually refers to a number which contains a decimal point. A decimal number may have digits only to the right of the decimal point, or it may have digits on both sides of the decimal point. A **digit** is a single figure: 0, 1, 2, 3, 4, 5, 6, 7, 8, or 9.

Actually, the word *decimal* can have a much broader meaning, since it comes from the Latin word for ten, **decem.** The actual value of a digit in a number, whether whole or decimal, depends upon the column of the number in which the digit is located. The columns change values by multiples of ten, as illustrated in Parts A and C of this section.

PART A Reading and Writing Whole Numbers

Digits are put together to form a number, but the value of the number depends on the order of the digits. For example, the number 456 is read and written as *four hundred fifty-six*. It consists of:

$$4 \times 100 = 400$$
$$5 \times 10 = 50$$
$$6 \times 1 = 6$$
$$\overline{456}$$

Now take the same three digits and form another number, 645. This number is read and written as *six hundred forty-five*. Next, write 645 to the right of 456. The new number is 456,645. It is read and written as *four hundred fifty-six thousand six hundred forty-five*. Note that the word *thousand* is used after the *four hundred fifty-six* to show the value of the new number, but 456 and 645 are read the same as if they were separate numbers.

If you can read a number in the hundreds correctly, then you can read any whole number by learning the names of the three-digit groups into which any whole number can be separated. Study the chart below in which the blank spaces represent digits. Each group is read and written as if it were a number by itself and then is followed by the group name. The last group of three has no group name.

$$\underbrace{\_\,\_\,\_\,}_{\text{billion}}\,,\,\underbrace{\_\,\_\,\_\,}_{\text{million}}\,,\,\underbrace{\_\,\_\,\_\,}_{\text{thousand}}\,,\,\underbrace{\_\,\_\,\_}_{\text{(no name)}}$$

Chapter 1 Working with Decimals

The group names to the left of billion are, respectively, trillion, quadrillion, quintillion, etc.

To read and write a large whole number, follow these four steps:

1. Starting at the right, use commas to separate the number into groups of three. The last group marked off at the extreme left may have less than three digits. For example, 53275819 is separated as follows: 53,275,819.
2. Starting at the left, read the first group as if it were alone and then say the group name. Continue in the same manner to the end, reading each group of three and saying the group name.

EXAMPLE:

53,275,819 is read and written as *fifty-three million two hundred seventy-five thousand eight hundred nineteen*.

3. If a three-digit group starts with one or two zeros, read only the number which follows the zeros and give the group name. The last group, of course, does not have a group name.

EXAMPLE:

40,003,006,051 is read and written as *forty billion three million six thousand fifty-one*.

4. If a three-digit group consists of three zeros, omit the group name entirely.

EXAMPLE:

4,000,006,000 is read and written as *four billion six thousand*.

Numbers in the hundreds group, such as 819 in the whole number 53,275,819, are read aloud without the word *and* after the word *hundred*. When writing a number in words, use a hyphen to combine two words to form numbers from twenty-one through ninety-nine.

PART B Writing Numbers from Words

In order to form a number from words, you must know the position of each group from its name and then be able to write correctly the number you hear or read. Follow these four steps:

1. Write the first number you hear or read as if it were by itself. If no group name is given, it means that the number is in the hundreds or less.

EXAMPLE:

Two hundred one is 201. *Twenty-nine* is 29.

2. If group names are given, there must be commas separating the various groups. When the first group name is *thousand*, there must be one group of three digits after the comma.

EXAMPLE:

Seven hundred fifty thousand two hundred six is written

$$\underline{7}\ \underline{5}\ \underline{0},\ \underline{2}\ \underline{0}\ \underline{6}$$

When the first group name is *million*, there must be two groups of three digits each after the first comma.

EXAMPLE:

Three million seven hundred fifty thousand two hundred six is written

$$\_\ \_\ \underline{3},\ \underline{7}\ \underline{5}\ \underline{0},\ \underline{2}\ \underline{0}\ \underline{6}$$

When the first group name is *billion*, there must be three groups of three digits each after the first comma.

EXAMPLE:

Sixty-eight billion four hundred three million seven hundred fifty thousand two hundred six is written

$$\_\ \underline{6}\ \underline{8},\ \underline{4}\ \underline{0}\ \underline{3},\ \underline{7}\ \underline{5}\ \underline{0},\ \underline{2}\ \underline{0}\ \underline{6}$$

3. After the first group has been written, if any other group after it has only *two* digits, write a zero before the digits to make a group of three digits.

EXAMPLE:

Four hundred three million fifty thousand two hundred six is written

$$\underline{4}\ \underline{0}\ \underline{3},\ \underline{\mathbf{0}}\ \underline{5}\ \underline{0},\ \underline{2}\ \underline{0}\ \underline{6}$$

FIGURE 1-1 *A Chart for Reading Decimals*

0. _ tenths	1 digit after the point
0. _ _ hundredths	2 digits after the point
0. _ _ _ thousandths	3 digits after the point
0. _ _ _ _ ten-thousandths	4 digits after the point
0. _ _ _ _ _ hundred-thousandths	5 digits after the point
0. _ _ _ _ _ _ millionths	6 digits after the point
0. _ _ _ _ _ _ _ ten-millionths	7 digits after the point

After the first group has been written, if any other group after it has only *one* digit, write *two zeros* before the digits to make a group of three digits.

EXAMPLE:

Four hundred three million fifty thousand six is written

$$4\,0\,3,\,0\,5\,0,\,0\,0\,6$$

After the first group has been written and if a *following group name is omitted entirely,* write *three zeros* for that group.

EXAMPLE:

Four hundred three million two hundred six is written

$$4\,0\,3,\,0\,0\,0,\,2\,0\,6$$

4. If the *last* word you hear or read is a group name (billion, million, or thousand), the groups after it must have three zeros each.

EXAMPLES:

Fifteen billion is written

$$1\,5,\,0\,0\,0,\,0\,0\,0,\,0\,0\,0$$

Fifteen billion five million is written

$$1\,5,\,0\,0\,5,\,0\,0\,0,\,0\,0\,0$$

Fifteen billion five million five thousand is written

$$1\,5,\,0\,0\,5,\,0\,0\,5,\,0\,0\,0$$

It is important to make neat, readable figures to prevent errors when someone else reads them. Follow this sample for both size and shape:

$$0,1,2,3,4,5,6,7,8,9$$

Note that the 0, 6, 8, and 9 are closed; the 2, 3, and 5 have no loops.

PART C Reading and Writing Decimals

If a decimal has a whole number to the left of the decimal point, the whole number is read and written as described in Part A of this section; the decimal point itself is read and written as *and*.

The decimal part of the number must always end with its decimal place name. This name depends on the number of digits after the decimal point. Study Figure 1-1, in which each blank space represents a digit. If a decimal has no whole-number part, it is common practice to write one zero to the left of the decimal point. You will notice that most calculators do this automatically because it helps the user notice exactly where the decimal point is located.

To read and write the decimal part of a given number, follow these two steps:

1. Read and write the number after the decimal point as if it were a whole number, as described in Part A.
2. Count the number of digits after the decimal point. Include all zeros after the decimal point as digits in order to get the correct decimal place name. Add the decimal place name as given in Figure 1-1.

Chapter 1   Working with Decimals

EXAMPLES:

a. 0.5 is read and written as *five tenths*.

b. 0.01 is read and written as *one hundredth*.

c. 0.138 is read and written as *one hundred thirty-eight thousandths*.

d. 0.2111 is read and written as *two thousand one hundred eleven ten-thousandths*.

e. 0.01815 is read and written as *one thousand eight hundred fifteen hundred-thousandths*.

f. 0.000042 is read and written as *forty-two millionths*.

g. 101.083 is read and written as *one hundred one and eighty-three thousandths*.

Another way to determine the correct decimal place name is to count out the places digit by digit, starting with the first one after the decimal point: tenths, hundredths, thousandths, ten-thousandths, etc. Be sure to count zeros as digits.

In the above examples, note that a hyphen is used in writing two-word place values, such as ten-thousandths and hundred-thousandths. Sometimes, shortcuts are taken when numbers are read aloud for checking purposes. Group names may be omitted and the word *comma* used instead; the number may be read as a series of single digits; or the word *point* may be used instead of *and*. When *point* is used, the decimal part of the number is read as a series of single digits without the decimal place names.

EXAMPLE:

For checking purposes, the number 286,375.40285 may be read as *two hundred eighty-six comma three hundred seventy-five point four zero two eight five* or *two eight six comma three seven five point four zero two eight five*.

PART D Writing Decimals from Words

In writing decimals, the key word is *and*. The *and* represents the decimal point. Write the whole-number part before the word *and* and the decimal part of the number after the word *and*.

EXAMPLE:

Forty-seven and thirty-nine hundredths is written as 47.39.

To write a decimal number from words, follow these three steps:

1. Write the whole-number part (47) of the decimal number as described in Part A.
2. Write the decimal point represented by the word *and*.
3. Write the decimal part of the number to the right of the decimal point. If the decimal part of the number has fewer digits than the decimal place name requires, write enough zeros between the decimal point and the decimal part of the number to give the correct number of decimal places.

EXAMPLE:

Four thousand three and nine thousandths is written as 4,003.009. To represent the nine thousandths, write the 9, and then write two zeros (00) between the decimal point and the 9.

There is another way of placing the decimal point correctly. Write the number by starting with the last digit of the decimal part and say its decimal place name. Going to the left, say each decimal place name through tenths, write enough zeros to give the correct number of places, and then place the decimal point before the number.

EXAMPLE:

Write *six hundred thirty-nine hundred-thousandths* as a decimal number. To begin, write 639. Start with the 9 and say *hundred-thousandths*. Go to the 3 and say *ten-thousandths*. Then go to the 6 and say *thousandths*. Since there are no more digits in 639, write a zero in front of the 6 (0639) and say *hundredths;* then write a zero in front of the 0639 (00639) and say *tenths*. Place the decimal point before the two zeros. Finish by writing a zero to the left of the decimal point. Thus, the example is correctly written as follows:

0.00639

PART E Comparing Decimals

In most cases, it is easy to select the larger of two whole numbers. You compare the numbers column by column or place by place *from the left*.

EXAMPLE:

51,321 is larger than 51,231, because the 3 in the hundreds place of 51,*3*21 is larger than the 2 in 51,*2*31. Note that the first two digits on the left, 51, are equal in both numbers.

However, be careful when the numbers being compared are decimals. To find which of two decimals is larger, follow these four steps:

1. If the numbers have whole-number parts, look at *only* the whole-number part in each one. The number that has the larger whole-number part is the larger number, regardless of its decimal part.

EXAMPLE:

639.4 is larger than 638.9912, because 639 is larger than 638 (the whole-number parts).

2. If the whole-number parts are exactly the same or if the numbers have only decimal parts, then look at *only* the first digit after the decimal point (the tenths place) in each. The number that has the larger digit in the tenths place is the larger number, regardless of the remaining decimal places.

EXAMPLE:

420.3 is larger than 420.289, because 420.*3* has a 3 in the tenths place, while 420.*2*89 has only a 2 in the same place. In this example, the digits in the tenths place determine which is the larger number.

3. If the numbers are exactly the same through the first decimal place, then look at *only* the second decimal place (the hundredths place) in each. The number that has the larger digit in the hundredths place is the larger number.

EXAMPLE:

420.32 is larger than 420.31985, even though the second number has more digits.

Continue doing this until one digit is found to be larger than another in the same decimal place. If two numbers are exactly the same through the last decimal place of one of them but the other still has at least one more non-zero decimal place, then the number with more decimal places is the larger number.

EXAMPLE:

420.32981 is larger than 420.3298.

4. Zeros can be written at the end of any decimal without changing the value of the number.

EXAMPLE:

0.6, 0.60, and 0.600 are all equal in value, even though they are read and written differently and have a different number of decimal places. Similarly, any zeros at the end of a decimal can be crossed out without changing the value of the number:

$$0.6 = 0.6\emptyset = 0.6\emptyset\emptyset$$

Remember that this applies *only* to zeros at the end of a *decimal, never* to zeros at the end of a *whole number.* Using this rule for the example in Step 2, you could write 420.3 as 420.300. Now, it is clear that 420.300 is larger than 420.289. Likewise, for the example in Step 3, 420.32 is equal to 420.32000, and 420.32000 is larger than 420.31985.

It is often necessary to arrange a series of numbers according to their size or value. Most of the time, the numbers are arranged in a column with the largest number first and the smallest number last. This arrangement is called **descending order.** Many of the same rules for doing this are used for finding the larger of two numbers. First, look at the whole-number parts and arrange them according to size. If two or more whole numbers are the same, then compare the decimal parts, place by place.

When decimals are written in a column, they are written so that the decimal points are lined up vertically. The digits in the same positions will then be in vertical columns; that is, the tenths digits are in one column and the hundredths digits are in the next column, and so on. When this is done, the numbers are said to be aligned or in **alignment.**

Chapter 1   Working with Decimals

When there are many numbers to be arranged according to size, as you put a number in a column it is a good idea to put a check mark after it or to draw a light line through it in the original list.

EXAMPLES:

a. Arrange these numbers in a column in descending order: 971.56, 975.9, 95.1589, 971.84, 975.93, 95.518, 979.3, 95.54, 975.8994, 917.68. Looking at only the whole-number parts, we see that 979 is the largest, so 979.3 will be the first number in the column.

　　　　979.3   (first)

The next largest number is 975, but there are three numbers which have 975 as the whole-number part. These are: 975.9, 975.93, and 975.8994. Looking at the first decimal place, we see that there are two numbers which have 9 in the tenths place. Since 975.93 has a digit in the hundredths place, it becomes the second number, and 975.9 is third. The fourth number is 975.8994. Our column so far looks like this:

　　　979.3
　　　975.93   (second)
　　　975.9   (third)
　　　975.8994   (fourth)

The next largest whole number is 971. There are two numbers with 971 as the whole-number part: 971.56 and 971.84. Since the 8 in 971.84 is larger than the 5 in 971.56, 971.84 becomes the fifth number, and 971.56 is the sixth. The next whole number is 917, so 917.68 becomes the seventh number. Our column now looks like this:

　　　979.3
　　　975.93
　　　975.9
　　　975.8994
　　　971.84   (fifth)
　　　971.56   (sixth)
　　　917.68   (seventh)

Finally, there are three numbers with 95 as the whole-number part: 95.1589, 95.518, and 95.54. After comparison, 95.54 becomes the eighth number and 95.518 is ninth. The number 95.1589 has only 1 in the tenths place and is the last and smallest number. The completed column shows the following:

　　　979.3
　　　975.93
　　　975.9
　　　975.8994
　　　971.84
　　　971.56
　　　917.68
　　　95.54   (eighth)
　　　95.518   (ninth)
　　　95.1589   (tenth)

b. Check the column below to see if the numbers are in correct descending order.

　　　91.00281
　　　91.02
　　　91.1

The numbers are in the wrong order. The correct descending order is:

　　　91.1
　　　91.02
　　　91.00281

PART F   The Decimal: The Number Used by the Calculator

While there are many different types of calculators, they all share common characteristics. All calculators have keys located on their faces (see Figure 1-2). These keys allow you to enter three types of information: **data** (numbers), operations, and functions. The process of entering this information is called **key-entering.** To key-enter information to the calculator, use the following steps:

1. Turn on the calculator by pressing the [ON] key. A 0 should appear on the **display,** or window, of your calculator. This prepares the calculator to receive information. Many modern calculators are solar powered. Simply exposing the face of the calculator to light will prepare the calculator to receive information.

2. Press the key(s) representing the number, operation, and/or function necessary for your calculation.

The differences between types of calculators result from the specifications for certain models. You should familiarize yourself with these specifications, which appear in the owner's manual accompanying your particular calculator.

All numbers key-entered on a calculator are decimal numbers. To key-enter a decimal number, key-enter each digit from left to right and include the decimal point. When key-entering decimals, enter all digits except trailing zeros and single zeros to the left of the decimal point (indicating that *zero* is the whole-number part of a decimal number).

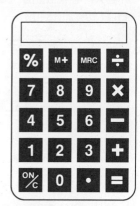

FIGURE 1-2 *A Standard Hand-Held Calculator*

EXAMPLES:

a. To key-enter the decimal number 7.3, press the 7 key, the decimal key, and the 3 key.

        7   .   3

b. Key-enter the decimal 7.03.

        7   .   0   3

c. Key-enter the decimal 1.600.

        1   .   6

d. Key-enter the decimal 0.523.

        .   5   2   3

While writing zeros is important when aligning decimals in columns, it is not necessary to use leading or trailing zeros when key-entering decimals. For example, the decimal 1.6 can be written as 1.6 or 1.60 or 1.600, depending on the number of columns needed for alignment. The reverse logic can be used for key-entering data to a calculator. Since 1.600 = 1.60 = 1.6, it is only necessary to key-enter 1.6.

The four basic operations of arithmetic are addition, subtraction, multiplication, and division. These operations are represented by the following keys on the calculator:

Operation	Key
addition	[+]
subtraction	[−]
multiplication	[×]
division	[÷]

When an operation or sequence of operations is necessary to perform a calculation, key-enter the equals key [=] to show the result of the calculation on the display of the calculator.

EXAMPLE:

$$7 + 4 = ?$$

Key-enter	Display
7	7.
[+]	7.
4	4.
[=]	11.

Notice that numbers appear in the display as decimals (11.). To erase the calculation, press the [C], or *clear,* key. This will prepare the calculator for the next calculation. Double-check each entry to make sure it is correct.

One final note about using a calculator. The calculator does not replace the problem-solving skills necessary to perform a calculation. Even with the use of a calculator, you must still interpret a problem, plan a solution, perform the calculation, and evaluate the correctness of your answer. Like all machines invented by human beings, the calculator is simply a time-saving tool.

COMPLETE ASSIGNMENT 1

Name _____

Date _____

ASSIGNMENT 1
SECTION 1.1

The Decimal Number System

	Perfect Score	Student's Score
PART A	10	
PART B	10	
PART C	10	
PART D	10	
PART E	40	
PART F	20	
TOTAL	100	

PART A Reading and Writing Whole Numbers

DIRECTIONS: On the lines provided, write the following numbers in words (2 points for each correct answer).

1. 857 _____

2. 4932 _____

3. 10047 _____

4. 280001 _____

5. 7000077 _____

STUDENT'S SCORE _____

PART B Writing Numbers from Words

DIRECTIONS: On the lines provided, write the following words in numbers (2 points for each correct answer).

6. Three thousand two hundred twenty-three _____

7. Eighty-nine thousand one hundred _____

8. Four hundred thousand five _____

9. Three million seventy _____

10. Six hundred million four thousand three _____

STUDENT'S SCORE _____

Unit 1   Fundamentals

PART C Reading and Writing Decimals

DIRECTIONS: On the lines provided, write the following decimals in words (2 points for each correct answer).

11. 122.845 _____

12. 608.005 _____

13. 4000.0004 _____

14. 10035.00056 _____

15. 907.047 _____

STUDENT'S SCORE _____

PART D Writing Decimals from Words

DIRECTIONS: On the lines provided, write the following words as decimals (2 points for each correct answer).

16. Two thousand six and five thousandths   _____

17. Four hundred thousand forty-five and four hundred four ten-thousandths   _____

18. Eighty-nine and five hundredths   _____

19. Two hundred million six hundred thousand two and four hundredths   _____

20. Twenty million twenty thousand twenty and two hundred-thousandths   _____

STUDENT'S SCORE _____

PART E Comparing Decimals

DIRECTIONS: For each pair of numbers, underline the one which is larger in value. If the numbers are equal, underline both (1 point for each correct answer).

21.	0.078	0.78	25.	0.01	0.0091
22.	839.06	839.060	26.	57.58	55.78000
23.	0.32	0.325	27.	0.5	0.500
24.	596.670	596.67	28.	845,902	845,092

Chapter 1   Working with Decimals   11

29.	0.330	0.033	**40.**	19.497	19.749
30.	0.07388	0.070883	**41.**	80.06	80.60
31.	0.2	0.1634	**42.**	10,829	10,982
32.	9,245.3	9,345.2	**43.**	6.177	61.77
33.	0.802	0.80100	**44.**	244	242.00
34.	160,094	106,094	**45.**	43.67	43.76
35.	0.751	0.7501	**46.**	5.12	51.2600
36.	370.9	370.90	**47.**	712,590	712,509
37.	0.675	0.675001	**48.**	0.633	0.6330
38.	29.314	29.134	**49.**	5.41825	5.4183
39.	300.125	301.025	**50.**	21.5	21.50

DIRECTIONS: On the lines provided, arrange the following groups of numbers in descending order. Vertically align the numbers by the decimal points (2 points for each correct answer).

51.   44.4
      3.44
      444
      4.440
      400.40
      0.44
      4.4
      344

53.   520
      0.4146
      173.08
      980.3
      520.5
      0.4164
      520.05
      980.29
      520.15
      173.008

52.   827.94
      872.94
      870.49
      827.49
      782.54
      784.29
      728.54
      782.49

54.   63.7
      6.37
      637
      6.307
      60.37
      673
      6.0370
      63.7370

Unit 1 Fundamentals

55. 483.801 _____

72.77 _____

504.712 _____

772 _____

0.9255 _____

771.99 _____

483.81 _____

0.92505 _____

504.721 _____

483.000 _____

STUDENT'S SCORE _____

PART F The Decimal: The Number Used by the Calculator

DIRECTIONS: On the lines provided, write the keys necessary to key-enter the following decimals on a calculator (2 points for each correct answer).

56. 370.90 _____
57. 12,000 _____
58. 28.124 _____
59. 9,245.300 _____
60. 938,467 _____

61. 0.0092 _____
62. 0.0851 _____
63. 0.675 _____
64. 0.66667 _____
65. 8,399 _____

STUDENT'S SCORE _____

Chapter 1 Working with Decimals

SECTION 1.2 Addition of Decimals

Most arithmetic calculations in business consist of some form of **addition,** or finding the total of two or more numbers. The numbers being added are called **addends.** The result of the addition is the **sum** or **total.**

Businesses add up all of their sales to find total sales. They add up their salaries, administrative costs, and other expenses to find total expenses.

PART A Adding Decimals Vertically

To add two or more decimal numbers in a column (vertical addition), follow these three steps:

1. Place the numbers in a column with the decimal points aligned.
2. Start at the top and add down; then check by starting at the bottom and adding up.
3. Place the decimal point in the sum directly under the decimal points in the column.

If all the addends do not have the same number of decimal places, you may write as many zeros as needed to the right of a number to give all addends the same number of decimal places.

EXAMPLES:

a. Add these numbers vertically and check: 0.97, 45.2, 3.001, 38, 0.0005, and 3.1875. Place a decimal point after any whole number. Place the numbers in a column with the decimal points aligned.

```
   0.97              0.9700  ⎫
  45.2              45.2000  ⎪
   3.001             3.0010  ⎬
  38.         or    38.0000  ⎨  addends
   0.0005            0.0005  ⎪
   3.1875            3.1875  ⎭
  ──────           ────────
  90.3590           90.3590      total
```

b. Add these numbers and check:

```
  0.487                    0.4870
  0.395                    0.3950
  0.0023       or          0.0023
  0.5                      0.5000
  ──────                   ──────
  1.3843    Carry the 1 to the    1.3843
            left of the decimal
            point.
```

You can total columns of numbers more quickly by following these steps: combine the numbers into groups of two, find the sum of each group, and add the sums of these groups. This shortcut is even faster if some of the numbers can be grouped into combinations that add up to 10, such as 1 and 9, 2 and 8, 3 and 7, 4 and 6, and 5 and 5. These shortcuts are illustrated in the following example.

EXAMPLE:

Find the sum of 8, 2, 3, 7, 4, 5, 2, 6, 1, and 8.

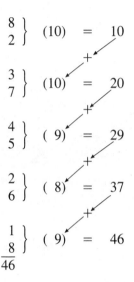

In review, note these points when vertically adding decimal numbers:

1. Place a decimal point at the end of a whole number so that it can be correctly positioned in the column. Thus, in *Example a*, 38 is written as 38. or 38.0000.
2. Correctly align each number so that the decimal points are under each other.
3. Check the sum of a column by adding in the opposite direction.
4. If the sum of any column, either to the right or to the left of the decimal point, is 10 or more, add the *carry-over* to the next column to the left. In particular, if there is a *carry-over* from the tenths column, add it to the ones column.
5. The total must have as many decimal places as the greatest number of places in any one of the addends. In *Example b,* the number 0.0023 has the most places; therefore, the total must also have four decimal places, including any trailing zeros that can be dropped.

PART B Adding Long Columns of Decimals

Long columns of numbers may be separated into groups. These groups can be added to find **subtotals.** The subtotals are then added to find the **grand total.** This is an especially valuable method when using a calculator that has no paper tape printout. Errors can be found more easily by checking the subtotals.

When amounts of money are added, the *first* number and the *total* should have dollar signs.

EXAMPLE:

Add by using subtotals.

$ 934.96		
829.17		
100.84		
873.12		
257.90	2,995.99	*subtotal*
698.75		
210.59		
542.20		
214.08		
764.95	2,430.57	*subtotal*
1,347.62		
764.00		
2,089.37		
481.63	4,682.62	*subtotal*
	$10,109.18	*total*

PART C Adding Decimals Horizontally

In a business report that consists of a table of numbers, it is quite common to show both *horizontal* (row) and *vertical* (column) totals. When this is done, the numbers are said to be **recapitulated,** or **recapped.**

To add a row of numbers horizontally, simply add across the page, being sure to keep track of the decimal places as the digits are being added.

When doing a *recap,* add the same numbers across and down. The sum of the row totals should equal the sum of the column totals. This sum is called the *grand total* and serves as a check on the row and column totals.

In the sales recap shown in Figure 1-3, find the total for Clerk 1 by adding across. Add the following: 3 (of 56.23) + 6 (of 43.76) + 3 (of 34.13) + 5 (of 88.25) + 0 (of 75.40) + 4 (of 58.94) = 21. Write the 1, carry the 2 to the tenths place, and add the following: 2 (carry over) + 2 (of 56.23) + 7 (of 43.76) + 1 (of 34.13) + 2 (of 88.25) + 4 (of 75.40) + 9 (of 58.94) = 27. Write the 7 in the tenths place and the decimal point. The result (.71) is the decimal part of the sum. Carry the 2 to the ones place. In the same way, add the digits in the *ones* place and then the digits in the *tens* place to find the first row total ($356.71). Using the same procedure, determine the remaining row totals. Find the grand total by adding the row totals vertically.

Find the column totals by adding down in the usual way. Find the grand total of these sums by adding horizontally. It should be the same as the grand total found by adding the row totals vertically. If it is not, this means that there are one or more errors in the row totals, column totals, or grand total. All totals must be checked. Do not forget to place a dollar sign before the first amount in each column and before each column total.

PART D Adding Decimals with the Calculator

When adding whole numbers or decimals with a calculator, follow these steps:

1. Key-enter each number.
2. Key-enter the addition key $\boxed{+}$ after each addend (except the final one).
3. After key-entering the final addend, use the equals key $\boxed{=}$ to find the final total, which will appear on the display.

EXAMPLE:

Find the sum of 3.01 and 6.23.

Key-enter	**Display**
3.01	3.01
$\boxed{+}$	3.01
6.23	6.23
$\boxed{=}$	9.24

Chapter 1 Working with Decimals

Clerk	Dept. A	Dept. B	Dept. C	Dept. D	Dept. E	Dept. F	Totals
1	$ 56.23	$ 43.76	$ 34.13	$ 88.25	$ 75.40	$ 58.94	$ 356.71
2	29.98	70.31	88.27	31.72	30.24	75.25	325.77
3	80.55	69.42	75.20	56.47	46.32	80.20	408.16
4	75.97	48.09	83.10	76.38	74.24	61.22	419.00
	$242.73	$231.58	$280.70	$252.82	$226.20	$275.61	$1,509.64

FIGURE 1-3 *A Sales Recap*

Key-enter the decimal 3.01, followed by the addition key [+]. Then key-enter the decimal 6.23, followed by the equals key [=]. If you key-enter both the decimals and operations correctly, the display will show the sum 9.24.

When adding decimal numbers with a calculator, you may add them vertically (in columns) or horizontally (in rows).

EXAMPLES:

a. Add the following column of numbers:

0.97
45.2
3.001
39.
0.0005
3.1875

Key-enter	Display
.97	0.97
[+]	0.97
45.2	45.2
[+]	46.17
3.001	3.001
[+]	49.171
39	39.
[+]	88.171
.0005	0.0005
[+]	88.1715
3.1875	3.1875
[=]	91.359 *total*

b. Add the following row of numbers:

8, 4, 3, 7

Key-enter	Display
8	8.
[+]	8.
4	4.
[+]	12.
3	3.
[+]	15.
7	7.
[=]	22. *total*

Checking all sums when adding with a calculator is a necessity. It is easy to press a wrong key and have an incorrect total. Checking sums with a calculator is the same as checking sums without a calculator. Instead of key-entering the numbers from top to bottom, change the order of the addition and key-enter the numbers from bottom to top.

EXAMPLE:

The sum of 8, 4, 3, and 7 is 22. Check this total by adding the numbers in reverse order.

Key-enter	Display
7	7.
[+]	7.
3	3.
[+]	10.
4	4.
[+]	14.
8	8.
[=]	22. *total*

COMPLETE ASSIGNMENT 2

Name _____

Date _____

ASSIGNMENT
SECTION 1.2

Addition of Decimals

	Perfect Score	Student's Score
PART A	24	
PART B	25	
PART C	36	
PART D	15	
TOTAL	100	

PART A Adding Decimals Vertically

DIRECTIONS: Write each set of numbers in a column. Find the sums, and write your answers on the lines provided. Check by adding in the opposite direction (4 points for each correct answer).

1.   0.72, 0.42, 0.27, 0.81   _____

3.   0.576, 0.5, 0.76, 0.03612, 0.95, 0.0121   _____

2.   0.51, 0.88, 0.47, 0.3, 0.1   _____

4.   0.102, 0.02057, 0.9, 0.638, 0.8218, 0.69   _____

5. 238.08, 71.0, 9.72, 67.449, 5.83, 54.0016

6. 6.08, 2.0, 381.85, 7.654, 8.31, 0.652, 0.00437

STUDENT'S SCORE _____

PART B Adding Long Columns of Decimals

DIRECTIONS: Add by using subtotals as indicated. Write the totals on the double-ruled lines. Check your work with a calculator (2 points for each correct subtotal and 3 points for each correct total).

7. $667.54
353.56
76.12
65.48   **a.** _____
341.23
689.74
901.43
54.17   **b.** _____   **c.** _____

8. $ 326.50
875.46
2,645.64
16.67
230.15   **a.** _____
66.55
543.30
11.22
5,873.38
767.41   **b.** _____
9.99
7,450.30
631.36
275.24   **c.** _____   **d.** _____

9. $ 92.90
261.25
79.80
359.87   **a.** _____
7,405.43
10.10
876.54
2,426.46   **b.** _____
1,780.27
73.73
437.02
260.35   **c.** _____   **d.** _____

STUDENT'S SCORE _____

Chapter 1 Working with Decimals

PART C Adding Decimals Horizontally

DIRECTIONS: Add horizontally and vertically. The grand total of the rows must equal the grand total of the columns. Check your work with a calculator (3 points for each correct answer).

780.04	728.47	88.88	700.25	363.49		**10.**
195.68	201.75	837.15	540.88	482.95		**11.**
251.76	160.43	205.94	149.69	187.32		**12.**
679.87	157.44	473.86	863.32	500.37		**13.**
894.77	494.73	548.76	677.74	294.93		**14.**
48.03	168.25	550.79	78.48	200.04		**15.**
16.	**17.**	**18.**	**19.**	**20.**	**21.**	

STUDENT'S SCORE _____

PART D Adding Decimals with the Calculator

DIRECTIONS: Add by using a calculator. Write your answers on the lines provided (3 points for each correct answer).

22. $84.383 + 21.48 + 3.7 =$ _____

25. 3.7812
    45.07
    0.038
    _____

23. $\$21.20 + \$81.35 + \$27.75 =$ _____

26. 8.7667
    3.0012
    5.8321
    _____

24. $1.2 + 0.0008 + 75.021 =$ _____

STUDENT'S SCORE _____

Chapter 1 Working with Decimals

SECTION 1.3 Subtraction of Decimals

Business reports and records usually require some subtraction. Subtraction is used to find such items as the amount of increase or decrease, the amount of net pay, the net amount after discount, and the amount of net profit. Subtraction is also used to compare numbers, such as the seating capacities of two auditoriums or the price of a skirt at two different stores.

Subtraction is the operation of finding the difference between two numbers. This is done by deducting one number from another. Subtraction can be expressed in the following ways:

1. Subtract $6.25 from $47.60.
2. $47.60 − $6.25
3. Deduct $6.25 from $47.60.
4. Take $6.25 from $47.60.
5. Find the difference between $47.60 and $6.25.
6. By how much is $47.60 greater than $6.25?

In these statements, $6.25 is the number being subtracted and is called the **subtrahend.** The number from which it is being subtracted is $47.60 and is called the **minuend.** The answer, or result, of the subtraction is called the **difference.**

Subtraction is checked by adding the difference and the subtrahend. The total of the difference and the subtrahend should equal the minuend. It is not necessary to recopy the numbers in order to add them for the check.

PART A Subtracting One Decimal from Another

To find the difference between two decimals which have the same number of places, write the larger number (the minuend) on top. Write the smaller number (the subtrahend) under the minuend with the decimal points aligned. Subtract as you would with whole numbers. Write the decimal point in the answer (the difference) directly under the other decimal points. Check by adding the answer (difference) to the subtrahend. Their sum is the minuend.

EXAMPLE:

```
  137.59   (minuend)      Check:    123.56
−  14.03   (subtrahend)           +  14.03
  123.56   (difference)             137.59
```

When a digit in the subtrahend is larger than the digit above it in the minuend, it is necessary to borrow a 1 from the digit to the left. Draw a line through the minuend digit from which the 1 is borrowed and reduce that digit by 1. Write the 1 that has been borrowed in front of the digit in the minuend. In the following example, the minuend is 567.28. In the tenths place, you must subtract 4 from 2. Borrow a 1 from the 7 to the left. The 7 becomes a 6, and the borrowed 1 is written next to the 2 in the tenths place. Now the 4 can be subtracted from 12.

EXAMPLE:

$$
\begin{array}{r} 6 \\ 56\cancel{7}.^{1}28 \\ -\ 252.43 \\ \hline 314.85 \end{array}
$$

$$
\begin{array}{r} 567.28 \\ -\ 252.43 \end{array}
$$

Sometimes it is necessary to go two or more places to the left to borrow a 1.

EXAMPLE:

$$
\begin{array}{r} 100.04 \\ -\ \ \ 6.05 \end{array}
\qquad
\begin{array}{r} 9\ 9\ \ \ 9 \\ \cancel{1}\,\cancel{0}\,\cancel{0}.\,\cancel{0}\,{}^{1}4 \\ -\ \ \ 6\ .\ 0\ 5 \\ \hline 9\ 3\ .\ 9\ 9 \end{array}
$$

These examples demonstrate the significance of a calculator. You do not need to borrow when you use a calculator, because it is done automatically by the calculator. Using a calculator to subtract is discussed in Part D of this section.

PART B Subtracting a Decimal from a Whole Number

Errors in subtraction may be caused by not correctly aligning the minuend and subtrahend. This often occurs when one of the numbers is a whole number written without a decimal point after it.

When the minuend is a whole number and the subtrahend is a decimal, place a decimal point after the whole number and write as many zeros to the right as needed to match the number of decimal places in the subtrahend. Write the subtrahend directly under the minuend with the decimal points aligned. Since the minuend has one or more zeros after the point, borrowing from the ones digit is necessary in order to subtract the decimal part of the subtrahend.

EXAMPLE:

Deduct 82.01 from 116.

```
  116          116.00      (write two zeros)
-  82.01     -  82.01
               ───────
                33.99
```

When the difference between two numbers is required, the larger number is considered the minuend and the smaller number is considered the subtrahend.

EXAMPLE:

Find the difference between 0.153 and 15. In this case, 15 is the minuend, since it is larger than 0.153.

```
  15           15.000
-  0.153     -  0.153
               ───────
                14.847
```

PART C Subtracting a Whole Number from a Decimal

When the minuend is a decimal and the subtrahend is a whole number, place a decimal point after the whole number and write as many zeros to the right as needed to match the decimal places in the minuend. Subtract and check as usual.

EXAMPLE:

Find the difference between 48 and 53.821.

```
  53.821       53.821
-    48      -  48.000       (write three zeros)
              ───────
                5.821
```

PART D Subtracting Decimals with the Calculator

When subtracting whole numbers or decimals with a calculator, follow these steps:

1. Key-enter the minuend.
2. Key-enter the subtraction key $\boxed{-}$.
3. Key-enter the subtrahend.
4. Use the equals key $\boxed{=}$ to find the difference, which will appear on the display.

EXAMPLES:

a. Find the difference between 125.48 and 14.03.

Key-enter	Display	
125.48	125.48	
$\boxed{-}$	125.48	
14.03	14.03	
$\boxed{=}$	111.45	*difference*

Key-enter the minuend, 125.48, followed by the subtraction key $\boxed{-}$. Then key-enter the subtrahend, followed by the equals key $\boxed{=}$. The result is the difference, 111.45.

b. Subtract 258.43 from 567.28.

Key-enter	Display	
567.28	567.28	
$\boxed{-}$	567.28	
258.43	258.43	
$\boxed{=}$	308.85	*difference*

c. By how much is $100.04 greater than $6.05?

Key-enter	Display	
100.04	100.04	
$\boxed{-}$	100.04	
6.05	6.05	
$\boxed{=}$	93.99	*difference*

Since there is no dollar sign on calculators, 93.99 is understood to represent $93.99.

Errors can easily be caused by incorrect key-entering. Therefore, checking the difference when subtracting with a calculator is advised. Checking a difference with a calculator is the same as checking the difference without a calculator. If the difference is correct, the sum of the subtrahend and the

Chapter 1   Working with Decimals

difference will be the minuend. Another method for checking a calculation is simply to repeat the calculation.

EXAMPLE:

Is $100.04 greater than $6.05 by a difference of $93.99?

	Key-enter	Display	
	100.04	100.04	
	[−]	100.04	
	6.05	6.05	
	[=]	93.99	*difference*
Check:	[+]	93.99	
	6.05	6.05	
	[=]	100.04	*minuend*

As you can see, using a calculator to subtract eliminates your need for borrowing, writing zeros to the right of a number, and aligning decimal points. The calculator does these tasks for you.

PART E Checking Calculator Tapes and Adding Machine Tapes

Almost all businesses use calculators and/or adding machines to do much of their arithmetic. Because it is so easy to press an incorrect key on a calculator, you must know how to check your work. Unlike the familiar hand-held calculator, adding machines usually produce a paper tape output that serves as a record of any calculations.

When using a calculator or an adding machine, the first step is always to clear the machine of any numbers that have been previously entered. The exact key to use will vary with the model and manufacturer. For example, an adding machine often has a symbol, such as *, which indicates a cleared machine.

After the total has been found, the numbers on the tape are checked against the original numbers to make sure that the numbers have been correctly key-entered. One person may do this, or the original numbers may be read aloud to someone checking the tape.

To check tape numbers and correct any errors, follow these steps, which are illustrated in the example that follows.

1. If the tape number is correct, place a small dot or check mark after it. (In the example, note the dots beside the numbers which are the same in the "Original Numbers" column and the "Tape" column.)

2. If the tape number is wrong, circle the number to show that it is wrong, and write the correct number to the right. Write the change to the adding machine total next to the corrected number. The change to the total is the difference between the correct number and the incorrect number.

   If the incorrect number is *larger* than the correct number, the tape total is too large by the amount of the difference. Since this must be *subtracted* from the tape total, place a minus sign before the written correction. The second number in the example was key-entered incorrectly. Circle the 5,555.50 and write 555.50 at the right. The tape total is too large by 5,000.00 (or 5,555.50 − 555.50), so write −5,000.00 to the right of 555.50 to show that 5,000.00 must be subtracted from the tape total.

   If the incorrect number is *smaller* than the correct number, the tape total is too small by the amount of the difference. Since this must be *added* to the tape total, place a plus sign before the written correction. The sixth number in the example was key-entered as 4.00 instead of 40.00. Circle the 4.00 and write 40.00 at the right. The tape total is short by 36.00 (or 40.00 − 4.00), so write +36.00 to the right of 40.00 to show that 36.00 must be added to the tape total.

   When only the last two digits of the tape number are wrong, the entire number does not have to be circled. In the example, the last two digits of the tenth number were key-entered as 98 instead of 89. Circle the 98 and write .89 to the right. The tape total is too large by 0.09 (or 0.98 − 0.89), so write −0.09 to the right of .89 to show that 0.09 must be subtracted from the tape total.

3. If a number in the original set of numbers was not key-entered, that number must be inserted in the tape and an arrow put in to show where it belongs. In the example, the last number in the column of original numbers was omitted. Write 300.00 between the 12.53 and the total, and insert an arrow to show where 300.00 belongs.

The tape total is short by 300.00, so write +300.00 at the right to show that it must be added to the tape total.

EXAMPLE:

Check and correct the tape.

Original Numbers	Tape	Corrections To Be Made
	*	
467.75	467.75 •	
555.50	(5555.50)	555.50 − 5,000.00
3,020.41	3020.41 •	
678.82	678.82 •	
607.10	607.10 •	
40.00	(4.00)	40.00+ 36.00
270.33	270.33 •	
3,476.98	3476.98 •	
561.55	561.55 •	
5,768.89	5768.(98)	.89− 0.09
12.53	12.53 •	300.00+ 300.00
300.00	20423.95 *	

```
     + 336.00
      20,759.95
      − 5,000.09
      15,759.86 (CORRECT TOTAL)
```

4. After the last tape number has been checked, add all the numbers with a plus sign in the "Corrections To Be Made" column. Then, add their sum to the tape total, as follows:

```
          300.00
           36.00
          336.00
       20,423.95   (tape total)
       +  336.00
       20,759.95
```

5. The last step is to find the sum of all the numbers with a minus sign. Then, subtract their sum from the new tape total found in Step 4, as follows:

```
           0.09
       + 5,000.00
         5,000.09

        20,759.95
       −  5,000.09
        15,759.86   (correct total)
```

Make corrections in the tape total after *all* the numbers have been checked. Some companies will allow one or two errors if the corrections are made clearly; otherwise, a new tape must be run.

This example shows more errors than are usually allowed. It was given as an example to show how corrections are made. Three common types of errors are reversing (or transposing) two digits, such as entering 98 instead of 89; entering too many or too few digits when the same digit occurs several times in a row, such as entering 5,555.50 or 55.50 instead of 555.50; and omitting a number entirely.

In the next example, there are six errors in the first tape. See if you can find the errors and make the corrections as shown in the previous example. The second tape was run correctly to give you the correct total of the original numbers.

EXAMPLE:

Check and correct the first tape.

Original Numbers	First Tape
	*
354.35	354.35
855.89	855.89
766.90	766.09
7,844.99	7894.99
90.76	90.76
130.45	130.54
8,604.29	8604.29
200.25	200.25
222.20	22.20
998.60	998.60
60.00	138.15
138.15	643.88
643.88	620.00
620.00	88.89
88.89	962.75
926.75	22371.63 *

Chapter 1 Working with Decimals

Second Tape

```
              *
      354.35
      855.89
      766.90
     7844.99
       90.76
      130.45
     8604.29
      200.25
      222.20
      998.60
       60.00
      138.15
      643.88
      620.00
       88.89
      926.75
    22546.35*
```

PART F Applications

In business, it is often necessary to find the increase or decrease of an amount from one period to another. This is done by subtracting the smaller amount from the larger one. When the later amount is larger than the earlier amount, there is an increase. When the later amount is smaller than the earlier amount, there is a decrease.

In business reports, generally there are columns called "Increase" and "Decrease." The amount of increase or decrease is written in the correct column.

EXAMPLES:

Find the amount of increase or decrease. Write the amount in the correct column.

	Last Month	This Month	Increase	Decrease
a.	701.91	862.74	_____	_____
b.	677.41	529.48	_____	_____

```
a.    862.74    (later period is larger)
    − 701.91    (earlier period is smaller)
      160.83    Increase

b.    677.41    (earlier period is larger)
    − 529.48    (later period is smaller)
      147.93    Decrease
```

Write the increase or decrease in the correct column, as follows:

	Last Month	This Month	Increase	Decrease
a.	701.91	862.74	160.83	
b.	677.41	529.48		147.93

In some reports, only the amount of change needs to be shown. When the change is an increase, a plus sign is written before the amount; when the change is a decrease, a minus sign is written before the amount. The previous examples are used below.

	Last Month	This Month	Change
a.	701.91	862.74	+160.83
b.	677.41	529.48	−147.93

COMPLETE ASSIGNMENT 3

Name _____

Date _____

ASSIGNMENT 3
SECTION 1.3

Subtraction of Decimals

	Perfect Score	Student's Score
PART A	20	
PART B	10	
PART C	10	
PART D	10	
PART E	30	
PART F	20	
TOTAL	100	

PART A Subtracting One Decimal from Another

DIRECTIONS: Subtract and check (2 points for each correct answer).

1.   63.09   2.   582.43   3.   5,665.04   4.   8,099.86   5.   15,000.25
    −38.24   − 25.99   −1,486.75   −5,188.70   − 1,421.01
    _____   _____   _____   _____   _____

DIRECTIONS: Find the difference between the two numbers. Check your answers (2 points for each correct answer).

6.   41.75 and 4.23   _____   9.   2,300.97 and 10,235.90   _____

7.   19.82 and 110.85   _____   10.   25,610.54 and 26,913.8   _____

8.   526.74 and 713.80   _____

STUDENT'S SCORE _____

PART B Subtracting a Decimal from a Whole Number

DIRECTIONS: Place a decimal point in the minuend and write zeros as needed. Write the subtrahend under the minuend and subtract. Write your answers on the lines provided (2 points for each correct answer).

11. 4 − 0.75 = _____   14. 635 − 52.618 = _____

12. 3 − 2.4 = _____   15. 384 − 3.7296 = _____

13. 14 − 7.83 = _____

STUDENT'S SCORE _____

PART C Subtracting a Whole Number from a Decimal

DIRECTIONS: Place a decimal point in the subtrahend and write zeros as needed. Write the subtrahend under the minuend and subtract. Write your answers on the lines provided (2 points for each correct answer).

16. 83.47 − 28 = _____   19. 5,668.84 − 5,082 = _____

17. 65.72 − 37 = _____   20. 12,012.29 − 8,998 = _____

18. 843.27 − 603 = _____

STUDENT'S SCORE _____

Chapter 1 Working with Decimals

PART D Subtracting Decimals with the Calculator

DIRECTIONS: Use a calculator to subtract the following numbers. Write your answers on the lines provided (2 points for each correct answer).

21. 4.383 − 3.7 = _____

22. 6,957.75 − 87.1 = _____

23. 91,001 − 15,482.85 = _____

24. 3.9512
      −0.4820   _____

25. 8.7667
      −1.8321   _____

STUDENT'S SCORE _____

PART E Checking Calculator Tapes and Adding Machine Tapes

DIRECTIONS: Check the calculator tape against the original numbers. Make the corrections on the tape and find the correct total (4 points for each correct answer).

26.

Original	Tape
574.99	574.99 *
142.15	142.15
6.87	6.87
3.00	30.00
356.57	356.57
987.48	987.48
1,246.88	1246.88
1,456.10	1456.10
524.58	524.50
12.08	0.08
67.52	67.52
3,005.50	3005.50
62.10	62.10
450.60	450.60
6,080.06	6080.60
	14991.94 *

_____ Correct Total

27.

```
  19.15        19.15  *
  42.06        42.06
 433.98       433.98
  31.42        31.42
  55.18        55.18
  26.72        26.27
 676.33       676.33
  28.47        28.47
 579.19      5779.19
   8.00       800.00
 349.86       349.86
 254.97       254.97
 316.54       316.54
 260.78       198.75
 198.75      9012.17 *
```

_____ Correct Total

28.

```
 815.90       815.90  *
 742.15       742.15
   6.33         6.33
 424.68       424.86
 525.36       525.36
 417.84       417.84
   7.95       268.05
 268.05       987.19
 987.19      1300.27
1,030.27      319.04
 319.04        22.85
  22.85       331.42
 331.42       944.16
 944.16        55.50
 555.50      7160.92 *
```

_____ Correct Total

Chapter 1   Working with Decimals   **31**

29.

```
                          *
   676.04       6 7 6 . 0 4
   778.13       7 7 8 . 1 3
   388.96       3 8 8 . 9 6
   270.86       2 7 0 . 8 6
    79.90       7 9 9 . 9 0
   435.68       4 3 5 . 6 8
   124.02       1 2 4 . 0 2
 1,389.15       1 3 8 9 . 1 5
   450.00               4 . 5 0
   426.25       4 2 6 . 2 5
 6,045.78           4 5 . 7 8
    12.63           1 2 . 6 3
   455.87       4 5 5 . 8 7
   153.13       1 5 3 . 1 3
   445.56       4 4 5 . 6 6
                6 4 0 6 . 5 6 *
```

_____ Correct Total

DIRECTIONS: If the numbers in the two columns are the same, put a check mark (✔) in the blank. If the numbers are different, put an (X) in the blank (1 point for each correct answer).

30.	683.93	638.93	_____	**37.**	23,303.09	23,303.09	_____
31.	3,889.68	3,389.68	_____	**38.**	56,930.14	56,930.14	_____
32.	75,358.00	7,358.00	_____	**39.**	18,880.05	1,880.05	_____
33.	1,241.76	1,241.76	_____	**40.**	7,777.50	77,777.50	_____
34.	4,688.40	6,486.40	_____	**41.**	900.09	910.09	_____
35.	5,962.50	55,962.50	_____	**42.**	9,091.00	9,091.00	_____
36.	13,810.45	13,810.45	_____	**43.**	4,520.82	4,520.82	_____

STUDENT'S SCORE _____

PART F Applications

DIRECTIONS: Find the amount of increase or decrease and write it in the correct column. Use a calculator to check your work (2 points for each correct answer).

	Department	Last Month	This Month	Increase	Decrease
44.	A	$ 5,421.29	$ 3,941.29		
45.	B	844.52	111.11		
46.	C	7,358.90	5,499.85		
47.	D	25,895.06	27,892.25		
48.	E	42,009.11	34,762.07		

DIRECTIONS: Find the amount of change and write it in the "Change" column. Use a plus sign to show an increase and a minus sign to show a decrease. Use a calculator to check your work (2 points for each correct answer).

	Department	Last Month	This Month	Change
49.	A	$ 600.18	$ 567.37	
50.	B	12,175.23	15,175.43	
51.	C	9,000.55	10,670.78	
52.	D	64,319.47	60,525.50	
53.	E	30,003.20	35,892.66	

STUDENT'S SCORE _____

Chapter 1 Working with Decimals

SECTION 1.4 Multiplication of Decimals

In business, addition is used to calculate such items as total revenue and total expenses; subtraction is used to find such items as net profit, net pay, increases, and decreases. **Multiplication,** which is adding a number to itself a specified number of times, is also an important arithmetic operation. It is used to calculate such items as sales tax, interest, discounts, and total cost.

The first number in a multiplication problem is called the **multiplicand.** The number by which it is multiplied is called the **multiplier.** The result of multiplication is called the **product.** Since $6 \times 4 = 24$ and $4 \times 6 = 24$, it is clear that the multiplicand and the multiplier can be interchanged. Thus, it is easier to say that each of the numbers is a **factor** of 24. Other factors of 24 are 8 and 3, since $8 \times 3 = 24$. Sometimes a number can be a product of more than two factors; for example, $30 = 2 \times 3 \times 5$.

One method of multiplication is to write the *partial products* and indent them. For example, find the product of 3,687 and 463 as shown below.

EXAMPLE:

```
    3,687     (multiplicand)
  ×   463     (multiplier)
   11 061  ⎫
   221 22   ⎬ (partial products)
 1 474 8   ⎭
 1,707,081   (final product)
```

Another method of multiplication is to take the digits of the multiplier at their full place value. In this second method, the following shortcuts are used:

1. To multiply by a number ending in one zero, multiply without the zero and then write one zero to the right of the product.
2. To multiply by a number ending in two zeros, multiply without the zeros and then write two zeros to the right of the product, and so on.

For example, taking the multiplication problem illustrated previously, the 6 of 463 is in the tens place and stands for 60. Thus, the product of 3,687 and 60 ($3,687 \times 60 = 221,220$) is a partial product. The 4 of 463 is in the hundreds place and stands for 400. Thus, the product of 3,687 and 400 ($3,687 \times 400 = 1,474,800$) is a partial product. With this method, all the partial products are right aligned and in their correct positions, as shown in the following example:

EXAMPLE:

```
        3,687
      ×   463
       11 061
      221 220
    1 474 800
    1,707,081
```

Three steps are used in the second method:

1. Find the first partial product as you would in the first method: $3,687 \times 3 = 11,061$.
2. Find the second partial product by multiplying 3,687 by 60. The partial product is $3,687 \times 60 = 221,220$.
3. Find the third partial product by multiplying 3,687 by 400. The partial product is $3,687 \times 400 = 1,474,800$.

When the multiplier has one or more zeros and the first method is used, errors may arise because the partial products are not indented correctly. This will not happen with the second method, since there is no indenting and the partial products are correctly aligned. Compare the two methods in the following example.

EXAMPLE:

$9,036 \times 3,704$

First Method	Second Method	
9,036	9,036	
× 3,704	× 3,704	
36 144	36 144	($9,036 \times 4$)
6 325 20	6 325 200	($9,036 \times 700$)
27 108	27 108 000	($9,036 \times 3,000$)
33,469,344	33,469,344	

PART A Multiplying a Decimal by a Decimal or a Whole Number

In many applications, decimals and whole numbers are multiplied. It is important that the decimal point is in the correct place. To find the product of two numbers, one or both of which are decimals, follow these four steps:

1. Right align both numbers. Do *not* line up the decimal points. Find their product.
2. Count the total number of decimal places (the digits to the right of the decimal point) in the multiplicand and the multiplier. The product should have the same number of decimal places.
3. In the product, begin with the digit on the right and count, to the left, the total number of decimal places found in Step 2. The decimal point goes to the left of this digit. After placing the decimal point in the product, check the decimal places to the right of the decimal point. Remember that, although the multiplication itself is correct, the answer is wrong if the decimal point is in the wrong position.
4. Zeros at the right end of a product are counted as part of the correct number of decimal places in the product. These zeros may be dropped *after* the decimal point has been placed. Zeros to the left of the decimal point are never dropped, since they belong to the whole-number part of the product.

The first method of multiplication illustrated previously is used in all of the following examples.

EXAMPLES:

a. 502 × 4.35

```
        502      (0 decimal places)
     × 4.35      (2 decimal places)
      25 10
     150 6
     2 008
     2,183.70    total = 2 decimal places
```

The zero on the right may be dropped to give 2,183.7 (unless the answer represents money).

b. 111.11 × 0.26005

```
       111.11    (2 decimal places)
    × 0.26005    (5 decimal places)
       55555
      6 6666
     22 222
     28.8941555  total = 7 decimal places
```

c. 412.5 × 0.64

```
       412.5     (1 decimal place)
     × 0.64      (2 decimal places)
      16 500
     247 50
     264.000     total = 3 decimal places
```

The three zeros after the decimal point may be dropped to give 264.

PART B Writing Zeros to the Left of a Product

The decimal point in a product is placed by starting with the right-hand digit and counting to the left. When the product does not have enough decimal places to correctly position the decimal point, write as many zeros as needed to the left of the number to give the correct number of decimal places. Remember that the end zeros in the product must be counted.

EXAMPLES:

a.
```
        0.785    (3 decimal places)
     × 0.1004    (4 decimal places)
        3140
        785
     0.0788140
```
Write one zero to the left of 7 for 7 decimal places.

b.
```
       0.1632    (4 decimal places)
     × 0.0078    (4 decimal places)
       13056
       11424
    0.00127296
```
Write two zeros to the left of 1 for 8 decimal places.

Chapter 1   Working with Decimals

PART C   Rounding Off to a Given Number of Decimal Places

Often a calculation results in more decimal places than are needed. For example, an answer may be 56.3875, but only two decimal places are required. The 56.3875 is **rounded off** to 56.39; that is, it is expressed correctly to two decimal places.

Generally, in rounding off a decimal to a given number of places, the whole-number part to the left of the decimal point remains the same (see Step 5, which follows, for an exception). In the decimal part, the digit at the required place may stay the same or it may be increased by one.

To round off a decimal to a given number of places, follow these five steps:

1. The whole-number part (if there is one) of the decimal number remains the same.
2. In the decimal part, look only at the digit to the right of the required place. You may wish to draw a vertical line after the required place and underline the next digit.

EXAMPLE:

Round off 31.58215 to two decimal places.

$$31.58|215$$

3. If the digit after the required place is less than 5, make no change in the required place and drop all places after it. In the example in Step 2:

$$31.58|215 = 31.58$$

4. If the digit after the required place is 5 or more, add 1 to the required place and drop all places after it.

EXAMPLE:

Round off 31.5858 to two decimal places.

$$31.58|58 = 31.58|58 = 31.59$$

5. If the digit at the required place is a 9 and the digit to its right is 5 or greater, the required place is increased by 1 as usual. However, this makes the required place a 10, which simply means that you should write 0 in the required place and carry 1 to the left.

EXAMPLE:

Round off 195.98952 to three decimal places.

$$195.989|52 = 195.989|52 = 195.990$$

To summarize, in rounding off a number, look only at the digit to the right of the required decimal place. The rest of the decimal places are always dropped.

EXAMPLES:

Round off as indicated.

a. 8.3499 to one decimal place

$$8.3|499 = 8.3|499 = 8.3$$

b. 21.8599 to one decimal place

$$21.8|599 = 21.8|599 = 21.9$$

c. 21.8549 to two decimal places

$$21.85|49 = 21.85|49 = 21.85$$

d. 21.8559 to two decimal places

$$21.85|59 = 21.85|59 = 21.86$$

e. 32.6492 to three decimal places

$$32.649|2 = 32.649|2 = 32.649$$

f. 32.64992 to three decimal places

$$32.649|92 = 32.649|92 = 32.650$$

g. 8.995061 to one, two, three, and four decimal places

One: $8.9|95061 = 8.9|95061 = 9.0$

Two: $8.99|5061 = 8.99|5061 = 9.00$

Three: $8.995|061 = 8.995|061 = 8.995$

Four: $8.9950|61 = 8.9950|61 = 8.9951$

PART D   Multiplying by Ten and Its Multiples

Since any number multiplied by 0 is always 0, there are some shortcuts for multiplying by 10, 100, 1,000, and other multiples of ten.

To multiply any whole number by 10, write one zero to the right of the number. To multiply any whole number by 100, write two zeros to the right of the number. To multiply any whole number by 1,000, write three zeros to the right of the number, and so on. The number of zeros written is always the same as the number of zeros in the multiplier. Thus, the long method of multiplication is unnecessary.

EXAMPLES:

a. 48 × 10 = 480 (write one zero to the right of 48)

b. 21 × 100 = 2,100 (write two zeros to the right of 21)

c. 90 × 1,000 = 90,000 (write three zeros to the right of 90)

To multiply a decimal by 10, 100, 1,000, etc., move the decimal point one place to the right for each zero in the multiplier.

EXAMPLES:

Multiply by moving the decimal point to the right.

a. 3.189 × 10 = 31.89 (move the decimal point one place to the right)

b. 0.0136 × 1,000 = 13.6 (move the decimal point three places to the right)

If the decimal has fewer places than the multiplier has zeros, write zeros to the right of the decimal to reach the correct number of decimal places.

EXAMPLES:

a. 200.1 × 100 = 20010 = 20,010 (write one zero and move the decimal point two places to the right)

b. 7.39 × 10,000 = 73900 = 73,900 (write two zeros and move the decimal point four places to the right)

A **multiple** of a number contains that number as a factor. Some multiples of 10 are 20, 30, 110, 350, 900, 4,600, etc. All multiples of 10 end in at least one zero; all multiples of 100 end in at least two zeros, and so on.

When multiplying any number by a multiple of 10, write the multiplier so that the end zeros are to the right of the last digit of the multiplicand. Multiply and write the same number of zeros in the multiplier to the right of the product.

EXAMPLE:

106 × 400

$$\begin{array}{r} 106 \\ \times\ \ 400 \\ \hline 42400 \end{array} = 42,400$$

Note that the 4 is written under the 6, with the two zeros written to the right of 4. Then 106 is multiplied by 4 (106 × 4 = 424) and two zeros are written to the right of 424 (42,400).

If the multiplicand is a decimal, place the decimal point after the zeros have been written.

EXAMPLE:

32.47 × 46,000

$$\begin{array}{r} 32.47 \\ \times\ 46000 \\ \hline 19482 \\ 12988 \\ \hline 149362000. \end{array} = 1,493,620.00$$

Note that the 6 (46,000) is written under the 7, with the three zeros to the right of 6. Then 3247 is multiplied by 46 and three zeros are written to the right of the product, 149362, to give 149362000. Then, move the decimal point two places to the left to give 1493620.00. The two end zeros are dropped to give the answer: 1,493,620.

EXAMPLE:

0.897 × 5,900

$$\begin{array}{r} 0.897 \\ \times\ 5900 \\ \hline 8073 \\ 4485 \\ \hline 5292300. \end{array} = 5,292.300$$

Note that the 9 (5900) is written under the 7, with the two zeros to the right. Then, 897 is multiplied by 59 and two zeros are written to the right of the

Chapter 1   Working with Decimals

product, 52923, to give 5292300. Move the decimal point three places to the left to give the answer: 5292.300 or 5,292.3.

There is another shortcut that can be used when multiplying. Move the decimal point of the top number (the multiplicand) one place to the right for each zero in the multiplier, cross out the zeros in the multiplier, and then multiply as usual.

In this shortcut, the multiplier is written as two factors, one of which is a multiple of 10. Since the order of multiplication does not matter, the multiplicand is multiplied first by the multiple of 10 and then by the other factor.

EXAMPLE:

Multiply 3.65 by 2,500. First, write 2,500 as 25 × 100. Since the order of multiplication does not matter, change 25 × 100 to 100 × 25.

$$3.65 \times 100 \times 25$$

Next, multiply 3.65 by 100 by moving the decimal point two places to the right:

$$3.65 \times 100 = 3.65 = 365$$

Then, multiply 365 by 25.

Second Shortcut	**First Shortcut**
365	3.65
× 25	× 2500
1 825	1825
7 30	730
9,125	912500. = 9,125.00

EXAMPLE:

Multiply 4.892 by 6,700. Change this to 489.2 × 67, then proceed as follows:

Second Shortcut	**First Shortcut**
489.2	4.892
× 67	× 6700
3 424 4	34244
29 352	29352
32,776.4	32776400. = 32,776.4

PART E Multiplying Decimals with the Calculator

When multiplying whole numbers or decimals with a calculator, follow these four steps:

1. Key-enter the multiplicand.
2. Key-enter the multiplication key, ×.
3. Key-enter the multiplier.
4. Use the equals key = to find the product, which will appear on the display.

EXAMPLES:

a. Find the product of 3,687 and 463.

Key-enter	**Display**	
3687	3687.	
×	3687.	
463	463.	
=	1707081.	*product*

Key-enter the multiplicand (3,687), followed by the multiplication key ×. Then key-enter the multiplier (463), followed by the equals key =. The result is the product 1707081. In this example, notice that you must be able to interpret the number 1707081 as 1,707,081. Not all calculators and computers place commas for us, so we must be able to interpret numbers correctly.

b. Multiply 502 and 3.65.

Key-enter	**Display**	
502.	502.	
×	502.	
3.65	3.65	
=	1832.3	*product*

c. Find the product of 148.92 and 0.26005.

Key-enter	**Display**	
148.92	148.92	
×	148.92	
.26005	0.26005	
=	38.726646	*product*

Rounding off to a desired number of decimal places is done the same way, with or without a calculator.

To review, use the following steps when rounding off:

1. In the rounded answer, the whole-number part of the decimal number (if there is one) usually remains the same.
2. In the decimal part, look only at the digit *to the right of* the required place. You may wish to draw a vertical line after the required place and underline the digit after it.
3. If the digit to the right of the required place is less than 5, make no change in the required place and drop all places after it.
4. If the digit to the right of the required place is 5 or more, add 1 to the required place and drop all places after it.
5. If the digit at the required place is a 9 and the digit to its right is 5 or greater, increase the required place by 1 as usual. However, this makes the required place a 10, which simply means that you should write 0 in the required place and carry 1 to the left.

EXAMPLE:

Find the product of 8.29 and 0.05. Round off the answer to two decimal places.

Key-enter	Display
8.29	8.29
×	8.29
.05	0.05
=	0.4145 *product*

0.41|4̲5 = 0.41

Key-enter 8.29, followed by the multiplication key ×. Key-enter .05, followed by the equals key =. The product to four decimal places is 0.4145. To round off to two decimal places, look at the digit 4 in the third decimal place. Since 4 is less than 5, make no change in the required place and drop all places after it.

PART F Applications

Once or twice a year, many businesses take a **physical inventory,** in which all items on hand are counted and recorded. This is usually done by departments. The inventory sheets generally show the stock number of each item and have columns for the quantity, the unit cost, and the extension. While the multiplication may be done by a computer, it is often done on a calculator or adding machine.

Most retail stores now have their cash registers linked to a computer to keep a running inventory. By the end of six months or a year, this inventory may not be accurate. The physical inventory provides a check on the computer's inventory.

When the unit cost is given in cents, it is usually changed to dollars before multiplying. Normally a cost in cents does not have a decimal point, but it is true that 26¢ = 26.¢. To change the cost in cents to dollars, find the decimal point in the cost and move it two places to the left. Then, drop the cents sign and write a dollar sign in front.

EXAMPLES:

Change the cost in cents to dollars.

a. 74¢ = 74.¢ = $0.74
b. 8¢ = 08.¢ = $0.08
c. 36.75¢ = 36.75¢ = $0.3675
d. 0.25¢ = 00.25¢ = $0.0025

After the quantities and the unit costs have been recorded, multiply each quantity by its unit cost. This is called **extending** the inventory. (An **extension** is a multiplication.) When all the extensions have been completed, add them together to get the entire inventory for the department. As a last step, add all the department inventories together to get the complete inventory for the business.

Because a department inventory can be very long, the following example shows only a few typical extensions.

Item	Quantity	Unit Cost	Extension
Y-264	324	14.5¢ ($0.145)	$ 46.98
4603	855	3.5¢ ($0.035)	29.925
M-4802	19	$7.61	144.59
C-510	4,791	0.25¢ ($0.0025)	11.9775

Notice that the extensions are not always rounded to two decimal places. This allows for more accuracy. Only the final answer should be rounded to two decimal places.

COMPLETE ASSIGNMENT 4

ASSIGNMENT 4
SECTION 1.4

Name _____

Date _____

Multiplication of Decimals

	Perfect Score	Student's Score
PART A	20	
PART B	10	
PART C	20	
PART D	30	
PART E	10	
PART F	10	
TOTAL	100	

PART A Multiplying a Decimal by a Decimal or a Whole Number

DIRECTIONS: Multiply. Check your answers with a calculator. Write your answers on the lines provided (2 points for each correct answer).

1. 8
   × 0.3
   _____

2. 3.213
   × 0.5
   _____

3. 7.2
   × 0.22
   _____

4. 5.8
   × 3.6
   _____

5. 8.816
   × 0.46
   _____

6. 0.071
   × 3.09
   _____

7. 29.6
   × 3.55
   _____

8. 23.8
   × 925
   _____

9. 1,009.8
   × 30.05
   _____

10. 7.401
    × 8.007
    _____

STUDENT'S SCORE _____

PART B Writing Zeros to the Left of a Product

DIRECTIONS: Multiply. Check your answers with a calculator. Write your answers on the lines provided (2 points for each correct answer).

11. 7
    × 0.007
    _____

12. 0.91
    × 0.04
    _____

13. 0.213
    × 0.047
    _____

14. 0.1074
    × 0.11
    _____

15. 1.009
    × 0.0068
    _____

STUDENT'S SCORE _____

PART C Rounding Off to a Given Number of Decimal Places

DIRECTIONS: Round off to the number of places indicated (1 point for each correct answer).

		One Place	Two Places			Three Places	Four Places
16.	4.7291	_____		26.	42.63763	_____	
17.	16.8626	_____		27.	54.6135	_____	
18.	0.05098	_____		28.	0.00117	_____	
19.	99.995	_____		29.	0.1486	_____	
20.	10.9949	_____		30.	10.999549	_____	
21.	8.66666		_____	31.	11.985931		_____
22.	0.3891		_____	32.	11.999949		_____
23.	0.03191		_____	33.	100.99999		_____
24.	321.65395		_____	34.	89.9999499		_____
25.	0.005555		_____	35.	29.292929		_____

STUDENT'S SCORE _____

PART D Multiplying by Ten and Its Multiples

DIRECTIONS: Multiply by moving the decimal point only. Write your answers on the lines provided (1 point for each correct answer).

36.	31.23 × 10 =	_____	44.	100 × 9.80	_____
37.	100 × 10.012 =	_____	45.	1,400 × 100 =	_____
38.	1.245 × 1,000 =	_____	46.	2.47 × 1,000 =	_____
39.	31 × 100 =	_____	47.	9.6351 × 1,000 =	_____
40.	0.004 × 1,000 =	_____	48.	10 × 0.029 =	_____
41.	19.9 × 1,000 =	_____	49.	11.0083 × 10 =	_____
42.	0.001 × 10,000 =	_____	50.	7.3 × 1,000 =	_____
43.	10,000 × 0.0002 =	_____	51.	1,000 × 643.25 =	_____

Chapter 1   Working with Decimals   43

52. 10,000 × 0.03 = _____   **54.** 21.5 × 100,000 = _____

53. 37.04 × 10,000 = _____   **55.** 100,000 × 0.0475 = _____

DIRECTIONS: Multiply by using shortcuts (1 point for each correct answer).

56. 4.5 × 320 = _____   **61.** 0.506 × 12,500 = _____

57. 30 × 200 = _____   **62.** 0.252 × 75,000 = _____

58. 0.002 × 1,200 = _____   **63.** 0.042 × 7,200 = _____

59. 1.01 × 6,000 = _____   **64.** 1.906 × 250,000 = _____

60. 0.0028 × 4,000 = _____   **65.** 68.75 × 4,900 = _____

STUDENT'S SCORE _____

PART E   Multiplying Decimals with the Calculator

DIRECTIONS: Multiply by using a calculator. Write the product as shown on the display, then round off the answer to two decimal places (1 point for each correct answer).

	Product		Two Places
66. 4.383 × 3.7 =	_____	**71.**	_____
67. 957.75 × 87.1 =	_____	**72.**	_____
68. 31.1 × 2,482.85 =	_____	**73.**	_____
69. 2.34 × 8.100 =	_____	**74.**	_____
70. 4.16 × 0.0057 =	_____	**75.**	_____

STUDENT'S SCORE _____

PART F Applications

DIRECTIONS: Extend the inventory and find the total. Abbreviations: cs. = case; lb = pound; bbl. = barrel; a unit cost followed by one of the abbreviations means the cost of one unit per item (1 point for each correct answer).

	Quantity	Unit Cost	Extension	Work Space
76.	41 cs.	$3.75/cs.	_____	
77.	409 lb	3.75¢/lb	_____	
78.	200 lb	7.5¢/lb	_____	
79.	55 cs.	$2.375/cs.	_____	
80.	2,100 lb	6.5 ¢/lb	_____	
81.	3,550 lb	5.25 ¢/lb	_____	
82.	20 cs.	$11.98/cs.	_____	
83.	17 bbl.	$6.95/bbl.	_____	
84.	120 lb	8.5 ¢/lb	_____	
85.		TOTAL	_____	

STUDENT'S SCORE _____

Chapter 1 Working with Decimals

SECTION 1.5 Division of Decimals

Of the four arithmetic operations, division usually takes more time to do than any of the others. Division is used to calculate important items such as averages, proportions, percentages, and expense distributions.

Division is the operation of finding how many times one number is contained in another. It is the **inverse** of multiplication. For example, the multiplication problem of 5 × 7 = 35 becomes a division problem written as 35 ÷ 7 = 5. In the division problem, 35 is the **dividend,** 7 is the **divisor,** and 5 is the **quotient.** We say that "7 goes into 35 five times," or "there are five 7's in 35," or "35 divided by 7 is 5." The division sign (÷) is read as *divided by.* To do division calculations, the problem may be written in this form:

$$\begin{array}{r} 5 \text{ (quotient)} \\ \text{(divisor) } 7\overline{)35} \text{ (dividend)} \end{array}$$

If the division is not exact, then there is a **remainder.** For example, 38 ÷ 7 has a quotient of 5 and a remainder of 3.

Division is checked by multiplying the divisor by the quotient and then adding the remainder, if any. The result will be the dividend. If the check is done with a rounded-off quotient, the result will not be exactly equal to the original dividend.

Here are some division facts you should know:

1. No number can be divided by zero. Division by zero is impossible.
2. Zero divided by any nonzero number is zero.
3. Any number divided by 1 is the same number.
4. Any nonzero number divided by itself is 1.

PART A Review of Division with Whole Numbers

In division, it is important to place the digits in the quotient in their correct positions. Place the *first* digit in the quotient directly above the *last* digit in the first *partial dividend.* In the following example, 40 is the first partial dividend, so the 1 in the quotient is directly above the 0. Other partial dividends are 87, 238, 147, and 192.

Each of the remaining digits in the quotient must be placed directly above the correct digit in the dividend. Errors can result from carelessly writing the digits in the quotient.

EXAMPLE:

407,872 ÷ 32

$$\begin{array}{r} 12{,}746 \\ 32\overline{)407{,}872} \\ \underline{32\phantom{0{,}000}} \\ 87\phantom{{,}000} \\ \underline{64\phantom{{,}000}} \\ 23\,8 \\ \underline{22\,4} \\ 1\,47 \\ \underline{1\,28} \\ 192 \\ \underline{192} \\ 0 \end{array}$$

Check:
$$\begin{array}{r} 12{,}746 \\ \times\ 32 \\ \hline 25\,492 \\ 382\,38 \\ \hline 407{,}872 \end{array}$$

When a digit in the dividend has been brought down to form the next partial dividend, the divisor may still be larger than the partial dividend. In this case, write a zero above that digit in the dividend to show that the divisor went into the partial dividend zero times.

EXAMPLE:

$$\begin{array}{r} 4{,}009 \\ 246\overline{)986{,}357} \\ \underline{984\phantom{{,}000}} \\ 2\,357 \\ \underline{2\,214} \\ 143 \\ \text{(remainder)} \end{array}$$

Check:
$$\begin{array}{r} 4{,}009 \\ \times\ 246 \\ \hline 24\,054 \\ 160\,36 \\ 801\,8 \\ \hline 986{,}214 \\ +\ \ \ 143 \\ \hline 986{,}357 \end{array}$$

In this example, 2 is the first difference. Bring 3 down to make a partial dividend of 23. The divisor, 246, is larger than 23. Therefore, write a 0 above the 3 in the quotient. Bring down the 5 to make a new partial dividend of 235. The divisor is still larger. Write another 0 in the dividend, this time above the 5. Bring down the 7 to make a new partial dividend of 2,357; 246 goes into 2,357 nine times, with a remainder of 143. Be sure that you place the zeros over the digits that were brought down. Otherwise, you may carelessly omit one or both of the zeros.

There is one step in division that is often omitted and may cause errors. After subtracting and before bringing down the next dividend digit, always check the remainder of the subtraction just performed. If it is equal to or larger than the actual divisor, it means that the quotient digit used is not large enough and must be increased by at least one.

EXAMPLE:

$$\begin{array}{r} 2 \\ 62{\overline{\smash{\big)}\,18{,}982}} \\ \underline{12\ 4} \\ 6\ 5 \end{array}$$

Note that 65 is larger than 62. This means that the quotient digit 2 is too small. Change the 2 to a 3 before continuing the division, as follows:

$$\begin{array}{r} 3 \\ 62{\overline{\smash{\big)}\,18{,}982}} \\ \underline{18\ 6} \\ 3 \end{array}$$

Short division may be used when the divisor is a single digit. In this process the subtraction and multiplication are done mentally. Only the quotient is written. However, some persons may choose to write down the remainders next to the digits in the dividend to illustrate the new partial dividends.

EXAMPLE:

$$\begin{array}{r} 23{,}756 \\ 6{\overline{\smash{\big)}\,142{,}536}} \end{array} \quad \text{or} \quad \begin{array}{r} 2\ \ 3\ \ 7\ \ 5\ \ 6 \\ 6{\overline{\smash{\big)}\,14\ {}^2\!2\ {}^4\!5\ {}^3\!3\ {}^3\!6}} \end{array}$$

When writing the remainders to illustrate the new partial dividends, the five steps are:

1. $14 \div 6 = 2$; $2 \times 6 = 12$; $14 - 12 = 2$.

2. Carry the 2 to make the next partial dividend 22. $22 \div 6 = 3$; $3 \times 6 = 18$; $22 - 18 = 4$.

3. Carry the 4 to make the next partial dividend 45. $45 \div 6 = 7$; $7 \times 6 = 42$; $45 - 42 = 3$.

4. Carry the 3 to make the next partial dividend 33. $33 \div 6 = 5$; $5 \times 6 = 30$; $33 - 30 = 3$.

5. Carry the next 3 to make the next partial dividend 36. $36 \div 6 = 6$.

PART B Dividing a Decimal by a Whole Number

When the dividend is a decimal and the divisor is a whole number, the division is the same as it is for whole numbers, except that a decimal point must first be placed *directly above* the decimal point in the dividend. After placing the decimal point in the quotient, divide as usual.

EXAMPLES:

a. $52.5 \div 15$

$$\begin{array}{r} 3.5 \\ 15{\overline{\smash{\big)}\,52.5}} \\ \underline{45} \\ 7\ 5 \\ \underline{7\ 5} \\ 0 \end{array} \qquad \textbf{Check:} \quad \begin{array}{r} 3.5 \\ \times\ 15 \\ \hline 17\ 5 \\ 35 \\ \hline 52.5 \end{array}$$

b. $3.3463 \div 307$

$$\begin{array}{r} 0.0109 \\ 307{\overline{\smash{\big)}\,3.3463}} \\ \underline{3\ 07} \\ 2763 \\ \underline{2763} \\ 0 \end{array} \qquad \textbf{Check} \quad \begin{array}{r} 0.0109 \\ \times\ 307 \\ \hline 763 \\ 3\ 270 \\ \hline 3.3463 \end{array}$$

In these examples, the final remainder is 0. Sometimes this will not happen, and you may need to carry the division to several more decimal places. To do this, simply write as many zeros as needed to the right of the dividend; then divide as usual.

EXAMPLE:

$0.09 \div 12$

$$\begin{array}{r} 0.0075 \\ 12{\overline{\smash{\big)}\,0.0900}} \\ \underline{0\ 084} \\ 60 \\ \underline{60} \\ 0 \end{array} \qquad \textbf{Check:} \quad \begin{array}{r} 0.0075 \\ \times\ 12 \\ \hline 150 \\ 75 \\ \hline 0.0900 \end{array}$$

Chapter 1 Working with Decimals

PART C Dividing a Whole Number by a Decimal

When the dividend is a whole number and the divisor is a decimal, a decimal point must be placed in the quotient before any division takes place. The actual division is performed with a whole-number divisor. To move the decimal point in the divisor, use the following steps:

1. First, since the dividend is a whole number, place a decimal point to the right of the last digit.
2. Count the number of decimal places in the divisor. Write the same number of zeros to the right of the decimal point in the dividend.
3. Move the decimal point in the divisor to the right of the last digit in the divisor.
4. Move the decimal point in the dividend the same number of places to the right as in Step 3.
5. Place a decimal point in the quotient directly above the new decimal point in the dividend.

Note that the new decimal point in the dividend is to the right of the last zero in the dividend.

EXAMPLE:

Divide and check: 750 ÷ 0.25.

$$0.25\overline{)750.00}$$

Place a decimal point after the 750. Count the number of decimal places in the divisor, 0.25. Write two zeros to the right of the decimal point: 750.00. Move the decimal point in the divisor two places to the right. Then move the decimal point in the dividend two places to the right. Write a decimal point in the quotient directly above the new decimal point in the dividend, and then divide.

```
         3,000.         Check:    3,000
    25)75,000.                  × 0.25
       75                        150 00
       ──                        600 0
        0                        ──────
                                 750.00
```

If the divisor is larger than the dividend, or if the division is not exact, you may wish to carry the division out to several decimal places. Simply write additional zeros to the right of the dividend and divide as usual.

EXAMPLE:

Divide to four decimal places and check: 9 ÷ 42.78.

$$42.78\overline{)9.00}$$

```
         0.2103      Check:    42.78
  4,278)900.0000            × 0.2103
        855 6                 12834
        ─────                 42780
         44 40               8 5560
         42 78              ─────────
         ─────              8.996634
          1 6200          + 0.003366
          1 2834          ─────────
          ─────            9.000000
            3366
       (remainder)
```

Note that the first two zeros were needed to place the decimal point in the quotient. The next four zeros were written after the decimal point so that the division could be carried out to four decimal places. In checking the problem, notice that the remainder is actually a decimal value. It will contain the number of decimal places indicated by the multiplication problem.

PART D Dividing a Decimal by a Decimal

When both the divisor and the dividend are decimals, the division is similar to dividing a whole number by a decimal (see Part C). First, change the divisor into a whole number by moving its decimal point to the extreme right. Then move the decimal point in the dividend the same number of places to the right. It may be necessary to write a few zeros to the right of the dividend to have enough decimal places. Finally, place a decimal point in the quotient directly above the new decimal point in the dividend.

As also illustrated in Part C, if you wish to carry out the division to a certain number of places, it may be necessary to write more zeros after the decimal point.

EXAMPLE:

$63.9 \div 2.457$. Divide to four decimal places and round off the quotient to three places.

$$2.457 \overline{)63.900}$$

```
              26.0073  = 26.007
2,457)63,900.0000
      49 14
      14 760
      14 742
          18 000
          17 199
             8010
             7371
              639
```

Check: 26.0073   26.007
            × 2.457   × 2.457
            ─────── or   ───────
            1820511   182049
            1 300365   1 30035
           10 40292   10 4028
           52 0146   52 014
           ─────────   ─────────
           63.8999361   63.899199
          + 0.0000639
           ─────────
           63.9000000

In this example, note that you first write two zeros in order to move the decimal point three decimal places to the right. Then you write four more zeros to extend the division to four decimal places. The rounded quotient is less than the actual quotient. Therefore, when you use 26.007 for the check, the product is less than the actual dividend.

Note that moving the decimal point in the divisor to the right is the same as multiplying the divisor by 10, or 100, or 1,000, etc. To maintain an equivalent problem, multiply the dividend by the same number (10, 100, 1,000, etc.) by moving its decimal point the same number of places to the right. In the preceding example, multiply both the divisor and the dividend by 1,000. Check to see if the quotient of $63.9 \div 2.457$ is the same as the quotient of $63,900 \div 2,457$.

PART E Dividing One Whole Number by Another to Several Places

As already mentioned, when the quotient is not exact, you may want to carry out the division to several decimal places. Simply place a decimal point to the right of the dividend, write as many zeros as needed to the right of the decimal point, and place a decimal point in the quotient directly above the decimal point in the dividend. Divide as usual.

EXAMPLE:

Divide 527 by 6 and round the quotient to four decimal places.

```
    87.83333  = 87.8333    Check:    87.8333
6)527.00000                         ×      6
                                    ────────
                                    526.9998
                                   = 527
```

This is also the method used to find the decimal form of a common fraction. The numerator is divided by the denominator to as many places as are required.

EXAMPLE:

$3 \div 8$

```
    0.375     Check:   0.375
8)3.000              ×     8
  2 4                ───────
  ───                  3.000
    60
    56
    ──
    40
    40
    ──
     0
```

Note that this division is the same that would be done to find the decimal form of $\frac{3}{8}$. Divide the numerator, 3, by the denominator, 8. The decimal form of a common fraction is called the **decimal equivalent** of that fraction. Fractions will be discussed in Chapter 2.

Chapter 1   Working with Decimals

PART F Dividing by Ten and Its Multiples

To divide a number by 10, 100, 1000, etc., simply move the decimal point one place to the left for each zero in the divisor. Move the decimal point one place when dividing by 10, two places when dividing by 100, three places when dividing by 1,000, and so on. It may be necessary to write zeros to the left of the dividend if the divisor is much larger than the dividend.

If the dividend is a whole number, first place a decimal point to the right of the dividend. Then proceed as described.

EXAMPLES:

Divide by moving the decimal point in the dividend.

a. $21.12 \div 10 = 21.12 = 2.112$

b. $8,495.2 \div 1,000 = 8,495.2 = 8.4952$

c. $7.25 \div 100 = 07.25 = 0.0725$

d. $25 \div 10 = 25. = 2.5$

e. $398 \div 10,000 = 0398. = 0.0398$

When the divisor is a multiple of 10, other than 100, 1,000, etc., it can be expressed as some number times 10, 100, 1,000, etc. For example, the divisor 5,700 can be expressed as 57×100. The division is then performed in two steps, as shown in the following example.

EXAMPLE:

$2,013 \div 1,500$ (1,500 is the same as 15×100.)

Step 1.   Divide by 100 by moving the decimal point of the dividend to the left. In this example,

$$2,013 \div 100 = 20.13$$

Step 2.   Divide by the other factor of the original divisor. In this example,

$$\begin{array}{r} 1.342 \\ 15\overline{)20.130} \\ \underline{15} \\ 5\ 1 \\ \underline{4\ 5} \\ 63 \\ \underline{60} \\ 30 \\ \underline{30} \\ 0 \end{array}$$

The advantage of using this method is that the zeros in the original divisor are eliminated, thus simplifying the division.

EXAMPLE:

$2,750.85 \div 45,000$ (45,000 is the same as $45 \times 1,000$.)

$$2,750.85 \div 1,000 = 2.75085$$

$$\begin{array}{r} 0.06113 \\ 45\overline{)2.75085} \\ \underline{2\ 70} \\ 50 \\ \underline{45} \\ 58 \\ \underline{45} \\ 135 \\ \underline{135} \\ 0 \end{array}$$

PART G Dividing Decimals with the Calculator

When dividing whole numbers or decimals with a calculator, follow these four steps:

1. Key-enter the dividend.
2. Key-enter the division key ÷ .
3. Key-enter the divisor.
4. Use the equals key = to find the quotient, which will appear on the display.

EXAMPLES:

a. Find the quotient of 407,872 divided by 32.

Key-enter	Display	
407872	407872.	
÷	407872.	
32	32.	
=	12746.	*quotient*

Key-enter the dividend, 407872, followed by the division key ÷ . Then, key-enter the divisor, 32, followed by the equals key = . The result is the quotient 12746 or 12,746.

b. Divide 87.5 by 25.

Key-enter	Display	
87.5	87.5	
÷	87.5	
25	25.	
=	3.5	*quotient*

c. 750 ÷ 0.125

Key-enter	Display	
750	750.	
÷	750.	
.125	0.125	
=	6000.	*quotient*

Checking quotients with a calculator is the same as checking quotients without a calculator. If the quotient is correct, the product of the quotient and the divisor will be the dividend.

EXAMPLE:

Divide and check: 3 ÷ 62.5.

	Key-enter	Display
	3	3.
	÷	3.
	62.5	62.5
	=	0.048
Check:	×	0.048
	62.5	62.5
	=	3.

PART H Applications

An average is often used to analyze such things as sales and costs. An **average** is a reasonable estimate of the size of each number in a given group of numbers. It is found by adding all the numbers in a group together and dividing the sum by the number of numbers in the group. An average can also represent an *overall* performance.

EXAMPLES:

a. Eight persons worked the following hours in a week: 40, 42, 25, 40, 50, 35, 20, and 39. Find the average hours worked that week.

$$40 + 42 + 25 + 40 + 50 + 35 + 20 + 39 = 291$$

$$\begin{array}{r} 36.375 \\ 8\overline{)291.000} \end{array}$$

The average hours worked that week were 36.375 hours.

b. Smith's Bookstore was open 25 days in April and had total sales of $50,382.30. Find the average daily sales for April to the nearest cent and to the nearest dollar.

The total sales for all 25 days are already given. Therefore,

$$\begin{array}{r} \$\ 2{,}015.292 \\ 25\overline{)\$50{,}382.300} \end{array}$$

To the nearest cent, the average daily sales were $2,015.29; to the nearest dollar, the average daily sales were $2,015.

c. Bill traveled a total of 255 miles. If the trip took 5 hours, find his average speed.

The total miles traveled are already given. Therefore,

$$\begin{array}{r} 51 \\ 5\overline{)255} \end{array}$$

Bill's average speed was 51 miles per hour.

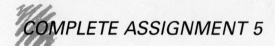

COMPLETE ASSIGNMENT 5

Name _____

Date _____

ASSIGNMENT 5
SECTION 1.5

Division of Decimals

	Perfect Score	Student's Score
PART A	8	
PART B	15	
PART C	15	
PART D	18	
PART E	12	
PART F	16	
PART G	12	
PART H	4	
TOTAL	100	

PART A Review of Division with Whole Numbers

DIRECTIONS: Find the quotient by short division. Check your answer with a calculator and write it on the line provided (2 points for each correct answer).

1. 3)654 _____

2. 6)528 _____

3. 3)276 _____

4. 2)48,316 _____

STUDENT'S SCORE _____

PART B Dividing a Decimal by a Whole Number

DIRECTIONS: Divide and round off the answer to three decimal places where necessary. Check your answer with a calculator and write it on the line provided (3 points for each correct answer).

5. 5)4.2 _____

6. 15)25.575 _____

7. 62)10.31 _____

8. 23.2 ÷ 56 _____

9. 5.428 ÷ 301 _____

STUDENT'S SCORE _____

PART C Dividing a Whole Number by a Decimal

DIRECTIONS: Divide and round off the answer to three decimal places where necessary. Check your answer with a calculator and write it on the line provided (3 points for each correct answer).

10. 4.8)̄16 _____   13. 80 ÷ 0.625 _____

11. 0.47)̄329 _____   14. 2,601 ÷ 4.25 _____

12. 0.095)̄525 _____

STUDENT'S SCORE _____

PART D Dividing a Decimal by a Decimal

DIRECTIONS: Divide and round off the answer to three decimal places where necessary. Check your answer with a calculator and write it on the line provided (3 points for each correct answer).

15. 2.3)̄0.5 _____   18. 0.086)̄0.0488 _____

16. 2.31)̄0.5 _____   19. 1.414 ÷ 1.22 _____

17. 1.9)̄0.038 _____   20. 2.5 ÷ 0.125 _____

STUDENT'S SCORE _____

Chapter 1   Working with Decimals   53

PART E Dividing One Whole Number by Another to Several Places

DIRECTIONS: Divide and round off the answer to four decimal places where necessary. Check your answer with a calculator and write it on the line provided (3 points for each correct answer).

21.   39 ÷ 7 = _____   **23.**   55 ÷ 125 = _____

22.   125 ÷ 45 = _____   **24.**   75 ÷ 210 = _____

STUDENT'S SCORE _____

PART F Dividing by Ten and Its Multiples

DIRECTIONS: Divide by only moving the decimal point (1 point for each correct answer).

25.   627 ÷ 10 = _____   **31.**   29.1 ÷ 10,000 = _____
26.   65.3 ÷ 10 = _____   **32.**   385 ÷ 10,000 = _____
27   28.4 ÷ 100 = _____   **33.**   0.101 ÷ 10 = _____
28.   3,840 ÷ 100 = _____   **34.**   0.1633 ÷ 100 = _____
29.   8.44 ÷ 1,000 = _____   **35.**   16,522 ÷ 1,000 = _____
30.   75 ÷ 1,000 = _____   **36.**   0.01 ÷ 10,000 = _____

DIRECTIONS: Divide by using a shortcut. Check your answer (1 point for each correct answer).

37.   28.35 ÷ 700 = _____   **39.**   65.4 ÷ 3,000 = _____

38.   28.2 ÷ 1,410 = _____   **40.**   539.4 ÷ 62,000 = _____

STUDENT'S SCORE _____

PART G Dividing Decimals with the Calculator

DIRECTIONS: Divide by using a calculator. Write the quotient as shown on the display, then round off the answer to two decimal places (1 point for each correct answer).

	Quotient		Two Places
41. 7.75 ÷ 0.4 =	_____	47.	_____
42. 7.5 ÷ 0.05 =	_____	48.	_____
43. 2.09375 ÷ 0.5625 =	_____	49.	_____
44. 34.65 ÷ 5.775 =	_____	50.	_____
45. 18 ÷ 57 =	_____	51.	_____
46. 2.538 ÷ 100.1 =	_____	52.	_____

STUDENT'S SCORE _____

PART H Applications

DIRECTIONS: Solve the problem as indicated and write your answer in the space provided (2 points for each correct answer).

53. The following are the weights (in pounds) of chickens sold in a grocery store: 2.6, 3.5, 2.7, 3.0, 3.2, 2.8, 2.7, and 3.1. Find the average weight of these chickens.

54. If Mary rode her bike 42 miles in 10 hours, what was her average speed?

STUDENT'S SCORE _____

CHAPTER 2

Working with Fractions

Fractions are used to represent the parts of a whole unit. A **fraction** is defined as one or more equal parts or divisions of a whole unit. The fraction $\frac{3}{4}$ means that a whole unit has been divided into 4 equal parts and $\frac{3}{4}$ represents 3 of these 4 equal parts. There are rules for working with fractions just as there are for decimals.

Fractions are used by businesses, stock brokerages, banks, and construction-related industries. Measurements are often given as fractional parts of inches, feet, yards, ounces, pounds, quarts, and gallons. Because calculators do not use fractions, a basic understanding of all types of fractions is important.

SECTION 2.1 Types of Fractions

In a **common fraction**, such as $\frac{3}{4}$, the top number, 3, is called the **numerator**; the bottom number, 4, is called the **denominator**.

The *fraction bar* can also be interpreted as a division sign. As will be explained more fully in Section 2.6 of this chapter, the common fraction $\frac{3}{4}$ can be changed into a decimal by dividing the 3 by the 4. Thus, the following notations are equal:

$$\tfrac{3}{4} = 3 \div 4 = 4\overline{)3.00} = 0.75$$

A common fraction may be proper or improper. In a **proper fraction,** the numerator is less than the denominator. Examples of proper fractions are:

$$\frac{3}{4}, \quad \frac{9}{10}, \quad \frac{11}{115}$$

In an **improper fraction**, the numerator is equal to or larger than the denominator. Examples of improper fractions are:

$$\frac{32}{30}, \quad \frac{7}{4}, \quad \frac{19}{12}, \quad \frac{12}{12}$$

A **mixed number** is a whole number followed by a proper fraction. Examples of mixed numbers are:

$$1\tfrac{3}{4}, \quad 15\tfrac{2}{3}, \quad 54\tfrac{7}{9}$$

Generally when the word "fraction" is used, it refers to a proper fraction.

PART A Changing a Fraction to Higher Terms

If both the numerator and the denominator of a fraction are multiplied by the same number, the value of the fraction remains the same. In effect, this is the same as multiplying the fraction by one. The new fraction is called an **equivalent fraction.** When both the numerator and the denominator are multiplied by the same number greater than 1, the fraction is said to be changed to **higher terms.**

EXAMPLES:

a. $\dfrac{2}{5} \times \dfrac{5}{5} = \dfrac{2 \times 5}{5 \times 5} = \dfrac{10}{25}$

b. $\dfrac{1}{6} \times \dfrac{7}{7} = \dfrac{1 \times 7}{6 \times 7} = \dfrac{7}{42}$

c. $\dfrac{5}{8} \times \dfrac{10}{10} = \dfrac{5 \times 10}{8 \times 10} = \dfrac{50}{80}$

To change a fraction to an equivalent fraction with a larger denominator, divide the denominator of the original fraction into the new denominator, and then multiply both the numerator and the denominator by that quotient.

EXAMPLES:

a. Change $\dfrac{1}{3}$ to 30ths.

$$\dfrac{1}{3} = \dfrac{?}{30}$$

$$30 \div 3 = 10$$

$$\dfrac{1}{3} = \dfrac{1 \times 10}{3 \times 10} = \dfrac{10}{30}$$

b. Change $\dfrac{3}{8}$ to 32nds.

$$\dfrac{3}{8} = \dfrac{?}{32}$$

$$32 \div 8 = 4$$

$$\dfrac{3}{8} = \dfrac{3 \times 4}{8 \times 4} = \dfrac{12}{32}$$

c. Change $\dfrac{1}{3}$ to 27ths.

$$\dfrac{1}{3} = \dfrac{?}{27}$$

$$27 \div 3 = 9$$

$$\dfrac{1}{3} = \dfrac{1 \times 9}{3 \times 9} = \dfrac{9}{27}$$

PART B   Reducing a Fraction to Lowest Terms

Reducing a fraction to lower terms is the opposite procedure of changing a fraction to higher terms. To reduce a fraction to lowest terms determine the largest number that will evenly divide into both the numerator and the denominator. This number is called the **greatest common divisor**, or G.C.D. Then divide the numerator and the denominator by the G.C.D.

A fraction is in its **lowest terms** when no whole number except 1 will divide evenly into both the numerator and the denominator.

EXAMPLE:

Reduce to lowest terms by using the G.C.D. The G.C.D. of 12 and 18 is 6.

$$\dfrac{12}{18} = \dfrac{12 \div 6}{18 \div 6} = \dfrac{2}{3}$$

Note that using the G.C.D. reduces a fraction to its lowest terms in one step. If you do not use the G.C.D., you will eventually get to the lowest terms by using any common divisor and repeating the process until there is no further common divisor.

EXAMPLE:

Reduce without using the G.C.D.

$$\dfrac{12}{18} = \dfrac{12 \div 2}{18 \div 2} = \dfrac{6}{9} = \dfrac{6 \div 3}{9 \div 3} = \dfrac{2}{3}$$

A shortcut used in reducing fractions is to mentally divide both the numerator and denominator by a common divisor. Cross out the former numerator and denominator, and write the new values. This is called *cancellation*.

EXAMPLES:

Reduce to lowest terms by using common divisors.

a. $\dfrac{12}{18} = \dfrac{\cancel{\cancel{12}}^{\,2}_{\,6}}{\cancel{\cancel{18}}^{\,9}_{\,3}} = \dfrac{2}{3}$

The common divisors used, in order, are 2 and 3.

Chapter 2   Working with Fractions

b. $\dfrac{36}{90} = \dfrac{\cancel{\cancel{\cancel{\cancel{36}}}}}{\cancel{\cancel{\cancel{\cancel{90}}}}} = \dfrac{2}{5}$

(successive reductions: 36→18→6→2, 90→45→15→5)

The common divisors used, in order, are 2, 3, and 3. This reduction could be shortened. After the first division by 2, the fraction is reduced to $\frac{18}{45}$. Using 9 as the next divisor gives the same result, $\frac{2}{5}$.

$\dfrac{36}{90} = \dfrac{\cancel{\cancel{36}}}{\cancel{\cancel{90}}} = \dfrac{2}{5}$

(18/45 then 2/5)

Even simpler, if the G.C.D. of 36 and 90 (18) is used, only one reduction is necessary.

$\dfrac{36}{90} = \dfrac{\cancel{36}}{\cancel{90}} = \dfrac{2}{5}$

c. $\dfrac{105}{420} = \dfrac{\cancel{\cancel{\cancel{105}}}}{\cancel{\cancel{\cancel{420}}}} = \dfrac{1}{4}$

(successive reductions: 105→35→7→1, 420→140→28→4)

The common divisors used, in order, are 3, 5, and 7. Note that the final numerator is a 1. It is very important to write the 1. If it is not written, the answer might be carelessly read as 4 instead of the correct fraction, $\frac{1}{4}$.

d. $\dfrac{800}{1{,}200}$

The two zeros in each term show that both the numerator and the denominator contain the factor 100. Divide by 100 by crossing out the two zeros in both the numerator and the denominator.

With a G.C.D. of 4, the fraction can then be reduced to $\frac{2}{3}$.

If you cannot easily find the G.C.D., use the following method to find the common divisors when reducing fractions. This method uses prime numbers. A **prime number** is a number greater than 1 whose only factors are 1 and itself. Examples of prime numbers are 2, 3, 5, 7, 11, 13, 17, 19, 23, and 29.

It is easy to see if numbers can be divided by the first three prime numbers, 2, 3, and 5. Just follow three very simple and helpful *divisibility rules:*

1. If a number is even, it is at least divisible by the prime number 2.

EXAMPLE:

Reduce $\frac{18}{28}$ to lowest terms.

$\dfrac{18}{28} = \dfrac{\cancel{18}}{\cancel{28}} = \dfrac{9}{14}$

2. If the sum of the digits in a number can be divided by 3, then the number is divisible by 3.

EXAMPLE:

Reduce $\frac{15}{54}$ to lowest terms. The sum of the digits in 15 is $1 + 5 = 6$. The sum of the digits in 54 is $5 + 4 = 9$. Both sums can be divided by 3. Therefore, 15 and 54 are both divisible by 3.

$\dfrac{15}{54} = \dfrac{\cancel{15}}{\cancel{54}} = \dfrac{5}{18}$

3. If the last digit of a number is a 0 or a 5, the number can be divided by 5.

EXAMPLE:

Reduce $\frac{60}{125}$ to the lowest terms.

$\dfrac{60}{125} = \dfrac{\cancel{60}}{\cancel{125}} = \dfrac{12}{25}$

EXAMPLES:

Reduce to lowest terms. Try each of the prime numbers as a common divisor, beginning with 2.

a. $\dfrac{33}{110}$

Step 1. 33 is odd, so 2 cannot be divided into the numerator.
Step 2. 33 is divisible by 3, but 110 is not (1 + 1 + 0 = 2, which is not divisible by 3).
Step 3. 33 does not end in either 0 or 5, so 5 cannot be a common divisor.
Step 4. The next prime number is 7, but 7 does not divide into 33. The next prime number is 11. 33 ÷ 11 = 3, and 110 ÷ 11 = 10. Thus, 11 is a common divisor.

$$\frac{33}{110} = \frac{\overset{3}{\cancel{33}}}{\underset{10}{\cancel{110}}} = \frac{3}{10}$$

b. $\dfrac{57}{133}$

By trying to divide 57 and 133 by each of the prime numbers, you can see that 2, 3, and 5 are not common divisors. Further, the prime numbers 7, 11, 13, and 17 are not common divisors. However, 57 = 3 x 19, and 133 = 7 x 19. Therefore, 19 is a common divisor.

$$\frac{57}{133} = \frac{\overset{3}{\cancel{57}}}{\underset{7}{\cancel{133}}} = \frac{3}{7}$$

Sometimes, a fraction looks as if it cannot be reduced any further. In the examples that follow, factoring the numerator and denominator separately shows a common divisor.

EXAMPLES:

a. $\dfrac{69}{184}$

Prime factors: 69 = 3 × 23
184 = $\underbrace{2 \times 2 \times 2}_{8} \times 23$

$$\frac{69}{184} = \frac{\overset{3}{\cancel{69}}}{\underset{8}{\cancel{184}}} = \frac{3}{8}$$

The common divisor of 69 and 184 is 23.

b. $\dfrac{26}{133}$

Prime factors: 26 = 2 × 13
133 = 7 × 19

There are no common factors in this example. Therefore, the fraction is in its lowest terms.

PART C Changing Mixed Numbers into Improper Fractions and Improper Fractions into Mixed Numbers

To change a mixed number to an improper fraction, multiply the whole number by the denominator of the fraction, add the numerator to this product, and then write that sum over the denominator to form the improper fraction.

EXAMPLE:

Change $7\tfrac{3}{8}$ to an improper fraction.

$7 \times 8 = 56$

$56 + 3 = 59$

$7\tfrac{3}{8} = \dfrac{59}{8}$

To change an improper fraction to a mixed number, divide the numerator by the denominator. The whole-number part of the quotient represents the whole-number part of the mixed number. Write the remainder from the division over the denominator to form the fractional part of the mixed

Chapter 2   Working with Fractions

number. Then reduce the fractional part to lowest terms. If there is no remainder, the answer is a whole number.

a. $\dfrac{250}{8} = 8\overline{)250}\begin{array}{r}31\text{ R}2\\\\24\\\hline 10\\8\\\hline 2\end{array} = 31\dfrac{2}{8} = 31\dfrac{1}{4}$

EXAMPLES:

Change to a mixed number or a whole number.

b. $\dfrac{135}{15} = 135 \div 15 = 9$

COMPLETE ASSIGNMENT 6

Name _____

Date _____

ASSIGNMENT 6
SECTION 2.1

Types of Fractions

	Perfect Score	Student's Score
PART A	40	
PART B	40	
PART C	20	
TOTAL	100	

PART A *Changing a Fraction to Higher Terms*

DIRECTIONS: Change to equivalent fractions with the indicated denominators (2 points for each correct answer).

1. $\frac{1}{3} = \frac{}{9}$
2. $\frac{2}{5} = \frac{}{10}$
3. $\frac{2}{3} = \frac{}{27}$
4. $\frac{3}{4} = \frac{}{36}$

5. $\frac{4}{7} = \frac{}{140}$
6. $\frac{5}{6} = \frac{}{96}$
7. $\frac{2}{7} = \frac{}{49}$
8. $\frac{5}{8} = \frac{}{112}$

9. $\frac{7}{9} = \frac{}{54}$
10. $\frac{9}{10} = \frac{}{150}$
11. $\frac{9}{11} = \frac{}{110}$
12. $\frac{11}{12} = \frac{}{156}$

13. $\frac{12}{13} = \frac{}{52}$
14. $\frac{9}{14} = \frac{}{84}$
15. $\frac{13}{15} = \frac{}{165}$
16. $\frac{15}{24} = \frac{}{72}$

17. $\frac{13}{20} = \frac{}{180}$
18. $\frac{21}{25} = \frac{}{150}$
19. $\frac{16}{27} = \frac{}{81}$
20. $\frac{23}{36} = \frac{}{288}$

STUDENT'S SCORE _____

PART B *Reducing a Fraction to Lowest Terms*

DIRECTIONS: Reduce the fractions to lowest terms. Write the G.C.D. of the numerator and denominator (2 points for each correct answer).

21. $\frac{6}{15} =$ G.C.D. = _____
22. $\frac{6}{10} =$ G.C.D. = _____
23. $\frac{8}{16} =$ G.C.D. = _____
24. $\frac{15}{28} =$ G.C.D. = _____
25. $\frac{16}{28} =$ G.C.D. = _____
26. $\frac{44}{55} =$ G.C.D. = _____

27. $\frac{10}{35} =$ G.C.D. = _____
28. $\frac{25}{30} =$ G.C.D. = _____
29. $\frac{21}{49} =$ G.C.D. = _____
30. $\frac{14}{91} =$ G.C.D. = _____
31. $\frac{85}{170} =$ G.C.D. = _____
32. $\frac{207}{299} =$ G.C.D. = _____

Unit 1   Fundamentals

33. $\frac{65}{115} =$ G.C.D. = _____   37. $\frac{112}{500} =$ G.C.D. = _____

34. $\frac{203}{259} =$ G.C.D. = _____   38. $\frac{148}{1,000} =$ G.C.D. = _____

35. $\frac{98}{161} =$ G.C.D. = _____   39. $\frac{64}{2,000} =$ G.C.D. = _____

36. $\frac{56}{112} =$ G.C.D. = _____   40. $\frac{75}{1,000} =$ G.C.D. = _____

STUDENT'S SCORE _____

PART C Changing Mixed Numbers into Improper Fractions and Improper Fractions into Mixed Numbers

DIRECTIONS: Change the mixed numbers to improper fractions (1 point for each correct answer).

41. $4\frac{2}{3} =$ _____   46. $10\frac{1}{6} =$ _____

42. $4\frac{3}{7} =$ _____   47. $3\frac{5}{8} =$ _____

43. $20\frac{2}{5} =$ _____   48. $7\frac{9}{10} =$ _____

44. $8\frac{5}{6} =$ _____   49. $10\frac{13}{16} =$ _____

45. $16\frac{3}{7} =$ _____   50. $4\frac{8}{25} =$ _____

DIRECTIONS: Change the improper fractions to mixed or whole numbers. Reduce your answers to lowest terms (1 point for each correct answer).

51. $\frac{19}{2} =$ _____   56. $\frac{250}{50} =$ _____

52. $\frac{43}{7} =$ _____   57. $\frac{1,000}{36} =$ _____

53. $\frac{96}{12} =$ _____   58. $\frac{2,050}{100} =$ _____

54. $\frac{120}{25} =$ _____   59. $\frac{3,900}{225} =$ _____

55. $\frac{210}{114} =$ _____   60. $\frac{3,036}{66} =$ _____

STUDENT'S SCORE _____

Chapter 2   Working with Fractions

SECTION 2.2 Addition of Fractions

Fractions that have the same denominator—a **common denominator**—are called **like terms**. Fractions that have different denominators are called **unlike terms**. Before you can add fractions with different denominators, you must change them to fractions having a common denominator.

Most of the fractions used in business are halves, thirds, fourths, fifths, sixths, eighths, tenths, twelfths, and sixteenths. When addition involves some of these, it is easy to change them to like terms.

PART A Adding a Fraction or a Mixed Number to a Whole Number

The sum of a whole number and a proper fraction is easy to find. Simply write the fraction after the whole number. The sum is a mixed number.

EXAMPLES:

a. $3 + \frac{2}{3} = 3\frac{2}{3}$

b. $\frac{1}{4} + 7 = 7\frac{1}{4}$

These sums are read as *three and two-thirds* and *seven and one-fourth,* respectively.

Finding the sum of a whole number and a mixed number is also easy. Add the two whole numbers, and then write the fraction after the sum. The answer is a mixed number.

EXAMPLES:

a. $14 + 7\frac{3}{10} = 21\frac{3}{10}$

b. $20\frac{1}{6} + 37 = 57\frac{1}{6}$

PART B Adding Fractions or Mixed Numbers with Common Denominators

To add fractions that have the same denominator (like terms), add the numerators and write the sum of the numerators over the common denominator. When the result is an improper fraction, change it to a whole or a mixed number. Reduce the fractional part of the mixed number to lowest terms.

EXAMPLES:

a. $\frac{2}{5} + \frac{3}{5} = \frac{2+3}{5} = \frac{5}{5} = 1$

b. $\frac{1}{8} + \frac{5}{8} = \frac{1+5}{8} = \frac{6}{8} = \frac{3}{4}$

c. $\frac{11}{12} + \frac{7}{12} = \frac{11+7}{12} = \frac{18}{12} = 1\frac{6}{12} = 1\frac{1}{2}$

To add mixed numbers, add the whole numbers together, and add the fractions together. Then, add these sums together.

EXAMPLES:

a. $4\frac{11}{12} + 5\frac{5}{12} + 11\frac{1}{12}$

$4 + 5 + 11 = 20$

$\frac{11}{12} + \frac{5}{12} + \frac{1}{12} = \frac{11+5+1}{12} = \frac{17}{12} = 1\frac{5}{12}$

$20 + 1\frac{5}{12} = 21\frac{5}{12}$

b. $8\frac{4}{15} + 6\frac{1}{15} + \frac{13}{15}$

$$\begin{array}{r|l} 8 & \frac{4}{15} \\ 6 & \frac{1}{15} \\ & \frac{13}{15} \\ \hline 14 & \frac{18}{15} = 1\frac{3}{15} = 1\frac{1}{5} \end{array}$$

$14 + 1\frac{1}{5} = 15\frac{1}{5}$

To find the sum of three or more mixed numbers, use either of the procedures shown in the preceding examples.

PART C Finding the Least Common Denominator

As stated earlier in this chapter, fractions with different denominators must be changed to fractions having a common denominator before they can be added. The common denominator is a number that is divisible by each of the different denominators. For example, 15 is divisible by 3 and 5. Therefore, 15 is the common denominator of thirds and fifths. When the common denominator is the smallest number possible, it is called the **least common denominator (L.C.D.)**. Therefore, 15 is the L.C.D. of the denominators 3 and 5.

When fractions have small denominators such as 2 and 3, the least common denominator can be easily recognized. The L.C.D. is said to be found *by inspection*.

EXAMPLES:

Find the L.C.D. by inspection.

a. $\dfrac{1}{3} + \dfrac{1}{2}$

By inspecting the denominators, you can see that 6 is the smallest number that is divisible by 2 and 3. Therefore, 6 is the L.C.D.

b. $\dfrac{2}{3} + \dfrac{3}{5} + \dfrac{1}{6}$

By inspecting the denominators, you can see that 30 is the smallest number divisible by 3, 5, and 6. Therefore, 30 is the L.C.D.

c. $\dfrac{1}{3} + \dfrac{5}{8} + \dfrac{7}{12}$

The smallest number divisible by 3, 8, and 12 is 24. Therefore, 24 is the L.C.D.

When inspection does not provide the L.C.D., you can find a common denominator by multiplying all the denominators. The resulting product is often not the L.C.D., but it can be used and is easy to find.

EXAMPLE:

Find a common denominator by multiplication.

$$\dfrac{2}{5} + \dfrac{7}{8} + \dfrac{5}{6}$$

$$5 \times 8 \times 6 = 240$$

The common denominator 240 can be used to add the fractions. However, it is not the smallest number divisible by 5, 8, and 6. The smallest number, or L.C.D., is 120.

To find the L.C.D., you can use the **repeated division method.** It may require a few more steps than some methods, but it is easy to use. The repeated division method is explained in detail in the following example.

EXAMPLES:

Find the L.C.D. by repeated division.

a. $\dfrac{1}{5} + \dfrac{1}{8} + \dfrac{5}{6}$

Step 1. Write the denominators in a row.

$$5 \quad 8 \quad 6$$

Step 2. If there is at least one denominator (such as 8) divisible by 2, divide 2 into it and into any other even denominators (such as 6). Write the quotients, 4 and 3, below the denominators. Bring down any denominator (such as 5) that is not divisible by 2.

$$2 \quad \underline{\lvert 5 \ 8 \ 6}$$
$$ 5 \ 4 \ 3$$

Step 3. Keep dividing by 2, as in Step 2, until there are only odd numbers left.

$$2 \quad \underline{\lvert 5 \ 4 \ 3}$$
$$2 \quad \underline{\lvert 5 \ 2 \ 3}$$
$$ 5 \ 1 \ 3$$

Step 4. If there is at least one number that is divisible by 3, divide 3 into it until you can no longer use 3 as a divisor.

$$3 \quad \underline{\lvert 5 \ 1 \ 3}$$
$$ 5 \ 1 \ 1$$

Step 5. Repeat this division with the prime numbers in turn until all the numbers brought down are 1. In this example, the prime number 5 is the last number you will use.

$$5 \quad \underline{\lvert 5 \ 1 \ 1}$$
$$ 1 \ 1 \ 1$$

Chapter 2   Working with Fractions

Step 6. Multiply all the prime-number divisors at the left. The product is the L.C.D.

$$\begin{array}{r|lll}2 & 5\ 8\ 6 \\ 2 & 5\ 4\ 3 \\ 2 & 5\ 2\ 3 \\ 3 & 5\ 1\ 3 \\ 5 & 5\ 1\ 1 \\ & 1\ 1\ 1\end{array}$$

L.C.D. $= 2 \times 2 \times 2 \times 3 \times 5 = 120$

b. $\dfrac{3}{4} + \dfrac{5}{12} + \dfrac{3}{5} + \dfrac{9}{20} + \dfrac{7}{15}$

$$\begin{array}{r|lllll}2 & 4\ 12\ 5\ 20\ 15 \\ 2 & 2\ \ 6\ 5\ 10\ 15 \\ 3 & 1\ \ 3\ 5\ \ 5\ 15 \\ 5 & 1\ \ 1\ 5\ \ 5\ \ 5 \\ & 1\ \ 1\ 1\ \ 1\ \ 1\end{array}$$

L.C.D. $= 2 \times 2 \times 3 \times 5 = 60$

PART D Adding Fractions with Unlike Denominators

After finding the L.C.D., change each fraction to an equivalent fraction with the L.C.D. as the new denominator. This makes all the fractions like terms. They can then be added as explained in Part B of this chapter.

EXAMPLES:

a. $\dfrac{1}{3} + \dfrac{1}{2}$

By inspection, the least common denominator is 6.

$$\dfrac{1}{3} + \dfrac{1}{2} = \dfrac{2}{6} + \dfrac{3}{6} = \dfrac{2+3}{6} = \dfrac{5}{6}$$

b. $\dfrac{2}{5} + \dfrac{7}{8} + \dfrac{5}{12}$

Use multiplication to find a common denominator. $5 \times 8 \times 12 = 480$

$$\dfrac{2}{5} = \dfrac{192}{480}$$
$$\dfrac{7}{8} = \dfrac{420}{480}$$
$$\dfrac{5}{12} = \dfrac{200}{480}$$
$$\dfrac{812}{480} = 1\dfrac{332}{480} = 1\dfrac{83}{120}$$

c. $\dfrac{2}{5} + \dfrac{7}{8} + \dfrac{5}{12}$

Use repeated division to find the L.C.D.

$$\begin{array}{r|lll}2 & 5\ 8\ 12 \\ 2 & 5\ 4\ 6 \\ 2 & 5\ 2\ 3 \\ 3 & 5\ 1\ 3 \\ 5 & 5\ 1\ 1 \\ & 1\ 1\ 1\end{array}$$

L.C.D. $= 2 \times 2 \times 2 \times 3 \times 5 = 120$

$$\dfrac{2}{5} = \dfrac{48}{120}$$
$$\dfrac{7}{8} = \dfrac{105}{120}$$
$$\dfrac{5}{12} = \dfrac{50}{120}$$
$$\dfrac{203}{120} = 1\dfrac{83}{120}$$

Using the L.C.D. as the common denominator saves work for two reasons. First, the smaller numbers make the equivalent fractions easier to work with. Second, the final sum will usually be in lowest terms.

PART E Adding Mixed Numbers with Unlike Denominators

To add two or more mixed numbers whose fractional parts have different denominators, first add the whole numbers together. Then, add the fractions as shown in Part D, and add this sum to the sum of the whole numbers.

EXAMPLES:

a. $5\frac{1}{3} + 2\frac{11}{12} + 8\frac{1}{6}$

$5 + 2 + 8 = 15$

$\frac{1}{3} + \frac{11}{12} + \frac{1}{6} = \frac{4}{12} + \frac{11}{12} + \frac{2}{12}$

$\frac{4 + 11 + 2}{12} = \frac{17}{12} = 1\frac{5}{12}$

$15 + 1\frac{5}{12} = 16\frac{5}{12}$

b. $7\frac{1}{4} + 2\frac{2}{5} + 9\frac{7}{8}$

7	$\frac{1}{4} = \frac{10}{40}$
2	$\frac{2}{5} = \frac{16}{40}$
9	$\frac{7}{8} = \frac{35}{40}$
18	$\frac{61}{40} = 1\frac{21}{40}$

$18 + 1\frac{21}{40} = 19\frac{21}{40}$

PART F Applications

Although much of the work in business and industry is done in decimals because of computers and calculators, fractions are still widely used in many different applications.

For example, time is often expressed in fractions, such as three-quarters or three-fourths of an hour. Two hours and forty minutes can be expressed as the mixed number $2\frac{40}{60}$ or $2\frac{2}{3}$ hours.

EXAMPLE:

Ann can walk to the train station in 35 minutes. The commuter train takes 1 hour and 20 minutes to get into the city. It takes Ann 20 minutes to walk from the train station in the city to her job. How long does it take Ann to get to work? Express your answer as a mixed number.

Step 1. Express each time period as a fraction.

$$\frac{35}{60}, 1\frac{20}{60}, \frac{20}{60}$$

Step 2. Add all fractions.

$$\frac{35}{60} + 1\frac{20}{60} + \frac{20}{60} = 1\frac{75}{60}$$

$$= 2\frac{15}{60}$$

Step 3. Reduce to lowest terms.

$$2\frac{15}{60} = 2\frac{1}{4} \text{ hours}$$

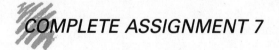

COMPLETE ASSIGNMENT 7

Name _____

Date _____

ASSIGNMENT 7
SECTION 2.2

Addition of Fractions

	Perfect Score	Student's Score
PART A	10	
PART B	10	
PART C	20	
PART D	25	
PART E	25	
PART F	10	
TOTAL	100	

PART A Adding a Fraction or a Mixed Number to a Whole Number

DIRECTIONS: Add. Reduce your answers to lowest terms (2 points for each correct answer).

1. $5 + \frac{2}{3} =$ _____

2. $2\frac{1}{2} + 7 =$ _____

3. $2\frac{1}{20} + 13 =$ _____

4. $18 + 4\frac{1}{5} =$ _____

5. $7\frac{3}{8} + 9 =$ _____

STUDENT'S SCORE _____

PART B Adding Fractions or Mixed Numbers with Common Denominators

DIRECTIONS: Add. Reduce your answers to lowest terms (2 points for each correct answer).

6. $\frac{5}{12} + \frac{11}{12} =$ _____

7. $\frac{7}{15} + \frac{11}{15} =$ _____

8. $\frac{5}{18} + \frac{3}{18} =$ _____

9. $5\frac{3}{4} + \frac{1}{4} + 3\frac{1}{4} + 9\frac{3}{4} =$ _____

10. $7\frac{2}{9} + 1\frac{8}{9} + \frac{5}{9} + 6\frac{7}{9} =$ _____

STUDENT'S SCORE _____

PART C Finding the Least Common Denominator

DIRECTIONS: Find the least common denominator (4 points for each correct answer).

11. $\frac{1}{6}, \frac{7}{10}, \frac{4}{15}$ L.C.D. = _____

12. $\frac{2}{5}, \frac{3}{4}, \frac{5}{12}$ L.C.D. = _____

13. $\frac{5}{12}, \frac{11}{28}, \frac{13}{42}$ L.C.D. = _____

14. $\frac{1}{5}, \frac{17}{24}, \frac{5}{6}, \frac{7}{8}$ L.C.D. = _____

15. $\frac{7}{36}, \frac{13}{54}, \frac{17}{63}, \frac{19}{84}$ L.C.D. = _____

STUDENT'S SCORE _____

PART D Adding Fractions with Unlike Denominators

DIRECTIONS: Add. Reduce your answers to lowest terms (5 points for each correct answer).

16. $\frac{1}{6} + \frac{5}{12} =$

17. $\frac{1}{5} + \frac{7}{15} =$

18. $\frac{2}{3} + \frac{3}{4} + \frac{5}{6} =$

19. $\frac{3}{16}$
     $\frac{1}{3}$
     $\frac{3}{8}$
     $+ \frac{7}{12}$
     _____

20. $\frac{1}{5}$
     $\frac{17}{24}$
     $\frac{5}{6}$
     $+ \frac{7}{8}$
     _____

STUDENT'S SCORE _____

Chapter 2 Working with Fractions

PART E Adding Mixed Numbers with Unlike Denominators

DIRECTIONS: Add. Reduce your answers to lowest terms (5 points for each correct answer).

21. $7\frac{5}{8}$
   $+ 5\frac{7}{12}$
   _____

22. $2\frac{7}{10}$
   $+ 7\frac{7}{20}$
   _____

23. $12\frac{1}{6}$
   $39\frac{2}{5}$
   $+ 27\frac{5}{12}$
   _____

24. $26\frac{2}{9}$
   $43\frac{1}{2}$
   $+ 29\frac{7}{15}$
   _____

25. $39\frac{1}{3}$
   $42\frac{5}{6}$
   $+ 10\frac{4}{15}$
   _____

STUDENT'S SCORE _____

PART F Applications

DIRECTIONS: Find the total time and give your answer in hours (5 points for each correct answer).

26. 1 hr., 23 min. + 3 hrs., 42 min.   _____

27. 3 hrs., 54 min.
   2 hrs., 48 min.
   + 5 hrs., 33 min.   _____

STUDENT'S SCORE _____

Chapter 2 Working with Fractions

SECTION 2.3 Subtraction of Fractions

The same general rules for adding fractions also apply to subtracting fractions. Before one fraction can be subtracted from another, both must have the same denominator.

Remember that subtraction is also indicated by such phrases as *find the difference between, deduct from, how much larger than, take from*, etc.

Recall from Chapter 1 that the subtrahend is the number being subtracted, and the minuend is the number from which it is subtracted.

PART A Subtracting Fractions Without Borrowing

To find the difference between two fractions with the same denominator, subtract the smaller numerator from the larger numerator. Write the difference over the common denominator. Reduce the answer to lowest terms.

EXAMPLES:

a. $\dfrac{11}{18} - \dfrac{7}{18}$

$$\dfrac{11}{18} - \dfrac{7}{18} = \dfrac{11-7}{18} = \dfrac{4}{18} = \dfrac{2}{9}$$

b. Find the difference between $\dfrac{2}{9}$ and $\dfrac{7}{9}$.

$$\dfrac{7}{9} - \dfrac{2}{9} = \dfrac{7-2}{9} = \dfrac{5}{9}$$

To find the difference between two mixed numbers whose fractional parts have the same denominator, find the difference between the whole numbers and then find the difference between the fractions. Reduce the answer to lowest terms.

EXAMPLE:

$14\dfrac{5}{6} - 2\dfrac{1}{6}$

$$\begin{array}{r} 14\dfrac{5}{6} \\ -\ 2\dfrac{1}{6} \\ \hline 12\dfrac{4}{6} = 12\dfrac{2}{3} \end{array}$$

To subtract a whole number from a mixed number, first find the difference between the whole numbers, and then write the fraction after that difference.

EXAMPLE:

Find the difference between 14 and $17\dfrac{1}{3}$.

$$\begin{array}{r} 17\dfrac{1}{3} \\ -\ 14 \\ \hline 3\dfrac{1}{3} \end{array}$$

To subtract fractions with different denominators, first change them to like terms. Then find the difference between the numerators and place this over the common denominator. Reduce the answer to lowest terms.

EXAMPLES:

a. How much smaller than $\dfrac{5}{6}$ is $\dfrac{1}{2}$?

$$\begin{array}{r} \dfrac{5}{6} = \dfrac{5}{6} \\ -\ \dfrac{1}{2} = \dfrac{3}{6} \\ \hline \dfrac{2}{6} = \dfrac{1}{3} \end{array}$$

b. Deduct $\dfrac{3}{20}$ from $\dfrac{2}{3}$.

$$\begin{array}{r} \dfrac{2}{3} = \dfrac{40}{60} \\ -\ \dfrac{3}{20} = \dfrac{9}{60} \\ \hline \dfrac{31}{60} \end{array}$$

To find the difference between two mixed numbers (or between a mixed number and a fraction) when the fractions have different denominators, first find the difference between the whole numbers. Then change the fractions to like terms and find the difference between the fractions. Write the fractional difference after the whole-number difference. Reduce the answer to lowest terms.

EXAMPLES:

a. $7\frac{3}{4} - 2\frac{1}{3}$

$$\begin{array}{r|l} 7 & \frac{3}{4} = \frac{9}{12} \\ -\;2 & \frac{1}{3} = \frac{4}{12} \\ \hline 5 & \frac{5}{12} = 5\frac{5}{12} \end{array}$$

b. Take $\frac{2}{15}$ from $4\frac{3}{8}$.

$$\begin{array}{r|l} 4 & \frac{3}{8} = \frac{45}{120} \\ - & \frac{2}{15} = \frac{16}{120} \\ \hline 4 & \frac{29}{120} = 4\frac{29}{120} \end{array}$$

PART B Subtracting Fractions by Borrowing

There is no difficulty subtracting $\frac{9}{5}$ from $\frac{22}{5}$ because the denominators are the same. The answer is $\frac{13}{5}$. However, if these improper fractions are written as mixed numbers, the problem becomes more complicated:

$$4\frac{2}{5} - 1\frac{4}{5}$$

The general rule in subtracting mixed numbers whose fractional parts have common denominators is to first subtract the whole-number parts and then to subtract the fractional parts. In this example, the fractional part of the minuend, $\frac{2}{5}$, has a smaller numerator than the fractional part of the subtrahend, $\frac{4}{5}$. To subtract, borrow 1 from the whole number 4 and change the 1 to $\frac{5}{5}$ ($\frac{5}{5} = 1$). Add $\frac{5}{5}$ to $\frac{2}{5}$ (of the minuend) to obtain $\frac{7}{5}$. Now subtract $\frac{4}{5}$ from $\frac{7}{5}$ to get $\frac{3}{5}$ as the fractional part of the answer. Remember that borrowing 1 changes the whole-number part of the minuend from 4 to 3. Subtract the whole-number parts.

$$\begin{array}{r} 3 \\ 4\frac{2}{5} \\ -1\frac{4}{5} \end{array} \qquad \begin{array}{r|l} 3 & \\ \cancel{4} & \frac{2}{5} + \frac{5}{5} = \frac{7}{5} \\ -\;1 & \frac{4}{5} = \frac{4}{5} \\ \hline 2 & \frac{3}{5} = 2\frac{3}{5} \end{array}$$

This method will work with any mixed numbers. Remember that the 1 that is borrowed is changed to a fraction whose *numerator* and *denominator* are equal to the common denominator of the fractions. For example, if the common denominator is 24, the 1 that is borrowed is changed to $\frac{24}{24}$. This fraction is added to the fractional part of the minuend.

EXAMPLES:

a. $9\frac{3}{10} - 6\frac{7}{10}$

The fraction $\frac{3}{10}$ is smaller than $\frac{7}{10}$. Therefore, borrow 1 from the 9; change the 1 to $\frac{10}{10}$ and add it to $\frac{3}{10}$. Then subtract the whole numbers and the fractions.

$$\begin{array}{r|l} 8 & \\ \cancel{9} & \frac{3}{10} + \frac{10}{10} = \frac{13}{10} \\ -\;6 & \frac{7}{10} = \frac{7}{10} \\ \hline 2 & \frac{6}{10} = 2\frac{6}{10} = 2\frac{3}{5} \end{array}$$

b. $40\frac{3}{5} - 21\frac{4}{5}$

Borrow 1 or $\frac{5}{5}$ from 40, making it 39. Add $\frac{5}{5}$ to $\frac{3}{5}$. Then subtract the whole numbers and the fractions.

$$\begin{array}{r|l} 39 & \\ \cancel{40} & \frac{3}{5} + \frac{5}{5} = \frac{8}{5} \\ -\;21 & \frac{4}{5} = \frac{4}{5} \\ \hline 18 & \frac{4}{5} = 18\frac{4}{5} \end{array}$$

c. $3\frac{1}{8} - \frac{5}{8}$

$$\begin{array}{r|l} 2 & \\ \cancel{3} & \frac{1}{8} + \frac{8}{8} = \frac{9}{8} \\ - & \frac{5}{8} = \frac{5}{8} \\ \hline 2 & \frac{4}{8} = 2\frac{4}{8} = 2\frac{1}{2} \end{array}$$

Chapter 2   Working with Fractions

Instead of a mixed number, the minuend may be a whole number without any fractional part. In this case, the minuend can be thought of as a mixed number whose fraction has a numerator of 0 and a denominator that is the same as the denominator of the fraction in the subtrahend. Borrowing in this case is always necessary.

EXAMPLES:

a. Find the difference between $12\frac{2}{5}$ and 20. In the subtraction, write 20 as $20\frac{0}{5}$. Then borrow 1 (or $\frac{5}{5}$) from 20, and add $\frac{5}{5}$ to $\frac{0}{5}$.

$$\begin{array}{r|l} 19 & \\ \cancel{20} & \frac{0}{5} + \frac{5}{5} = \frac{5}{5} \\ -12 & \frac{2}{5} \quad\quad = \frac{2}{5} \\ \hline 7 & \quad\quad\quad \frac{3}{5} = 7\frac{3}{5} \end{array}$$

b. Take $9\frac{5}{8}$ from 17. Write 17 as $17\frac{0}{8}$. Then borrow 1 (or $\frac{8}{8}$) from 17. Subtract the whole numbers and the fractions.

$$\begin{array}{r|l} 16 & \\ \cancel{17} & \frac{0}{8} + \frac{8}{8} = \frac{8}{8} \\ -9 & \frac{5}{8} \quad\quad = \frac{5}{8} \\ \hline 7 & \quad\quad\quad \frac{3}{8} = 7\frac{3}{8} \end{array}$$

To find the difference between two mixed numbers (or a mixed number and a fraction) with different denominators, change the fractions to like terms. Borrow when necessary.

EXAMPLES:

a. How much larger than $4\frac{5}{8}$ is $9\frac{2}{5}$? The least common denominator is found by multiplying the denominators: $8 \times 5 = 40$. Change the fractions to like terms as follows:

$$\begin{array}{r|l} 9 & \frac{2}{5} = \frac{16}{40} \\ -4 & \frac{5}{8} = \frac{25}{40} \\ \hline \end{array}$$

Borrow 1 (or $\frac{40}{40}$) from 9 and add it to $\frac{16}{40}$.

$$\begin{array}{r|l} 8 & \\ \cancel{9} & \frac{16}{40} + \frac{40}{40} = \frac{56}{40} \\ -4 & \frac{25}{40} \quad\quad\quad = \frac{25}{40} \\ \hline 4 & \quad\quad\quad\quad\quad \frac{31}{40} = 4\frac{31}{40} \end{array}$$

b. Deduct $\frac{5}{12}$ from $4\frac{3}{10}$. Change the fractions to like terms with the least common denominator of 60:

$$\begin{array}{r|l} 4 & \frac{3}{10} = \frac{18}{60} \\ - & \frac{5}{12} = \frac{25}{60} \\ \hline \end{array}$$

Borrow 1 (or $\frac{60}{60}$) from 4 and add it to $\frac{18}{60}$.

$$\begin{array}{r|l} 3 & \\ \cancel{4} & \frac{18}{60} + \frac{60}{60} = \frac{78}{60} \\ - & \frac{25}{60} \quad\quad\quad = \frac{25}{60} \\ \hline 3 & \quad\quad\quad\quad\quad = \frac{53}{60} = 3\frac{53}{60} \end{array}$$

PART C   Applications

A **stock record** is a card or sheet on which a record is kept of all the goods that are put into or taken out of a stockroom or warehouse. A separate stock record is kept for each item handled by the business. The quantity that comes in is recorded in a column called *Received;* the quantity that goes out is recorded in a column called *Issued.*

The stock record is kept up-to-date by the stock clerk. Any quantity that is issued is subtracted from the balance on hand; any quantity that is received is added to the balance. The information on the stock record is used for ordering purposes.

To keep from running out of stock on an item, most companies try to keep a minimum amount of stock on hand. When the balance falls below a desired amount, an order for more stock is placed. All orders are numbered for checking when the stock is received.

A request for stock to be taken out or issued is called a **requisition**. All requisitions are numbered for reference.

Today, a computer is often used to keep a running record of the stock on hand. The printout may look different, but the information on it is very similar to that on a stock record.

In the following example, stock-record amounts are recorded in gross. One gross equals 12 dozen. For example, if the requisition is for 185 dozen, the 185 dozen are changed to $15\frac{5}{12}$ gross (185 ÷ 12) on the stock record.

EXAMPLE:

Complete the stock record shown in Figure 2-1, based on the following facts:

1. May 8-Balance on hand is 26 gross.
2. May 12-Received $45\frac{1}{4}$ gross on Order No. 3796.
3. May 14-Issued $15\frac{5}{12}$ gross on Requisition No. 540.
4. May 16-Issued $17\frac{1}{3}$ gross on Requisition No. 593.
5. May 17-Received $48\frac{3}{4}$ gross on Order No. 4231.
6. May 18-Issued $16\frac{5}{6}$ gross on Requisition No. 612.

		Stock No. B-64			Minimum Amount 50 gross	
	Date	Received		Issued		Balance on hand
		Order No.	Quantity	Req. No.	Quantity	
	May					
1.	8					26 gross
2.	12	3796	$45\frac{1}{4}$			$71\frac{1}{4}$
3.	14			540	$15\frac{5}{12}$	$55\frac{5}{6}$
4.	16			593	$17\frac{1}{3}$	$38\frac{1}{2}$
5.	17	4231	$48\frac{3}{4}$			$87\frac{1}{4}$
6.	18			612	$16\frac{5}{6}$	$70\frac{5}{12}$

FIGURE 2-1 Stock Record

1. Balance on hand on May 8 is 26 gross.
2. On May 12, $45\frac{1}{4}$ gross are received. Add this to the 26 gross. The new balance is $71\frac{1}{4}$ gross. Write $71\frac{1}{4}$ in the Balance column under the 26 gross.
3. and 4. Subtract the issues of May 14 and May 16 to leave a current balance of $38\frac{1}{2}$ gross. Since this is below the desired minimum amount, an order is placed. The order quantity is not recorded until it is received.
5. On May 17, the order is received. Add $48\frac{3}{4}$ gross to the balance of $38\frac{1}{2}$ gross. The new balance is $87\frac{1}{4}$ gross.
6. On May 18, subtract the issue of $16\frac{5}{6}$ gross from $87\frac{1}{4}$ gross to give a balance on hand of $70\frac{5}{12}$ gross.

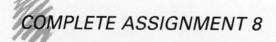

Name _____

Date _____

ASSIGNMENT 8
SECTION 2.3

Subtraction of Fractions

	Perfect Score	Student's Score
PART A	40	
PART B	40	
PART C	20	
TOTAL	100	

PART A Subtracting Fractions Without Borrowing

DIRECTIONS: Subtract. Reduce your answers to lowest terms (4 points for each correct answer).

1. $\frac{5}{6} - \frac{1}{6} =$ _____

2. $\frac{7}{10} - \frac{3}{10} =$ _____

3. $6\frac{4}{5} - 3\frac{1}{5} =$ _____

4. $9\frac{8}{9} - 3\frac{2}{9} =$ _____

5. Take 7 from $18\frac{7}{16}$. _____

6. Deduct 21 from $31\frac{3}{8}$. _____

7. Find the difference between $\frac{5}{6}$ and $\frac{11}{12}$. _____

8. How much larger than $\frac{2}{5}$ is $\frac{7}{10}$? _____

9. $6\frac{11}{12}$
   $-4\frac{3}{8}$ _____

10. $21\frac{2}{3}$
    $-16\frac{1}{5}$ _____

STUDENT'S SCORE _____

PART B Subtracting Fractions by Borrowing

DIRECTIONS: Subtract. Reduce your answers to lowest terms (4 points for each correct answer).

11. $3\frac{1}{6}$
    $-2\frac{5}{6}$ _____

16. $23\frac{5}{12}$
    $-12\frac{4}{5}$ _____

12. $15\frac{3}{8}$
    $- 8\frac{5}{8}$ _____

17. $45\frac{1}{12}$
    $-39\frac{2}{3}$ _____

13. 11
    $- 2\frac{5}{6}$ _____

18. $59\frac{7}{16}$
    $-58\frac{5}{6}$ _____

14. 27
    $- 3\frac{1}{3}$ _____

19. $93\frac{2}{15}$
    $-64\frac{7}{9}$ _____

15. $19\frac{3}{10}$
    $-16\frac{3}{4}$ _____

20. $101\frac{1}{16}$
    $- 99\frac{7}{20}$ _____

STUDENT'S SCORE _____

Chapter 2　Working with Fractions

PART C　Applications

DIRECTIONS:　Complete the "Balance on hand" column in the stock record below (4 points for each correct answer).

Stock No. M-18					Minimum Amount 50 gross	
Date	Received		Issued		Balance on hand	
	Order No.	Quantity	Req. No.	Quantity		
May 20					$95\frac{1}{3}$	
21			645	$34\frac{3}{4}$		21.
23			657	$35\frac{11}{12}$		22.
25	4258	$62\frac{1}{2}$				23.
27			669	$23\frac{1}{3}$		24.
28			680	$31\frac{5}{6}$		25.

STUDENT'S SCORE _____

Chapter 2   Working with Fractions

SECTION 2.4 Multiplication of Fractions

The multiplication of fractions and mixed numbers is faster than addition and subtraction, because there is no need to find a common denominator. With fractions, either proper or improper, there are two basic ways to multiply. You should know both methods so that you can determine which is easier for you.

1. Find the product of the numerators and place it over the product of the denominators. Reduce the answer to lowest terms.

EXAMPLE:

$$\frac{3}{5} \times \frac{5}{6} = \frac{3 \times 5}{5 \times 6} = \frac{15}{30} = \frac{1}{2}$$

2. It is usually easier to first divide any common divisors that occur in the numerators by those in the denominators. There are two advantages to working with this cancellation (reduction) method. First, smaller numbers make the work easier. Second, if all possible cancellations are performed, the answer will already be in its lowest terms.

EXAMPLE:

$$\frac{3}{5} \times \frac{5}{6} = \frac{\overset{1}{\cancel{3}}}{\underset{1}{\cancel{5}}} \times \frac{\overset{1}{\cancel{5}}}{\underset{2}{\cancel{6}}} = \frac{1}{2}$$

The common divisors are 3 and 5.

Remember, a common divisor must cancel out of both a numerator *and* a denominator. If the cancellation results in a 1 in either the numerator or the denominator, the 1 must be written down to prevent errors in the final product.

PART A Multiplying Two or More Fractions

Either of the two basic methods of multiplying fractions may be used to multiply two fractions. When the numbers are large, cancellation is easier to use.

EXAMPLE:

$$\frac{5}{18} \times \frac{14}{15} = \frac{5 \times 14}{18 \times 15} = \frac{70}{270} = \frac{7}{27}$$

or

$$\frac{5}{18} \times \frac{14}{15} = \frac{\overset{1}{\cancel{5}}}{\underset{9}{\cancel{18}}} \times \frac{\overset{7}{\cancel{14}}}{\underset{3}{\cancel{15}}} = \frac{7}{27}$$

The common divisors are 2 and 5.

To find the product of three or more fractions, cancellation is generally preferable. Only one numerator and denominator can be canceled at a time. Also, the canceled numbers need not be *adjacent* (next to each other).

EXAMPLE:

$$\frac{9}{14} \times \frac{2}{5} \times \frac{8}{9} \times \frac{35}{72}$$

Follow these steps when cancelling common factors:

Step 1. Divide 9 into the numerator 9 and the denominator 9.
Step 2. Divide 2 into the numerator 2 and the denominator 14.
Step 3. Divide 5 into the numerator 35 and the denominator 5.
Step 4. Divide 8 into the numerator 8 and the denominator 72.

$$\frac{\overset{1}{\cancel{9}}}{\underset{7}{\cancel{14}}} \times \frac{\overset{1}{\cancel{2}}}{\underset{1}{\cancel{5}}} \times \frac{\overset{1}{\cancel{8}}}{\underset{1}{\cancel{9}}} \times \frac{\overset{7}{\cancel{35}}}{\underset{9}{\cancel{72}}}$$

Step 5. Finally, divide 7 into the reduced numerator 7 and the reduced denominator 7.

$$\frac{\overset{1}{\cancel{9}}}{\underset{\underset{1}{7}}{\cancel{14}}} \times \frac{\overset{1}{\cancel{2}}}{\underset{1}{\cancel{5}}} \times \frac{\overset{1}{\cancel{8}}}{\underset{1}{\cancel{9}}} \times \frac{\overset{\overset{1}{\cancel{7}}}{\cancel{35}}}{\underset{9}{\cancel{72}}} = \frac{1}{9}$$

Note that all the numerators cancel to 1, resulting in the final product, $\frac{1}{9}$.

PART B Multiplying a Fraction by a Whole Number

Either of the two basic methods of multiplying fractions can be used to multiply a fraction by a whole number. It is important to remember to place the whole number over 1, so that it will have a numerator and a denominator.

EXAMPLES:

a. $\frac{3}{8} \times 44$ (often written as $\frac{3}{8}$ of 44)

$$\frac{3}{\cancel{8}_2} \times \frac{\cancel{44}^{11}}{1} = \frac{33}{2} = 16\frac{1}{2}$$

The common divisor is 4.

or

$$\frac{3 \times 44}{8 \times 1} = \frac{132}{8} = 16\frac{4}{8} = 16\frac{1}{2}$$

b. $3 \times \frac{1}{3}$

$$\frac{\cancel{3}^1}{1} \times \frac{1}{\cancel{3}_1} = \frac{1}{1} = 1$$

The common divisor is 3.

or

$$\frac{3}{1} \times \frac{1}{3} = \frac{3}{3} = 1$$

Note that when only 1's are left in the numerators and denominators, the answer is 1.

PART C Multiplying a Mixed Number by a Fraction, a Whole Number, or a Mixed Number

To find the product of a mixed number and a fraction, change the mixed number to an improper fraction and then multiply.

EXAMPLE:

$$3\frac{3}{10} \times \frac{5}{6}$$

$$3\frac{3}{10} = \frac{33}{10}$$

$$\frac{33}{10} \times \frac{5}{6} = \frac{\cancel{33}^{11}}{\cancel{10}_2} \times \frac{\cancel{5}^1}{\cancel{6}_2} = \frac{11}{4} = 2\frac{3}{4}$$

The common divisors are 3 and 5.

To find the product of a mixed number and a whole number, first change the mixed number to an improper fraction. Then place the whole number over 1 and multiply.

EXAMPLE:

$$3\frac{3}{4} \times 10$$

$$3\frac{3}{4} = \frac{15}{4}$$

$$\frac{15}{4} \times \frac{10}{1} = \frac{15}{\cancel{4}_2} \times \frac{\cancel{10}^5}{1} = \frac{75}{2} = 37\frac{1}{2}$$

The common divisor is 2.

To find the product of two mixed numbers, change both to improper fractions and then multiply.

EXAMPLE:

$$3\frac{4}{15} \times 2\frac{5}{14}$$

$$3\frac{4}{15} = \frac{49}{15} \text{ and } 2\frac{5}{14} = \frac{33}{14}$$

$$\frac{49}{15} \times \frac{33}{14} = \frac{\cancel{49}^7}{\cancel{15}_5} \times \frac{\cancel{33}^{11}}{\cancel{14}_2} = \frac{77}{10} = 7\frac{7}{10}$$

The common divisors are 3 and 7.

Follow the same procedure when multiplying three or more fractions or mixed numbers.

Chapter 2   Working with Fractions

EXAMPLE:

$$2\frac{2}{3} \times \frac{7}{10} \times \frac{3}{4} \times 1\frac{7}{8}$$

$$2\frac{2}{3} = \frac{8}{3} \text{ and } 1\frac{7}{8} = \frac{15}{8}$$

$$\frac{8}{3} \times \frac{7}{10} \times \frac{3}{4} \times \frac{15}{8} = \frac{\cancel{8}}{\cancel{3}} \times \frac{7}{\cancel{10}} \times \frac{\cancel{3}}{4} \times \frac{\cancel{15}}{\cancel{8}}$$

$$= \frac{21}{8} = 2\frac{5}{8}$$

The common divisors are 3, 5, and 8.

PART D Multiplying Mixed Numbers—An Alternate Method

When multiplying a mixed number by a whole number, it is not necessary to change the mixed number to an improper fraction before multiplying. To multiply a mixed number by a whole number, follow these three steps:

1. Multiply the fractional part of the mixed number by the whole number.
2. Multiply the whole-number part of the mixed number by the whole number.
3. Add the two products from Steps 1 and 2.

EXAMPLES:

a. $48 \times 2\frac{1}{2}$

Step 1. $\frac{1}{2} \times \frac{48}{1} = 24$

Step 2. $48 \times 2 = 96$

Step 3. $96 + 24 = 120$

b. $468\frac{2}{3} \times 91$

Step 1. $\frac{2}{3} \times \frac{91}{1} = \frac{182}{3} = 60\frac{2}{3}$

Step 2. $468 \times 91 = 42{,}588$

Step 3. $42{,}588 + 60\frac{2}{3} = 42{,}648\frac{2}{3}$

PART E Applications

When fractions or mixed numbers occur in a calculation, it may be simpler to work in fractions than to change the numbers to decimals (see Section 2.6, Part D, of this chapter).

EXAMPLES:

a. Find the cost of $2\frac{2}{3}$ yards of carpet @ $18.00 per yard. (The symbol @ stands for *at*.)

$$2\frac{2}{3} \times \$18 = \frac{8}{3} \times \$\frac{18}{1} = \frac{8}{\cancel{3}} \times \$\frac{\cancel{18}}{1} = \$48$$

The cost is $48.00.

b. Find the cost of $92\frac{5}{12}$ dozen @ 96¢ per dozen.

$$\frac{5}{12} \times \frac{96}{1} = \frac{5}{\cancel{12}} \times \frac{\cancel{96}}{1} = 40¢$$

$92 \times \$0.96 = \$88.32$

$\$88.32 + \$0.40 = \$88.72$

Today on the New York Stock Exchange, prices for stocks are quoted in dollars and fractional parts of a dollar. The fractional parts are $\frac{1}{8}, \frac{1}{4}, \frac{3}{8}, \frac{1}{2}, \frac{5}{8}, \frac{3}{4}$, and $\frac{7}{8}$. Thus, if a particular stock were selling for $17.50 per share, it would be quoted as $17\frac{1}{2}$.

EXAMPLES:

a. Find the cost of 20 shares of a stock priced at $12\frac{3}{8}$ per share.

$$20 \times 12\frac{3}{8} = \frac{\cancel{20}}{1} \times \frac{99}{\cancel{8}} = \frac{495}{2} = 247\frac{1}{2}$$

The cost of the 20 shares is $247\frac{1}{2}$ dollars, or $247.50.

b. What is the cost of 50 shares of a stock priced at $36\frac{5}{8}$ per share?

Step 1. $\frac{5}{\overset{8}{4}} \times \frac{\overset{25}{\cancel{50}}}{1} = \frac{125}{4} = 31\frac{1}{4}$

Step 2. $36 \times 50 = 1,800$

Step 3. $1,800 + 31\frac{1}{4} = 1,831\frac{1}{4}$

The cost of the 50 shares is $1,831\frac{1}{4}$ dollars, or $1,831.25.

COMPLETE ASSIGNMENT 9

Name _____

Date _____

ASSIGNMENT 9
SECTION 2.4

Multiplication of Fractions

	Perfect Score	Student's Score
PART A	20	
PART B	20	
PART C	20	
PART D	20	
PART E	20	
TOTAL	100	

PART A Multiplying Two or More Fractions

DIRECTIONS: Find the products. Reduce your answers to lowest terms (2 points for each correct answer).

1. $\dfrac{1}{5} \times \dfrac{5}{6} =$ _____

2. $\dfrac{1}{4} \times \dfrac{4}{5} =$ _____

3. $\dfrac{1}{10} \times \dfrac{3}{8} =$ _____

4. $\dfrac{9}{20} \times \dfrac{7}{8} =$ _____

5. $\dfrac{8}{9} \times \dfrac{9}{16} =$ _____

6. $\dfrac{4}{5} \times \dfrac{5}{16} =$ _____

7. $\dfrac{7}{9} \times \dfrac{3}{14} =$ _____

8. $\dfrac{5}{12} \times \dfrac{8}{15} =$ _____

9. $\dfrac{5}{6} \times \dfrac{6}{7} \times \dfrac{7}{10} =$ _____

10. $\dfrac{8}{9} \times \dfrac{3}{16} \times \dfrac{5}{8} \times \dfrac{2}{25} =$ _____

STUDENT'S SCORE _____

PART B Multiplying a Fraction by a Whole Number

DIRECTIONS: Find the products. Reduce your answers to lowest terms (2 points for each correct answer).

11. $\dfrac{1}{8} \times 8 =$ _____

15. $12 \times \dfrac{3}{5} =$ _____

12. $10 \times \dfrac{1}{10} =$ _____

16. $\dfrac{5}{16} \times 12 =$ _____

13. $36 \times \dfrac{2}{15} =$ _____

17. $\dfrac{3}{8} \times 30 =$ _____

14. $\dfrac{7}{30} \times 24 =$ _____

18. $\dfrac{1}{2} \times 72 =$ _____

Chapter 2   Working with Fractions   85

19. $\frac{3}{4} \times 26 =$ _____   **20.** $\frac{5}{6} \times 72 =$ _____

STUDENT'S SCORE _____

PART C   Multiplying a Mixed Number by a Fraction, a Whole Number, or a Mixed Number

DIRECTIONS: Find the products. Reduce your answers to lowest terms (2 points for each correct answer).

21. $4\frac{2}{3} \times \frac{1}{6} =$ _____   **25.** $12\frac{2}{5} \times 4\frac{1}{2} =$ _____

22. $\frac{3}{8} \times 1\frac{1}{3} =$ _____   **26.** $13\frac{1}{3} \times 3\frac{3}{5} =$ _____

23. $1\frac{1}{2} \times 6 =$ _____   **27.** $\frac{9}{10} \times 1\frac{1}{6} \times 2\frac{1}{3} =$ _____

24. $6\frac{3}{4} \times 4 =$ _____   **28.** $4\frac{1}{2} \times 2\frac{2}{15} \times 3\frac{1}{9} =$ _____

86 Unit 1 Fundamentals

29. $5\frac{5}{6} \times \frac{3}{10} \times 3\frac{1}{5} \times 3\frac{3}{4} =$ _____

30. $1\frac{4}{5} \times 2\frac{5}{8} \times \frac{16}{27} \times 3\frac{1}{3} =$ _____

STUDENT'S SCORE _____

PART D Multiplying Mixed Numbers—An Alternate Method

DIRECTIONS: Use the alternate method of multiplying mixed numbers to find the products. Reduce your answers to lowest terms (4 points for each correct answer).

31. $84 \times 5\frac{1}{2} =$ _____

34. $496 \times 20\frac{5}{8} =$ _____

32. $85\frac{2}{15} \times 30 =$ _____

35. $36\frac{3}{4} \times 25 =$ _____

33. $105 \times 6\frac{1}{3} =$ _____

STUDENT'S SCORE _____

PART E Applications

DIRECTIONS: Find the cost of each item. Show your work in the space provided (4 points for each correct answer).

36. $27\frac{1}{2}$ ft. @ 88¢ a foot = _____

37. $65\frac{1}{4}$ yd @ $2.00 a yard = _____

38. $26\frac{2}{3}$ doz @ 57¢ per doz = _____

Chapter 2   Working with Fractions

39. $10\frac{1}{3}$ shares @ $2\frac{1}{4}$ =   _____

40.   32 shares @ $3\frac{3}{8}$ =   _____

STUDENT'S SCORE _____

Chapter 2 Working with Fractions

SECTION 2.5 Division of Fractions

Division of fractions is less common than multiplication. Business applications typically will be found in such fields as construction, construction materials, and any area where formulas are mixed (paints, insecticides, foods, etc.).

Division is the inverse of multiplication. To divide by a fraction, invert the divisor (the number *after* the division sign) and multiply. To invert a fraction, turn it upside down. The inverted fraction is called the **reciprocal** of the original fraction.

EXAMPLE:

$$\frac{8}{15} \div \frac{2}{5}$$

The reciprocal of $\frac{2}{5}$ is $\frac{5}{2}$, so

$$\frac{8}{15} \div \frac{2}{5} = \frac{8}{15} \times \frac{5}{2} = \frac{\overset{4}{\cancel{8}}}{\underset{3}{\cancel{15}}} \times \frac{\overset{1}{\cancel{5}}}{\underset{1}{\cancel{2}}} = \frac{4}{3} = 1\frac{1}{3}$$

Another way of stating the rule is multiply the *dividend* (the number *before* the division sign) by the reciprocal of the divisor. Note that the dividend is the number being divided and it remains the same. Only the divisor is inverted.

PART A Dividing by a Fraction

To divide by a fraction, invert the divisor and multiply. The following examples are separated into three groups according to the form of the dividend: fraction, whole number, or mixed number.

When the Dividend is a Fraction.

EXAMPLES:

a. $\dfrac{3}{10} \div \dfrac{3}{5}$

$$\frac{3}{10} \div \frac{3}{5} = \frac{\overset{1}{\cancel{3}}}{\underset{2}{\cancel{10}}} \times \frac{\overset{1}{\cancel{5}}}{\underset{1}{\cancel{3}}} = \frac{1}{2}$$

b. $\dfrac{1}{6} \div \dfrac{1}{6}$

$$\frac{1}{6} \div \frac{1}{6} = \frac{1}{\underset{1}{\cancel{6}}} \times \frac{\overset{1}{\cancel{6}}}{1} = \frac{1}{1} = 1$$

Note from the previous examples that the quotient of two fractions might be less than 1, greater than 1, or equal to 1. Just as with whole numbers, if a fraction is divided by itself, the quotient is always equal to 1.

When the Dividend is a Whole Number. Before dividing, write the whole number over 1 to make it a fraction.

EXAMPLES:

a. $6 \div \dfrac{3}{4}$

$$6 \div \frac{3}{4} = \frac{6}{1} \div \frac{3}{4} = \frac{\overset{2}{\cancel{6}}}{1} \times \frac{4}{\underset{1}{\cancel{3}}} = \frac{8}{1} = 8$$

b. $18 \div \dfrac{2}{3}$

$$18 \div \frac{2}{3} = \frac{18}{1} \div \frac{2}{3} = \frac{\overset{9}{\cancel{18}}}{1} \times \frac{3}{\underset{1}{\cancel{2}}} = \frac{27}{1} = 27$$

When the Dividend is a Mixed Number. Change the mixed number to an improper fraction, then divide.

EXAMPLES:

a. $5\dfrac{3}{5} \div \dfrac{7}{8}$

$$5\frac{3}{5} \div \frac{7}{8} = \frac{28}{5} \div \frac{7}{8} = \frac{\overset{4}{\cancel{28}}}{5} \times \frac{8}{\underset{1}{\cancel{7}}} = \frac{32}{5}$$

$$= 6\frac{2}{5}$$

b. $15\frac{3}{4} \div \frac{3}{10}$

$$15\frac{3}{4} \div \frac{3}{10} = \frac{63}{4} \div \frac{3}{10} = \frac{\overset{21}{\cancel{63}}}{\underset{2}{\cancel{4}}} \times \frac{\overset{5}{\cancel{10}}}{\underset{1}{\cancel{3}}} = \frac{105}{2}$$

$$= 52\frac{1}{2}$$

PART B Dividing by a Whole Number

Before dividing a fraction by a whole number, write the whole number over 1. Follow the general rule of inverting the divisor and then multiplying. Again, the examples given here are separated into three groups, according to the form of the dividend.

When the Dividend is a Fraction.

EXAMPLES:

a. $\frac{8}{9} \div 4$

$$\frac{8}{9} \div 4 = \frac{8}{9} \div \frac{4}{1} = \frac{\overset{2}{\cancel{8}}}{9} \times \frac{1}{\underset{1}{\cancel{4}}} = \frac{2}{9}$$

b. $\frac{9}{11} \div 15$

$$\frac{9}{11} \div 15 = \frac{9}{11} \div \frac{15}{1} = \frac{\overset{3}{\cancel{9}}}{11} \times \frac{1}{\underset{5}{\cancel{15}}} = \frac{3}{55}$$

When the Dividend is a Whole Number.

EXAMPLES:

a. $8 \div 28$

$$8 \div 28 = \frac{8}{1} \div \frac{28}{1} = \frac{\overset{2}{\cancel{8}}}{1} \times \frac{1}{\underset{7}{\cancel{28}}} = \frac{2}{7}$$

Note that this example is actually one whole number divided by another. This problem is solved more quickly by writing the problem as a fraction, with the dividend as the numerator and the divisor as the denominator. Then reduce to lowest terms.

b. $8 \div 28$

$$8 \div 28 = \frac{8}{28} = \frac{\overset{2}{\cancel{8}}}{\underset{7}{\cancel{28}}} = \frac{2}{7}$$

When the Dividend is a Mixed Number. Change the mixed number to an improper fraction, then divide.

EXAMPLES:

a. $4\frac{1}{5} \div 3$

$$4\frac{1}{5} \div 3 = \frac{21}{5} \div \frac{3}{1} = \frac{\overset{7}{\cancel{21}}}{5} \times \frac{1}{\underset{1}{\cancel{3}}} = \frac{7}{5} = 1\frac{2}{5}$$

b. $15\frac{3}{4} \div 21$

$$15\frac{3}{4} \div 21 = \frac{63}{4} \div \frac{21}{1} = \frac{\overset{3}{\cancel{63}}}{4} \times \frac{1}{\underset{1}{\cancel{21}}} = \frac{3}{4}$$

PART C Dividing by a Mixed Number

To divide a fraction by a mixed number, first change the mixed number to an improper fraction. Then divide. The examples given here are again separated into three groups, according to the form of the dividend.

When the Dividend is a Fraction.

EXAMPLES:

a. $\frac{7}{8} \div 2\frac{1}{3}$

$$\frac{7}{8} \div 2\frac{1}{3} = \frac{7}{8} \div \frac{7}{3} = \frac{\overset{1}{\cancel{7}}}{8} \times \frac{3}{\underset{1}{\cancel{7}}} = \frac{3}{8}$$

Chapter 2 Working with Fractions

b. $\dfrac{7}{10} \div 1\dfrac{1}{3}$

$$\dfrac{7}{10} \div 1\dfrac{1}{3} = \dfrac{7}{10} \div \dfrac{4}{3} = \dfrac{7}{10} \times \dfrac{3}{4} = \dfrac{21}{40}$$

When the Dividend is a Whole Number.

EXAMPLES:

a. $5 \div 6\dfrac{2}{3}$

$$5 \div 6\dfrac{2}{3} = \dfrac{5}{1} \div \dfrac{20}{3} = \dfrac{\cancel{5}^{1}}{1} \times \dfrac{3}{\cancel{20}_{4}} = \dfrac{3}{4}$$

b. $24 \div 3\dfrac{3}{4}$

$$24 \div 3\dfrac{3}{4} = \dfrac{24}{1} \div \dfrac{15}{4} = \dfrac{\cancel{24}^{8}}{1} \times \dfrac{4}{\cancel{15}_{5}} = \dfrac{32}{5}$$

$$= 6\dfrac{2}{5}$$

When the Dividend is a Mixed Number.

EXAMPLES:

a. $4\dfrac{4}{5} \div 9\dfrac{1}{3}$

$$4\dfrac{4}{5} \div 9\dfrac{1}{3} = \dfrac{24}{5} \div \dfrac{28}{3} = \dfrac{\cancel{24}^{6}}{5} \times \dfrac{3}{\cancel{28}_{7}} = \dfrac{18}{35}$$

b. $234\dfrac{7}{8} \div 155\dfrac{1}{2}$

$$234\dfrac{7}{8} \div 155\dfrac{1}{2} = \dfrac{1{,}879}{8} \div \dfrac{311}{2}$$

$$= \dfrac{1{,}879}{\cancel{8}_{4}} \times \dfrac{\cancel{2}^{1}}{311}$$

$$= \dfrac{1{,}879}{1{,}244} = 1\dfrac{635}{1{,}244}$$

When the mixed numbers are large, as in *Example b*, many persons (especially when using a calculator) prefer to change the fractions to decimals and then divide. The problem would then become as follows:

$$234.875 \div 155.5 = 1.51045$$

This procedure will be discussed more fully in Section 2.6 of this chapter.

PART D Simplifying Complex Fractions

A **complex fraction** is a fraction whose numerator or denominator contains a fraction or a mixed number.

There are two different methods of simplifying complex fractions. The first, and most straightforward, is the **division method**. This method considers the complex fraction as a division problem. Recall that the fraction bar indicates division, with the numerator as the dividend and the denominator as the divisor. If the numerator or denominator is a mixed number, change it to an improper fraction and divide.

EXAMPLES:

a. $\dfrac{\tfrac{3}{5}}{\tfrac{3}{10}} = \dfrac{3}{5} \div \dfrac{3}{10}$

$$\dfrac{3}{5} \div \dfrac{3}{10} = \dfrac{\cancel{3}^{1}}{\cancel{5}_{1}} \times \dfrac{\cancel{10}^{2}}{\cancel{3}_{1}} = \dfrac{2}{1} = 2$$

b. $\dfrac{\tfrac{5}{6}}{10} = \dfrac{5}{6} \div 10$

$$\dfrac{5}{6} \div 10 = \dfrac{5}{6} \div \dfrac{10}{1} = \dfrac{\cancel{5}^{1}}{6} \times \dfrac{1}{\cancel{10}_{2}} = \dfrac{1}{12}$$

The second method, called the **common multiple method**, is longer. The steps for this method are:

1. Multiply the denominators of the two fractions. The product is a *common multiple* of the two denominators.

2. Multiply both the numerator and the denominator of the complex fraction by the common multiple of the two denominators.
3. Reduce the new fraction to lowest terms.

EXAMPLE:

$$\frac{8\frac{2}{3}}{4\frac{4}{5}} = \frac{\frac{26}{3}}{\frac{24}{5}}$$

Step 1. $3 \times 5 = 15$

Step 2. $\dfrac{\frac{26}{3} \times 15}{\frac{24}{5} \times 15} = \dfrac{\frac{26}{\cancel{3}} \times \frac{\cancel{15}^{5}}{1}}{\frac{24}{\cancel{5}} \times \frac{\cancel{15}^{3}}{1}} = \dfrac{130}{72}$

Step 3. $\dfrac{130}{72} = 1\dfrac{58}{72} = 1\dfrac{29}{36}$

PART E Applications

Civil service and other employment examinations usually have word problems to solve. A typical problem might be worded as follows: *What is the total net pay if $\frac{2}{3}$ of the pay is $150?* The word *of* stands for multiplication, and the word *is* stands for the equal sign. In such a problem, the whole amount must be found when only a fraction of it is given. In this case, you can change the words of a problem into the following form:

Fraction of Whole Amount is Number
(given)   (to be found)   (given)

The equation now becomes:

Fraction × Whole Amount = Number
(given)   (to be found)   (given)

Multiplication cannot be used to solve this problem, because there is only one known number (the given fraction) on the left side of the equal sign. However, this problem can be solved by dividing the given number by the given fraction, as follows:

$$\text{Whole Amount} = \frac{\text{Number (given)}}{\text{Fraction (given)}}$$

EXAMPLES:

a. Mary Ann Forbes spends $150 a week for living expenses. This is $\frac{2}{3}$ of her net pay for a week. What is her weekly net pay?

$\frac{2}{3}$ of weekly net pay is $150

$\frac{2}{3}$ × weekly net pay = $150

weekly net pay = $\dfrac{\$150}{\frac{2}{3}}$

$\$150 \div \dfrac{2}{3} = \dfrac{\cancel{\$150}^{75}}{1} \times \dfrac{3}{\cancel{2}_{1}} = \$225$

A good problem-solving strategy includes a check.

$\dfrac{2}{3} \times \$225 = \dfrac{2}{\cancel{3}} \times \dfrac{\cancel{\$225}^{75}}{1} = \dfrac{\$150}{1} = \$150$

b. A company spent $9,600 for advertising in one month. This was $\frac{3}{20}$ of the amount allowed for advertising for the entire year. What was the annual amount allowed?

$\frac{3}{20}$ of annual amount is $9,600

$\frac{3}{20}$ × annual amount = $9,600

annual amount = $\dfrac{\$9,600}{\frac{3}{20}}$

$\$9,600 \div \dfrac{3}{20} = \dfrac{\cancel{\$9,600}^{3,200}}{1} \times \dfrac{20}{\cancel{3}_{1}} = \dfrac{\$64,000}{1}$

$= \$64,000$

Check: $\dfrac{3}{20} \times \$64,000 = \dfrac{3}{\cancel{20}} \times \dfrac{\cancel{\$64,000}^{3,200}}{1}$

$= \dfrac{\$9,600}{1} = \$9,600$

If you are puzzled as to why you should divide, think of a very simple problem, like 4 × what number = 36. You know the answer is 9, because 4 × 9 = 36. Since division is the inverse of multiplication, to find the unknown factor in an indicated multiplication, divide the known product (36) by the one known factor (4). Thus, 36 ÷ 4 = 9.

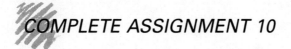
COMPLETE ASSIGNMENT 10

Name _____

Date _____

ASSIGNMENT 10
SECTION 2.5

Division of Fractions

	Perfect Score	Student's Score
PART A	30	
PART B	10	
PART C	40	
PART D	10	
PART E	10	
TOTAL	100	

PART A   Dividing by a Fraction

DIRECTIONS: Divide. Reduce your answers to lowest terms (2 points for each correct answer).

1. $\frac{3}{5} \div \frac{1}{2} =$ _____

2. $\frac{1}{2} \div \frac{1}{4} =$ _____

3. $\frac{5}{8} \div \frac{3}{4} =$ _____

4. $\frac{3}{4} \div \frac{3}{4} =$ _____

5. $\frac{14}{15} \div \frac{2}{5} =$ _____

6. $6 \div \frac{3}{5} =$ _____

7. $15 \div \frac{6}{7} =$ _____

8. $12 \div \frac{2}{3} =$ _____

9. $4 \div \dfrac{2}{5} =$ _____

10. $5 \div \dfrac{3}{10} =$ _____

11. $2\dfrac{5}{8} \div \dfrac{7}{24} =$ _____

12. $1\dfrac{3}{8} \div \dfrac{3}{10} =$ _____

13. $6\dfrac{2}{3} \div \dfrac{1}{12} =$ _____

14. $9\dfrac{3}{4} \div \dfrac{9}{16} =$ _____

15. $6\dfrac{2}{9} \div \dfrac{2}{9} =$ _____

STUDENT'S SCORE _____

PART B Dividing by a Whole Number

DIRECTIONS: Divide. Reduce your answers to lowest terms (2 points for each correct answer).

16. $\dfrac{3}{5} \div 9 =$ _____

17. $\dfrac{5}{8} \div 15 =$ _____

Chapter 2   Working with Fractions

18. $15 \div 33 =$ _____

19. $5\frac{5}{6} \div 25 =$ _____

20. $9\frac{1}{3} \div 4 =$ _____

STUDENT'S SCORE _____

PART C   Dividing by a Mixed Number

DIRECTIONS: Divide. Reduce your answers to lowest terms (2 points for each correct answer).

21. $\frac{3}{3} \div 1\frac{1}{3} =$ _____

22. $\frac{5}{6} \div 3\frac{1}{3} =$ _____

23. $\frac{5}{8} \div 1\frac{3}{4} =$ _____

24. $\frac{5}{6} \div 2\frac{4}{5} =$ _____

25. $\frac{11}{15} \div 1\frac{3}{8} =$ _____

26. $4 \div 13\frac{1}{2} =$ _____

27. $4 \div 2\frac{2}{5} =$ _____

28. $240 \div 6\frac{2}{5} =$ _____

29. $15 \div 2\frac{2}{3} =$ _____

30. $120 \div 5\frac{5}{6} =$ _____

31. $3\frac{2}{3} \div 5\frac{1}{2} =$ _____

32. $6\frac{2}{3} \div 2\frac{3}{4} =$ _____

33. $2\frac{3}{8} \div 1\frac{3}{4} =$ _____

34. $1\frac{2}{5} \div 2\frac{1}{2} =$ _____

35. $1\frac{1}{6} \div 2\frac{2}{3} =$ _____

36. $3\frac{1}{6} \div 3\frac{1}{6} =$ _____

37. $6\frac{4}{5} \div 3\frac{2}{5} =$ _____

38. $19\frac{1}{2} \div 1\frac{1}{2} =$ _____

Chapter 2   Working with Fractions

39. $72\frac{1}{2} \div 3\frac{3}{4} =$ _____

40. $429\frac{3}{8} \div 12\frac{3}{5} =$ _____

STUDENT'S SCORE _____

PART D   Simplifying Complex Fractions

DIRECTIONS: Simplify the complex fractions. Reduce all answers to lowest terms (2 points for each correct answer).

41. $\dfrac{\frac{9}{20}}{\frac{3}{5}} =$ _____

44. $\dfrac{\frac{3}{3}}{3\frac{3}{8}} =$ _____

42. $\dfrac{3\frac{2}{3}}{8\frac{5}{9}} =$ _____

45. $\dfrac{2\frac{1}{3}}{10} =$ _____

43. $\dfrac{3\frac{3}{10}}{5} =$ _____

STUDENT'S SCORE _____

PART E Applications

DIRECTIONS: Solve the problems and write your answers on the lines provided. Reduce your answers to lowest terms (5 points for each correct answer).

46. A firm allowed $\frac{7}{12}$ of its operations budget for selling expenses. If the selling expenses amounted to $84,000, what is the total amount in the operations budget?

_____

47. The cost of a new tire is $\frac{8}{15}$ of the selling price. If the cost of the tire is $20, what is the selling price?

_____

STUDENT'S SCORE _____

Chapter 2 Working with Fractions

SECTION 2.6 Equivalent Forms

There are two reasons why a person should be able to work with equivalent forms, and both reasons involve calculators. First, if a calculator is available, it is much easier to do the arithmetic in decimals on the calculator. If numbers are given in fractions, change the fractions to decimal equivalents for calculator use. Second, and just as important, it is very easy to incorrectly key-enter numbers. The ability to quickly estimate your calculator answer is valuable.

PART A Changing a Fraction to a Decimal

To find the decimal equivalent of any fraction, divide the numerator by the denominator. Follow these steps:

1. Use long division. Write the numerator as the dividend, and write the denominator as the divisor. Then place a decimal point after the dividend, and another one directly above it in the quotient. After the decimal point in the dividend, write one more zero than the required number of decimal places. For example, if four decimal places are required, write five zeros. When there is no whole number to the left of the decimal point in the quotient, write a zero there.

EXAMPLE:

Find the decimal equivalent of $\frac{4}{11}$. Round off your answer to four decimal places.

$$\begin{array}{r} 0.36363 \\ 11\overline{)4.00000} \\ \underline{3\;3} \\ 70 \\ \underline{66} \\ 40 \\ \underline{33} \\ 70 \\ \underline{66} \\ 40 \\ \underline{33} \\ 7 \end{array}$$

2. Round off the quotient to the required number of places. Four places are enough for most calculations.

$$0.36363 \text{ rounds to } 0.3636$$

The preceding example also illustrates a **repeating decimal,** with 36 as the repeating digits. When a number appears a second time as a remainder in the long division (in this example, the number is 7), the quotient will be a repeating decimal. It is not necessary to continue the division, since the rest of the quotient will be the same repeating digits.

The decimal equivalents of most fractions are repeating decimals. The only exception is when all of the prime factors of the denominator or divisor are 2's and 5's.

EXAMPLES:

a. $\frac{5}{16}$

$$\begin{array}{r} 0.3125 \\ 16\overline{)5.0000} \end{array} \quad (16 = 2 \times 2 \times 2 \times 2)$$

b. $\frac{657}{250}$

$$\begin{array}{r} 2.6280 \\ 250\overline{)657.0000} \end{array} \quad (250 = 2 \times 5 \times 5 \times 5)$$

As just mentioned, if the denominator has any prime factor other than 2 or 5, the decimal equivalent will eventually repeat. To indicate a repeating decimal, write a solid line above the repeating digits.

EXAMPLES:

a. $\frac{11}{6} = 1.833\ldots = 1.8\overline{3}$

b. $\frac{5}{37} = 0.135135\ldots = 0.\overline{135}$

To change a fraction with a denominator of 10, 100, 1,000, etc., to a decimal, write the numerator as a whole number. Then, starting at the extreme right of this number, move the decimal point one place to the left for each zero in the denominator. As always, it is wise to write a zero to the left of the decimal point (see Section 1.5, Part F, in Chapter 1).

EXAMPLES:

$$\frac{359}{1,000} = 0.359 \text{ and } \frac{6,245}{10,000} = 0.6245$$

If necessary, write as many zeros as needed to the left of the number to indicate the correct number of decimal places.

EXAMPLES:

a. $\dfrac{7}{100,000} = 0.00007$

Write 4 zeros to the left of the 7 to indicate 5 decimal places.

b. $\dfrac{1,001}{1,000,000} = 0.001001$

Write 2 zeros to the left of the 1 to indicate 6 decimal places.

Changing a fraction to a decimal becomes an easy task with a calculator. Simply think of the fraction as a division problem. Then divide the numerator by the denominator.

EXAMPLES:

Find the decimal equivalents by using a calculator.

a. $\dfrac{4}{11} = 11\overline{)4}$

Key-enter	Display	
4	4.	
÷	4.	
11	11.	
=	0.3636363	*quotient*

Key-enter the dividend 4, followed by the division key. Then key-enter the divisor 11, followed by the equals key. Remember to key-enter the numerator (the dividend) first. Your display will show the quotient 0.3636363. Rounded off to four decimal places, the quotient is 0.3636.

b. $\dfrac{5}{16} = 16\overline{)5}$

Key-enter	Display	
5	5.	
÷	5.	
16	16.	
=	0.3125	*quotient*

PART B Changing a Decimal to a Fraction in Its Lowest Terms

To change a decimal to a fraction, write the decimal as a fraction (with a numerator and a denominator). The digits to the right of the decimal point represent the numerator. The denominator is a multiple of 10, with the number of zeros in the denominator equal to the number of decimal places in the decimal (see Figure 1-1 in Chapter 1). Then reduce the fraction to its lowest terms.

EXAMPLES:

a. $0.135 = \dfrac{135}{1,000} = \dfrac{\cancel{135}^{27}}{\cancel{1,000}_{200}} = \dfrac{27}{200}$

b. $0.005 = \dfrac{5}{1,000} = \dfrac{\cancel{5}^{1}}{\cancel{1,000}_{200}} = \dfrac{1}{200}$

c. $0.479 = \dfrac{479}{1,000}$ (It cannot be reduced.)

Note that the decimal and fraction equivalent are read the same way. This provides a shortcut for changing a decimal to a fraction. Read the decimal, then write it as a fraction.

EXAMPLES:

a. 0.135 is read as *one hundred thirty-five thousandths*.

Equivalent fraction = $\dfrac{135}{1,000}$

$\dfrac{135}{1,000} = \dfrac{27}{200}$

b. 0.6 is read as *six tenths*.

Equivalent fraction = $\dfrac{6}{10}$

$\dfrac{6}{10} = \dfrac{3}{5}$

Chapter 2   Working with Fractions

c. 0.46 is read as *forty-six hundredths*.

Equivalent fraction $= \frac{46}{100}$

$\frac{46}{100} = \frac{23}{50}$

To change a decimal with a whole-number part to a mixed number, first find the fractional equivalent of the decimal part. Then replace the decimal with its fractional equivalent. The whole-number part remains the same.

EXAMPLE:

Change 5.0625 to a mixed number.

Step 1. $0.0625 = \frac{625}{10,000} = \frac{1}{16}$

Step 2. $5.0625 = 5\frac{1}{16}$

PART C Adding and Subtracting Fractions and Decimals

Sometimes, numbers to be added are in different forms. Some are decimals and others are fractions or mixed numbers. Before they can be added, all numbers must be in the same form; that is, all numbers must be decimals, fractions, or mixed numbers. The decimals can be changed to fractions, or the fractions can be changed to decimals, depending on which form is required for the answer.

EXAMPLES:

a. Find the sum of $\frac{3}{4}$ and 0.55.

Fractions

$0.55 = \frac{55}{100} = \frac{11}{20}$

$\frac{3}{4} = \frac{15}{20}$

$+\frac{11}{20} = \frac{11}{20}$

$\phantom{+\frac{11}{20} =\ }\frac{26}{20} = 1\frac{6}{20} = 1\frac{3}{10}$

Decimals

$\frac{3}{4} = 0.75$

$+0.55 = \frac{0.55}{1.30}$

$ = 1.3$

b. Find the difference between $\frac{3}{4}$ and 0.55.

Fractions

$0.55 = \frac{55}{100} = \frac{11}{20}$

$\frac{3}{4} = \frac{15}{20}$

$-\frac{11}{20} = \frac{11}{20}$

$\phantom{-\frac{11}{20} =\ }\frac{4}{20} = \frac{1}{5}$

Decimals

$\frac{3}{4} = 0.75$

$-0.55 = \frac{0.55}{0.20} = 0.2$

If you are using a calculator, work the problem in decimals. First, change the fraction to a decimal with the calculator. Second, add or subtract the two decimals as indicated.

EXAMPLES:

Add or subtract with a calculator.

a. Find the sum of $\frac{3}{4}$ and 0.55.

Key-enter	Display	
3	3.	
÷	3.	
4	4.	
=	0.75	
+	0.75	
.55	0.55	
=	1.3	*sum*

The decimal equivalent of $\frac{3}{4}$ is 0.75. Note that you do *not* re-enter this number. Once it is displayed, you may use it to find the sum (1.3).

b. Find the difference between $\frac{3}{4}$ and 0.55.

Key-enter	Display	
3	3.	
÷	3.	
4	4.	
=	0.75	
−	0.75	
.55	0.55	
=	0.2	*difference*

PART D Multiplying Fractions and Decimals

One method of multiplying fractions and decimals is to work the problem using fractions. Change the decimal to a fraction by writing the decimal over 1. Multiply the numerators. Multiply the denominators. Divide the product of the numerators by the product of the denominators.

EXAMPLES:

a. $\frac{3}{4} \times 2.76$

$$\frac{3}{4} \times \frac{2.76}{1} = \frac{3 \times 2.76}{4 \times 1} = \frac{8.28}{4} = 2.07$$

b. $\frac{7}{15} \times 478.95$

$$\frac{7}{15} \times \frac{478.95}{1} = \frac{7 \times 478.95}{15 \times 1} = \frac{3{,}352.65}{15}$$
$$= 223.51$$

c. $\frac{5}{12} \times 0.078$

$$\frac{5}{12} \times \frac{0.078}{1} = \frac{5 \times 0.078}{12 \times 1} = \frac{0.390}{12}$$
$$= 0.0325$$

You may cancel by using common divisors, as shown in Section 2.5. However, you must take care to retain the decimal point when canceling. For this reason, no cancellations are shown here.

When multiplying decimals and mixed numbers, use the procedure shown in the previous examples. Remember to change a mixed number to an improper fraction.

EXAMPLES:

a. $3\frac{3}{4} \times 76.8$

$$3\frac{3}{4} = \frac{15}{4}$$

$$\frac{15}{4} \times \frac{76.8}{1} = \frac{15 \times 76.8}{4 \times 1} = \frac{1{,}152.0}{4} = 288$$

b. $2\frac{1}{12} \times 110.46$

$$2\frac{1}{12} = \frac{25}{12}$$

$$\frac{25}{12} \times \frac{110.46}{1} = \frac{25 \times 110.46}{12 \times 1} = \frac{2{,}761.50}{12}$$
$$= 230.125$$

Multiplying Fractions and Decimals with a Calculator. Work the problem in decimals by changing the fraction to a decimal with the calculator, and then multiplying the two decimals.

EXAMPLES:

a. $\frac{3}{4} \times 2.76$

Key-enter	Display	
3	3.	
÷	3.	
4	4.	
=	0.75	
×	0.75	
2.76	2.76	
=	2.07	*product*

The decimal equivalent of $\frac{3}{4}$ is 0.75. Note that you do *not* re-enter this number. Once it is displayed, you may use it to find the product, 2.07.

b. $\frac{7}{15} \times 478.95$

Key-enter	Display	
7	7.	
÷	7.	
15	15.	
=	0.4666666	
×	0.4666666	
478.95	478.95	
=	223.50996	*product*

The decimal equivalent of $\frac{7}{15}$ is 0.4666666. It is a repeating decimal and has been rounded off to seven decimal places by the calculator. Rounded off to two decimal places, the final product is 223.51.

c. $\frac{5}{12} \times 0.078$

Key-enter	Display	
5	5.	
÷	5.	
12	12.	
=	0.4166666	
×	0.4166666	
.078	0.078	
=	0.0324999	*product*

Chapter 2 Working with Fractions

The decimal equivalent of $\frac{5}{12}$ is 0.4166666, a repeating decimal. Rounded off to four decimal places, the final product is 0.0325.

Multiplying Decimals and Mixed Numbers with a Calculator. Change the mixed number to a decimal, and then multiply. To change a mixed number to a decimal, change the fraction of the mixed number to its decimal equivalent, and then add this decimal equivalent to the whole-number part of the mixed number.

EXAMPLES:

a. $3\frac{3}{4} \times 76.8$

Key-enter	Display
3	3.
÷	3.
4	4.
=	0.75
+	0.75
3	3.
=	3.75
×	3.75
76.8	76.8
=	288 *product*

The decimal equivalent of $\frac{3}{4}$ is 0.75. Adding 3 to 0.75 gives 3.75, which is the decimal equivalent of $3\frac{3}{4}$. The final product is 288.

b. $2\frac{1}{12} \times 110.46$

Key-enter	Display
1	1.
÷	1.
12	12.
=	0.0833333
+	0.0833333
2	2.
=	2.0833333
×	2.0833333
110.46	110.46
=	230.12499 *product*

The decimal equivalent of $\frac{1}{12}$ is 0.0833333. Adding 2 to 0.0833333 gives 2.0833333, which is the decimal equivalent of $2\frac{1}{12}$. Rounded off to three decimal places, the final product is 230.125.

PART E Dividing Fractions and Decimals

One method of dividing fractions and decimals is to work the problem in fractions. Change the decimal to a fraction by using the methods described in Part B, and then invert the divisor and multiply. Reduce your answer to lowest terms.

EXAMPLES:

a. $\frac{1}{2} \div 0.76$

$$\frac{1}{2} \div \frac{76}{100} = \frac{1}{2} \times \frac{100}{76}$$

$$= \frac{1}{2} \times \frac{\cancel{100}^{25}}{\cancel{76}_{38}}$$

$$= \frac{25}{38}$$

The final quotient is $\frac{25}{38}$. The decimal equivalent of this answer is 0.6578947. Rounded off to four decimal places, the quotient is 0.6579.

b. $3\frac{3}{4} \div 0.6$

$$3\frac{3}{4} \div \frac{6}{10} = \frac{15}{4} \times \frac{10}{6}$$

$$= \frac{\cancel{15}^5}{\cancel{4}_2} \times \frac{\cancel{10}^5}{\cancel{6}_2}$$

$$= \frac{25}{4} = 6\frac{1}{4}$$

The final quotient is $6\frac{1}{4}$. The decimal equivalent of this answer is 6.25.

When dividing decimals and mixed numbers, use the procedure shown in the previous examples. Remember to change the mixed number to an improper fraction.

EXAMPLE:

$0.33 \div 3\frac{3}{10}$

$$\frac{33}{100} \div \frac{33}{10} = \frac{33}{100} \times \frac{10}{33}$$

$$= \frac{\cancel{33}}{\cancel{100}} \times \frac{\cancel{10}}{\cancel{33}} = \frac{1}{10}$$

The final quotient is $\frac{1}{10}$. The decimal equivalent of this answer is 0.1.

To divide decimals and fractions with a calculator, work the problem in decimals. Change the fraction to a decimal by using the calculator, and then divide the two decimals.

EXAMPLES:

a. $\frac{1}{2} \div 0.76$

Key-enter	Display
1	1.
÷	1.
2	2.
=	0.5
÷	0.5
.76	0.76
=	0.6578947 *quotient*

The decimal equivalent of $\frac{1}{2}$ is 0.5. The final quotient is 0.6578947. Rounded off to four decimal places, the quotient is 0.6579.

b. $3\frac{3}{4} \div 0.6$

Key-enter	Display
3	3.
÷	3.
4	4.
=	0.75
+	0.75
3	3.
=	3.75
÷	3.75
.6	0.6
=	6.25 *quotient*

The decimal equivalent of $3\frac{3}{4}$ is 3.75. The final quotient is 6.25. Its fractional equivalent is $6\frac{1}{4}$.

c. $0.33 \div 3\frac{3}{10}$

Key-enter	Display
3	3.
÷	3.
10	10.
=	0.3
+	0.3
3	3.
=	3.3
C	0.
.33	0.33
÷	0.33
3.3	3.3
=	0.1 *quotient*

The decimal equivalent of $3\frac{3}{10}$ is 3.3. The final quotient is 0.1. Its fractional equivalent is $\frac{1}{10}$.

When the divisor is a fraction or a mixed number, as in the preceding example, interpreting the problem becomes a difficult task for a standard calculator. The easiest method is to change the fraction to a decimal, write it down or remember it, and then enter the problem from left to right. If your calculator has a memory, place the decimal equivalent in the memory.

It is beneficial for you to learn the fractional and decimal equivalents, since they provide shortcuts with or without a calculator. These equivalents are given in Table 2-1. The "Cents Form" is provided for applications throughout this text. The "Percent Form" is included for use in Chapter 3.

Referring to Table 2-1, note the decimals containing fractions, such as $0.33\frac{1}{3}$, $0.16\frac{2}{3}$, and $0.58\frac{1}{3}$. A decimal with a fraction at the end can be simplified by changing it to a complex fraction with a numerator and a denominator. The numerator is the mixed number after the decimal point, and the denominator may be 10, 100, 1,000, etc., depending on the number of decimal places. This method is used when the fraction at the end of the decimal cannot be changed to an exact decimal and when an exact value is required. The fraction is an exact value, while the decimal equivalent is not.

Chapter 2   Working with Fractions

	Decimal Form	Cents Form	Percent Form	Fractional Equivalent		Decimal Form	Cents Form	Percent Form	Fractional Equivalent
Halves	0.5 (0.50)	50¢	50%	$\frac{1}{2}$	**Eighths**	$0.12\frac{1}{2}$ or 0.125	$12\frac{1}{2}$¢	$12\frac{1}{2}$%	$\frac{1}{8}$
Thirds	$0.33\frac{1}{3}$ or 0.333333	$33\frac{1}{3}$¢	$33\frac{1}{3}$%	$\frac{1}{3}$		$0.37\frac{1}{2}$ or 0.375	$37\frac{1}{2}$¢	$37\frac{1}{2}$%	$\frac{3}{8}$
	$0.66\frac{2}{3}$ or 0.666667	$66\frac{2}{3}$¢	$66\frac{2}{3}$%	$\frac{2}{3}$		$0.62\frac{1}{2}$ or 0.625	$62\frac{1}{2}$¢	$62\frac{1}{2}$%	$\frac{5}{8}$
Fourths	0.25	25¢	25%	$\frac{1}{4}$		$0.87\frac{1}{2}$ or 0.875	$87\frac{1}{2}$¢	$87\frac{1}{2}$%	$\frac{7}{8}$
	0.75	75¢	75%	$\frac{3}{4}$	**Twelfths**	$0.08\frac{1}{3}$ or 0.083333	$8\frac{1}{3}$¢	$8\frac{1}{3}$%	$\frac{1}{12}$
Fifths	0.2 (0.20)	20¢	20%	$\frac{1}{5}$		$0.41\frac{2}{3}$ or 0.416667	$41\frac{2}{3}$¢	$41\frac{2}{3}$%	$\frac{5}{12}$
	0.4 (0.40)	40¢	40%	$\frac{2}{5}$		$0.58\frac{1}{3}$ or 0.583333	$58\frac{1}{3}$¢	$58\frac{1}{3}$%	$\frac{7}{12}$
	0.6 (0.60)	60¢	60%	$\frac{3}{5}$		$0.91\frac{2}{3}$ or 0.916667	$91\frac{2}{3}$¢	$91\frac{2}{3}$%	$\frac{11}{12}$
	0.8 (0.80)	80¢	80%	$\frac{4}{5}$	**One Fifteenth**	$0.06\frac{2}{3}$ or 0.066667	$6\frac{2}{3}$¢	$6\frac{2}{3}$%	$\frac{1}{15}$
Sixths	$0.16\frac{2}{3}$ or 0.166667	$16\frac{2}{3}$¢	$16\frac{2}{3}$%	$\frac{1}{6}$	**One Sixteenth**	$0.06\frac{1}{4}$ or 0.0625	$6\frac{1}{4}$¢	$6\frac{1}{4}$%	$\frac{1}{16}$
	$0.83\frac{1}{3}$ or 0.833333	$83\frac{1}{3}$¢	$83\frac{1}{3}$%	$\frac{5}{6}$					

TABLE 2-1   *Fractional and Decimal Equivalents*

EXAMPLES:

a. $0.33\frac{1}{3}$

$$0.33\frac{1}{3} = \frac{33\frac{1}{3}}{100} = 33\frac{1}{3} \div 100$$

$$= \frac{100}{3} \div \frac{100}{1} = \frac{100}{3} \times \frac{1}{100}$$

$$= \frac{\cancel{100}^{1}}{3} \times \frac{1}{\cancel{100}_{1}} = \frac{1}{3}$$

b. $0.16\frac{2}{3}$

$$0.16\frac{2}{3} = \frac{16\frac{2}{3}}{100} = 16\frac{2}{3} \div 100$$

$$= \frac{50}{3} \div \frac{100}{1} = \frac{50}{3} \times \frac{1}{100}$$

$$= \frac{\cancel{50}^{1}}{3} \times \frac{1}{\cancel{100}_{2}} = \frac{1}{6}$$

c. $0.58\frac{1}{3}$

$$0.58\frac{1}{3} = \frac{58\frac{1}{3}}{100} = 58\frac{1}{3} \div 100$$

$$= \frac{175}{3} \div \frac{100}{1} = \frac{175}{3} \times \frac{1}{100}$$

$$= \frac{\cancel{175}^{7}}{3} \times \frac{1}{\cancel{100}_{4}} = \frac{7}{12}$$

Note in Table 2-1 that the thirds, sixths, and twelfths which have a $\frac{1}{3}$ or $\frac{2}{3}$ ending in the two-place decimal are changed to six-place decimals. Remember that you can extend these decimals to as many places as required, but the $\frac{1}{3}$ ending must be rounded to 3 in the last decimal place, and the $\frac{2}{3}$ ending must be rounded to 7 in the last decimal place. Keep in mind that the decimal ending in a fraction, such as $0.58\frac{1}{3}$, and its fractional equivalent, $\frac{7}{12}$, are exact values. However, the rounded decimal equivalent, 0.583333, is *not* an exact value; it is an *approximated* value.

PART F Applications

An amount of money less than one dollar can be written with a cents sign or a dollar sign. When the cents sign (¢) is used, it is written after the amount, as in 35¢, 12.5¢, and $7\frac{1}{4}$¢. Before these amounts are used in calculations, they are usually changed to decimal parts of dollars.

To change cents to dollars, locate the decimal point, move it two places to the left, drop the cents sign, and write the dollar sign in front. If there is no decimal point in the original amount, write a decimal point after the last whole digit. As usual, when there is no whole number to the left of the decimal point in the final answer, write a zero to the left of the decimal.

EXAMPLES:

a. 35¢ = 35.¢ = 35.¢ = $0.35

b. 12.5¢ = 12.5¢ = $0.125

If the amount in cents has a fractional ending, first change the fractional ending to a decimal and then write the amount in dollars. Remember to move the decimal point.

EXAMPLE:

$7\frac{1}{4}$¢ = 7.25¢

7.25¢ = 07.25¢ = $0.0725

Grocery stores list the price of items as *cost per unit of weight*. This allows a shopper to compare, for example, the price of a 12-ounce box of cereal to the price of a 15-ounce box of the same cereal. If the unit price of the 12-ounce box is 0.0725 cents per ounce and the unit price of the 15-ounce box is 0.08 cents per ounce, the smaller box is the better buy.

As mentioned several times already in this text, calculators are used extensively for many calculations in everyday living. Thus, when any of the numbers in the "Applications" sections of this book is a fraction or a mixed number and you are using a calculator, first change the number to its decimal form.

EXAMPLES:

Change to decimals (five places) for use on a calculator.

a. $25\frac{5}{6} + 17\frac{3}{4} = 25.83333 + 17.75$

b. $3{,}175 \ @ \ 5\frac{2}{3}¢ = 3{,}175 \times \$0.05667$

c. $6{,}250 \ @ \ 4\frac{1}{4}¢ = 6{,}250 \times \$0.0425$

d. $5\frac{7}{12} \div 0.07\frac{3}{8} = 5.58333 \div 0.07375$

COMPLETE ASSIGNMENT 11

ASSIGNMENT 11
SECTION 2.6

Name _____

Date _____

Equivalent Forms

	Perfect Score	Student's Score
PART A	20	
PART B	10	
PART C	20	
PART D	20	
PART E	20	
PART F	10	
TOTAL	100	

PART A Changing a Fraction to a Decimal

DIRECTIONS: Find the decimal equivalents. Round off your answers to four decimal places (2 points for each correct answer).

1. $\frac{11}{12} =$ _____

2. $\frac{13}{12} =$ _____

3. $\frac{13}{24} =$ _____

4. $\frac{89}{24} =$ _____

5. $\frac{23}{32} =$ _____

DIRECTIONS: Using a calculator, find the decimal equivalents. Round off your answers to four decimal places (2 points for each correct answer).

6. $\frac{155}{64} =$ _____

7. $\frac{8}{11} =$ _____

8. $\frac{427}{80} =$ _____

9. $\frac{35}{252} =$ _____

10. $\frac{260}{111} =$ _____

STUDENT'S SCORE _____

PART B Changing a Decimal to a Fraction in Its Lowest Terms

DIRECTIONS: Change the decimals to fractions reduced to lowest terms (2 points for each correct answer).

11.   0.06 = _____   14.   1.864 = _____

12.   0.15 = _____   15.   4.0425 = _____

13.   0.125 = _____

STUDENT'S SCORE _____

PART C Adding and Subtracting Fractions and Decimals

DIRECTIONS: Change the decimals to fractions, and then add or subtract as indicated. Reduce your answers to lowest terms (2 points for each correct answer).

16. $4\frac{1}{2} + 8.75 =$ _____   19. $9\frac{1}{8} + 65.05 =$ _____

17. $53.5 - 26\frac{1}{4} =$ _____   20. $8.875 + 3\frac{7}{20} =$ _____

18. $0.375 + 7\frac{1}{2} =$ _____

Chapter 2   Working with Fractions   111

DIRECTIONS: Change the fractions to decimals, and add or subtract as indicated (2 points for each correct answer).

21. $4\frac{1}{2}$ + 8.75 = _____

24. $9\frac{1}{8}$ + 65.05 = _____

22. 53.5 − $26\frac{1}{4}$ = _____

25. 8.875 + $3\frac{7}{20}$ = _____

23. 0.375 + $7\frac{1}{2}$ = _____

STUDENT'S SCORE _____

PART D   Multiplying Fractions and Decimals

DIRECTIONS: Change the decimals to fractions and multiply. Reduce your answers to lowest terms (4 points for each correct answer).

26. $\frac{3}{4}$ × 32.8 = _____

27. 62.5 × $\frac{3}{5}$ = _____

28. $2\frac{2}{5}$ × 0.24 = _____

Unit 1   Fundamentals

DIRECTIONS: With a calculator, change the fractions to decimals and multiply. Round off your answers to four decimal places (4 points for each correct answer).

29. $5.61 \times 1\frac{2}{3} =$ _____

30. $3\frac{1}{2} \times 0.25 =$ _____

STUDENT'S SCORE _____

PART E Dividing Fractions and Decimals

DIRECTIONS: Change the decimals to fractions and divide. Reduce your answers to lowest terms (4 points for each correct answer).

31. $24 \div 0.25 =$ _____   **32.** $243 \div 0.9 =$ _____   **33.** $870 \div 0.08 =$ _____

DIRECTIONS: With a calculator, change the fractions to decimals and divide. Round off your answers to four decimal places (4 points for each correct answer).

34. $84 \div 37\frac{1}{2} =$ _____

35. $5.30 \div 41\frac{2}{3} =$ _____

STUDENT'S SCORE _____

Chapter 2   Working with Fractions

PART F   Applications

DIRECTIONS: Change the amounts in cents to dollars (2 points for each correct answer).

36. $2.25¢ =$ _____   **37.** $43\frac{3}{4}¢ =$ _____   **38.** $2\frac{1}{2}¢ =$ _____

DIRECTIONS: With a calculator, change the fractions to decimals and perform the indicated operations (2 points for each correct answer).

39. $17\frac{3}{8} - 9.24 =$ _____   **40.** $3,250 @ 5\frac{1}{8}¢ =$ _____

STUDENT'S SCORE _____

Working with Percents

Percents have many uses; for example, they are used to represent interest rates, discounts, and changes in sales from one month to the next. Generally, it is easier to compare percents than it is to compare decimals or fractions.

The numbers that you have used so far have been whole numbers, decimals, fractions, or mixed numbers. In some cases, it was necessary to change one equivalent form to another, such as changing the fraction $\frac{5}{8}$ to its decimal equivalent, 0.625. Percents are another form in which numbers can be expressed.

You should understand percents, how to convert them to equivalent forms, and how to use them with or without a calculator. A percent key on a calculator can simplify operations with percents, but you must know how to use it correctly. On most calculators, the percent key is pressed *after* the multiplication or division key. Check with your manual to find out how to use the percent key on your calculator.

SECTION 3.1 Percent Conversions

A percent cannot be used directly in multiplication or division. It must first be changed, or **converted,** to a decimal or a fraction. However, percents can be added to or subtracted from other percents without being changed.

The term **percent** refers to a part of, or a fraction of, a hundred. A number followed by a percent sign means that number *per hundred*. Since *per hundred* is the same as "$\frac{}{100}$", the percent sign can be replaced by the word *hundredths*. This means that the percent can be changed to either a hundredths decimal (a decimal with two decimal places) or to a hundredths fraction (a fraction with a denominator of 100).

For example, $5\% = 0.05 = \frac{5}{100}$. The horizontal bar in a fraction indicates that the numerator is divided by the denominator. Therefore, a denominator of 100 means that the numerator is divided by 100. This leads to the following general rule for converting percents:

To convert any number in percent form to a decimal or a fraction, drop the percent sign and divide the number by 100.

The result can then be used in a calculation.

The rest of this section explains how a percent is changed to its decimal or fractional equivalent and how to change a decimal or a fraction to its percent form.

PART A Converting a Percent to a Decimal or a Whole Number

To change a percent to its decimal equivalent, follow the general rule for conversion. Use the shortcut for dividing by 100. The general rule can be restated as: *To change any number in a percent form to its decimal equivalent, drop the percent sign and move the decimal point in the number two places to the left.*

Converting Percents Between 1% and 100% (Inclusive) to Decimals or Whole Numbers. If there is no decimal point in a whole or mixed number, place one to the immediate right of the units digit. Change the fraction in any mixed number to its decimal equivalent. Then follow the shortcut conversion rule given earlier. As usual, write a zero to the left of the decimal point in the final answer if there is no digit there.

EXAMPLES:

a. $80\% = 80.\% = 80.\% = 0.80$ or 0.8
b. $6\% = 6.\% = 06.\% = 0.06$
c. $10\% = 10.\% = 10.\% = 0.10$ or 0.1
d. $16\frac{2}{3}\% = 16.67\% = 16.67\% = 0.1667$
e. $56.5\% = 56.5\% = 0.565$
f. $1\% = 1.\% = 01.\% = 0.01$
g. $100\% = 100.\% = 100.\% = 1.00$ or 1

Note that the decimal equivalents of percents between 1 and 100 cannot be less than 0.01 or greater than 1.00.

If the percent signs in each of the preceding examples were cent signs, amounts of money expressed in cents would be converted to their decimal equivalents. *Cents* and *percents* are very much alike. The same rules are used to convert both cents and percents to their decimal equivalents. Note that 1¢ and 1% both equal 0.01; and 100¢ and 100% both equal 1.00 (see Table 2-1 in Chapter 2).

EXAMPLES:

	Percents	Cents	Decimal Equivalents
a.	65%	65¢	0.65 or $0.65
b.	6%	6¢	0.06 or $0.06
c.	66.67%	66.67¢	0.6667 or $0.6667

Converting Percents Less Than 1% to Decimals. Percents that are less than 1% have either a fraction or a decimal less than 1 before the percent sign, such as 0.5%. To convert, change all fractions to their decimal equivalents. Write two zeros to the left of the decimal point. Follow the shortcut conversion rule. Drop the percent sign, and move the decimal point two places to the left. As usual, write a zero to the left of the decimal point in the final answer if there is no digit there.

EXAMPLES:

a. $0.5\% = 00.5\% = 00.5\% = 0.005$
$\frac{1}{2}\% = 0.5\% = 00.5\% = 0.005$

b. $0.25\% = 00.25\% = 00.25\% = 0.0025$
$\frac{1}{4}\% = 0.25\% = 00.25\% = 0.0025$
c. $0.33\frac{1}{3}\% = 00.3333\% = 00.3333\% = 0.003333$
$\frac{1}{3}\% = 0.3333\% = 00.3333\% = 0.003333$
d. $0.04\% = 00.04\% = 00.04\% = 0.0004$
$\frac{1}{25}\% = 0.04\% = 00.04\% = 0.0004$

Note that decimal equivalents of percents less than 1% must have at least two zeros to the immediate right of the decimal point.

Converting Percents Greater Than 100% to Decimals or Whole Numbers. If the number in front of the percent sign is a whole number, place a decimal point after the last digit. If the number is a mixed number, change the fraction to its decimal equivalent. Then follow the shortcut conversion rule. Drop the percent sign, and move the decimal point two places to the left.

EXAMPLES:

a. $120\% = 120.\% = 120.\% = 1.20$ or 1.2
b. $375\% = 375.\% = 375.\% = 3.75$
c. $200\% = 200.\% = 200.\% = 2.00$ or 2
d. $206\frac{2}{3}\% = 206.67\% = 206.67\% = 2.0667$

Be sure to move the decimal point *only two* places to the left. Note that decimal equivalents of percents greater than 100% must be either whole numbers or decimals greater than 1.

PART B Converting a Percent to a Fraction or a Mixed Number

There are several ways to convert a percent to a fraction or a mixed number. Only two methods will be given here. The first method is easier, but the second method is given for those who may prefer it.

To use the first method, change the percent to its decimal equivalent, as shown in Part A. Then change the decimal to its fractional or mixed-number equivalent.

As in Part A, the details of converting will be given in three groups: percents between 1% and 100% (inclusive), percents less than 1%, and percents greater than 100%. In the examples for each

Chapter 3   Working with Percents

group, the same percents illustrated in Part A are used, since their decimal equivalents have already been found. When the percent ends with a fraction, write the percent as a complex fraction with 100 as the denominator rather than rounding off the decimal.

Converting Percents Between 1% and 100% (Inclusive) to Fractions.   Since the decimal equivalents of these percents are at least 0.01 but not more than 1.00, their fractional equivalents must be at least $\frac{1}{100}$ but not more than 1.

EXAMPLES:

a. $80\% = 0.8 = \dfrac{\cancel{8}^{4}}{\cancel{10}_{5}} = \dfrac{4}{5}$

b. $6\% = 0.06 = \dfrac{\cancel{6}^{3}}{\cancel{100}_{50}} = \dfrac{3}{50}$

c. $10\% = 0.10 = 0.1 = \dfrac{1}{10}$

d. $16\tfrac{2}{3}\% = \dfrac{16\tfrac{2}{3}}{100} = 16\tfrac{2}{3} \div \dfrac{100}{1}$

$= \dfrac{\cancel{50}^{1}}{3} \times \dfrac{1}{\cancel{100}_{2}} = \dfrac{1}{6}$

e. $56.5\% = 0.565 = \dfrac{\cancel{565}^{113}}{\cancel{1,000}_{200}} = \dfrac{113}{200}$

f. $1\% = 0.01 = \dfrac{1}{100}$

g. $100\% = 1.00 = \dfrac{1}{1} = 1$

Converting Percents Less Than 1% to Fractions. The decimal equivalents of these percents have at least three decimal places and at least two zeros to the immediate right of the decimal point. Therefore, when the decimal is *first* changed to a fraction, make sure the denominator has at least *three zeros*.

A common error in converting percents less than 1% is to drop the percent sign without moving the decimal point two places to the left.

As in Part A, the decimal and fractional equivalents of the same percents are used in the examples that follow.

EXAMPLES:

a. $\left.\begin{array}{r}0.5\% \\ \tfrac{1}{2}\%\end{array}\right\} = 0.005 = \dfrac{\cancel{5}^{1}}{\cancel{1{,}000}_{200}} = \dfrac{1}{200}$

b. $\left.\begin{array}{r}0.25\% \\ \tfrac{1}{4}\%\end{array}\right\} = 0.0025 = \dfrac{\cancel{25}^{1}}{\cancel{10{,}000}_{400}} = \dfrac{1}{400}$

c. $\left.\begin{array}{r}0.33\tfrac{1}{3}\% \\ \tfrac{1}{3}\%\end{array}\right\} = 0.0033\tfrac{1}{3} = \dfrac{33\tfrac{1}{3}}{10{,}000} = 33\tfrac{1}{3} \div \dfrac{10{,}000}{1}$

$= \dfrac{\cancel{100}^{1}}{3} \times \dfrac{1}{\cancel{10{,}000}_{100}} = \dfrac{1}{300}$

d. $\left.\begin{array}{r}0.04\% \\ \tfrac{1}{25}\%\end{array}\right\} = 0.0004 = \dfrac{\cancel{4}^{1}}{\cancel{10{,}000}_{2{,}500}} = \dfrac{1}{2{,}500}$

Note that if the percent sign is dropped from the fractional form and two zeros are written to the right of the denominator of the fraction, the result is the fractional equivalent of the percent. For example, $\tfrac{1}{25}\%$ is equal to $\tfrac{1}{2{,}500}$. Try this shortcut method with the preceding examples.

Converting Percents Greater Than 100% to Mixed or Whole Numbers.   Since 100% equals 1, the whole-number equivalent of a percent that is greater than 100% must be more than 1. Therefore, the conversion must result in a mixed or a whole number.

EXAMPLES:

a. $120\% = 12\underset{\smile}{0.}\% = 1.2 = 1\frac{2}{10} = 1\frac{1}{5}$

b. $375\% = 37\underset{\smile}{5.}\% = 3.75 = 3\frac{3}{4}$

c. $200\% = 2\underset{\smile}{00.}\% = 2.00 = 2$

d. $206\frac{2}{3}\% = 20\underset{\smile}{6.}\frac{2}{3}\% = 2.06\frac{2}{3} = 2\frac{6\frac{2}{3}}{100} = 2\frac{1}{15}$

Note in *Example d* that when the fraction of a mixed number is a complex fraction, generally it is reduced separately as follows:

$$\frac{6\frac{2}{3}}{100} = 6\frac{2}{3} \div \frac{100}{1} = \frac{\overset{20}{\cancel{20}}}{3} \times \frac{1}{\underset{5}{\cancel{100}}} = \frac{1}{15}$$

You may recall from Table 2-1 in Chapter 2 that $0.06\frac{2}{3}$ has a fractional equivalent of $\frac{1}{15}$.

Multiplying by $\frac{1}{100}$: A Second Method. The second method of converting a percent to a fraction or a mixed number is based on the fact that dividing any number by 100 is the same as multiplying it by $\frac{1}{100}$. The conversion rule can be restated as: *To convert a percent to a fraction or a mixed number, drop the percent sign and multiply the number by $\frac{1}{100}$.* Any decimal should be changed to its fractional equivalent before multiplying.

While the arithmetic is simpler in this method, it requires learning another conversion rule. To illustrate this method, the same examples given for the first method are used here for comparison.

EXAMPLES:

Convert the percents (between 1% and 100%, inclusive) to their fractional equivalents.

a. $80\% = 80 \times \frac{1}{100} = \frac{\overset{4}{\cancel{80}}}{1} \times \frac{1}{\underset{5}{\cancel{100}}} = \frac{4}{5}$

b. $6\% = 6 \times \frac{1}{100} = \frac{\overset{3}{\cancel{6}}}{1} \times \frac{1}{\underset{50}{\cancel{100}}} = \frac{3}{50}$

c. $10\% = 10 \times \frac{1}{100} = \frac{\overset{1}{\cancel{10}}}{1} \times \frac{1}{\underset{10}{\cancel{100}}} = \frac{1}{10}$

d. $16\frac{2}{3}\% = 16\frac{2}{3} \times \frac{1}{100} = \frac{\overset{1}{\cancel{50}}}{3} \times \frac{1}{\underset{2}{\cancel{100}}} = \frac{1}{6}$

e. $56.5\% = 56\frac{1}{2}\% = 56\frac{1}{2} \times \frac{1}{100}$
$= \frac{113}{2} \times \frac{1}{100} = \frac{113}{200}$

f. $1\% = 1 \times \frac{1}{100} = \frac{1}{100}$

g. $100\% = 100 \times \frac{1}{100} = \frac{\overset{1}{\cancel{100}}}{1} \times \frac{1}{\underset{1}{\cancel{100}}} = \frac{1}{1} = 1$

EXAMPLES:

Convert the percents (less than 1%) to their fractional equivalents.

a. $0.5\% = \frac{1}{2}\% = \frac{1}{2} \times \frac{1}{100} = \frac{1}{200}$

b. $0.25\% = \frac{1}{4}\% = \frac{1}{4} \times \frac{1}{100} = \frac{1}{400}$

c. $0.33\frac{1}{3}\% = \frac{1}{3}\% = \frac{1}{3} \times \frac{1}{100} = \frac{1}{300}$

d. $0.04\% = \frac{1}{25}\% = \frac{1}{25} \times \frac{1}{100} = \frac{1}{2,500}$

EXAMPLES:

Convert the percents (greater than 100%) to their fractional equivalents.

a. $120\% = 120 \times \frac{1}{100} = \frac{\overset{6}{\cancel{120}}}{1} \times \frac{1}{\underset{5}{\cancel{100}}}$
$= \frac{6}{5} = 1\frac{1}{5}$

b. $375\% = 375 \times \dfrac{1}{100} = \dfrac{\overset{15}{\cancel{375}}}{1} \times \dfrac{1}{\underset{4}{\cancel{100}}}$

$= \dfrac{15}{4} = 3\dfrac{3}{4}$

c. $200\% = 200 \times \dfrac{1}{100} = \dfrac{\overset{2}{\cancel{200}}}{1} \times \dfrac{1}{\underset{1}{\cancel{100}}}$

$= \dfrac{2}{1} = 2$

d. $206\dfrac{2}{3}\% = 206\dfrac{2}{3} \times \dfrac{1}{100} = \dfrac{\overset{31}{\cancel{620}}}{3} \times \dfrac{1}{\underset{5}{\cancel{100}}}$

$= \dfrac{31}{15} = 2\dfrac{1}{15}$

PART C Converting a Decimal to Its Percent Equivalent

The general rule for converting any decimal to its percent equivalent is to multiply the number by 100 and add the percent sign. This is the reverse of what was done in changing a percent to a decimal. Since the shortcut to multiply a decimal by 100 is to move the decimal point two places to the right, this conversion rule can be restated as: *To change a decimal to a percent, move the decimal point two places to the right and add a percent sign.*

The decimal equivalents found in Part A are used here as examples for comparison.

Converting Decimals Between 0.01 and 1.00 (inclusive) to Percent Equivalents. If the number has only one decimal place, as in *Example a*, write a zero to the right of the number, and then move the decimal point two places to the right.

EXAMPLES:

a. $0.8 = 0.80 = 0.80 = 80.\%$ or 80%

b. $0.06 = 0.06 = 6.\%$ or 6%

c. $0.10 = 0.10 = 10.\%$ or 10%

d. $0.1667 = 0.1667 = 16.67\%$ or $16\dfrac{2}{3}\%$

e. $0.565 = 0.565 = 56.5\%$ or $56\dfrac{1}{2}\%$

f. $0.01 = 0.01 = 1.\%$ or 1%

g. $1.00 = 1.00 = 100.\%$ or 100%

Note that when the numbers have been converted to their percent equivalents, the decimal point is dropped if the result is a whole number, as in *Examples a, b, c, f,* and *g.*

Converting Decimals Less Than 0.01 to Percent Equivalents. Decimals less than 0.01 always have at least three decimal places and at least two zeros to the immediate right of the decimal point, such as 0.009 or 0.0009. Therefore, the percent equivalent must always be a decimal.

EXAMPLES:

a. $0.005 = 0.005 = 0.5\%$

b. $0.0025 = 0.0025 = 0.25\%$

c. $0.0033 = 0.0033 = 0.33\%$

d. $0.0004 = 0.0004 = 0.04\%$

Converting Decimals Greater Than 1 to Percent Equivalents. A decimal number must have two decimal places before it can be converted to its percent equivalent. Therefore, write one or two zeros to the right of the decimal point if necessary. In *Example a*, write one zero to the right of the decimal 1.2. In *Example c*, write two zeros to the right of the whole number 2.

EXAMPLE:

a. $1.2 = 1.20 = 1.20 = 120\%$

b. $3.75 = 3.75 = 375\%$

c. $2 = 2.00 = 2.00 = 200\%$

d. $2.0667 = 2.0667 = 206.67\%$

PART D Converting a Fraction or Mixed Number to Its Percent Equivalent

There are two ways to convert a fraction or mixed number to its percent equivalent. The first method is easier, but the second method is given for those who may prefer it.

For the first method, the rule may be stated as follows: *To convert a fraction or a mixed number to its percent equivalent, first change it to its decimal equivalent, then move the decimal point two places to the right and add a percent sign.*

EXAMPLES:

a. $\frac{3}{50} = 0.06 = 0.06 = 6\%$
b. $\frac{1}{4} = 0.25 = 0.25 = 25\%$
c. $\frac{3}{16} = 0.1875 = 0.1875 = 18.75\%$
d. $\frac{1}{2} = 0.50 = 0.50 = 50\%$
e. $2\frac{7}{8} = 2.875 = 2.875 = 287.5\%$
f. $4\frac{3}{4} = 4.75 = 4.75 = 475\%$

The rule for the second conversion method can be stated as follows: *To change a fraction or a mixed number to its percent equivalent, multiply the number by 100 and add a percent sign to the result.* The same examples used to illustrate the first method are also used here for comparison.

EXAMPLES:

a. $\frac{3}{50} = \frac{3}{\cancel{50}} \times \frac{\cancel{100}^{2}}{1} = 6\%$

b. $\frac{1}{4} = \frac{1}{\cancel{4}} \times \frac{\cancel{100}^{25}}{1} = 25\%$

c. $\frac{3}{16} = \frac{3}{\cancel{16}_{4}} \times \frac{\cancel{100}^{25}}{1} = \frac{75}{4}\% = 18\frac{3}{4}\%$

d. $\frac{1}{2} = \frac{1}{\cancel{2}_{1}} \times \frac{\cancel{100}^{50}}{1} = 50\%$

e. $2\frac{7}{8} = \frac{23}{\cancel{8}_{2}} \times \frac{\cancel{100}^{25}}{1} = \frac{575}{2}\% = 287\frac{1}{2}\%$

f. $4\frac{3}{4} = \frac{19}{\cancel{4}_{1}} \times \frac{\cancel{100}^{25}}{1} = 475\%$

Note that the percents can be written as mixed numbers, as in *Examples c* and *e*.

COMPLETE ASSIGNMENT 12

Name: _____

Date: _____

ASSIGNMENT 12
SECTION 3.1

Percent Conversions

	Perfect Score	Student's Score
PART A	30	
PART B	30	
PART C	20	
PART D	20	
TOTAL	100	

PART A Converting a Percent to a Decimal or a Whole Number

DIRECTIONS: Convert the percents to their decimal or whole-number equivalents. Round off your answers to four decimal places (1 point for each correct answer).

1. 50% = _____

2. 25% = _____

3. 4% = _____

4. 12% = _____

5. 20% = _____

6. 45% = _____

7. 47.5% = _____

8. $27\frac{1}{4}$% = _____

9. $3\frac{3}{4}$% = _____

10. 2.25% = _____

11. 41.67% = _____

12. $8\frac{1}{3}$% = _____

13. 81.5% = _____

14. 0.25% = _____

15. $\frac{2}{3}$% = _____

16. 0.1% = _____

17. 0.02% = _____

18. 0.05% = _____

19. 0.9% = _____

20. 110% = _____

21. 101% = _____

22. 400% = _____

23. 500% = _____

24. 1,000% = _____

25. $111\frac{1}{3}$% = _____

26. 100% = _____

27. 212% = _____

28. 75% = _____

29. 303% = _____

30. 0.09% = _____

STUDENT'S SCORE _____

Unit 1   Fundamentals

PART B   Converting a Percent to a Fraction or a Mixed Number

DIRECTIONS: Convert the percents to their fractional or mixed-number equivalents (2 points for each correct answer).

31. 25% = _____

32. 50% = _____

33. $\frac{3}{5}$% = _____

34. 15% = _____

35. $33\frac{1}{3}$% = _____

36. 600% = _____

37. $4\frac{1}{6}$% = _____

38. $6\frac{1}{4}$% = _____

39. $12\frac{1}{2}$% = _____

40. $8\frac{1}{3}$% = _____

41. $3\frac{1}{8}$% = _____

42. 140% = _____

43. 1,000% = _____

44. $\frac{1}{2}$% = _____

45. $\frac{1}{3}$% = _____

STUDENT'S SCORE _____

PART C   Converting a Decimal to Its Percent Equivalent

DIRECTIONS: Convert the numbers to their percent equivalents (1 point for each correct answer).

46.   0.5 = _____

53.   0.675 = _____

47.   0.25 = _____

54.   0.012 = _____

48.   3.25 = _____

55.   0.5667 = _____

49.   0.01 = _____

56.   0.0275 = _____

50.   10 = _____

57.   0.2 = _____

51.   1 = _____

58.   0.002 = _____

52.   0.175 = _____

59.   0.00275 = _____

Chapter 3 Working with Percents

60. 2.58 = _____   **63.** 1.01 = _____

61. 0.6 = _____   **64.** 1.25 = _____

62. 4 = _____   **65.** 1.5 = _____

STUDENT'S SCORE _____

PART D Converting a Fraction or Mixed Number to Its Percent Equivalent

DIRECTIONS: Convert the fractions or mixed numbers to their percent equivalents (2 points for each correct answer).

66. $\frac{1}{2}$ = _____   **71.** $\frac{7}{15}$ = _____

67. $3\frac{1}{2}$ = _____   **72.** $\frac{5}{8}$ = _____

68. $\frac{1}{3}$ = _____   **73.** $5\frac{2}{5}$ = _____

69. $\frac{3}{500}$ = _____   **74.** $\frac{7}{10}$ = _____

70. $\frac{1}{1,200}$ = _____   **75.** $1\frac{1}{2}$ = _____

STUDENT'S SCORE _____

SECTION 3.2 Finding the Percentage

The term **percentage** refers to a part of, or a fraction of, a whole. For example, 80% *of* the 90 students in the class passed the final exam, or the sales tax will be 5% *of* $4.50.

The given percent is called the **rate**; the given number is called the **base**. The following formula is used to find the percentage:

$$P = B \times R$$
Percentage = Base × Rate

In this formula, the percentage (P) is the product of the base (B) and the rate (R). On many calculators, the base must be key-entered *before* the rate.

It is a good idea to identify the parts of the formula before applying it in a problem. This procedure will prove useful in later sections of this chapter. The following is an example of a percentage problem.

Find 6% of $45. P = ? (percentage)
B = $45 (base)
R = 6% (rate)

In a percentage problem like this, the rate and the base are easily identified, since they are connected to each other by either the word *of* or a multiplication sign. In these problems, the word *of* indicates multiplication.

It was stated at the beginning of this chapter that a percent cannot be used directly in multiplication or division. To be used in multiplication or division, the percent must be converted to a decimal or a fraction. In general, a decimal is used, unless a fraction is easier to use. Therefore, to solve the preceding percentage problem, convert 6% (R) to its decimal equivalent, 0.06. Then apply the formula as follows:

$$P = B \times R$$
$$P = \$45 \times 0.06 = \$2.70$$

Note that the base in this problem is an amount of money. The percentage (P) found is always stated in the same terms as the base, whether it is money, a weight, or a measure. Sometimes, errors are made by adding the percent sign to the answer. Remember to drop the percent sign when the percent is converted to a decimal or a fraction.

The percent key $\boxed{\%}$ on a calculator looks like the percent sign %. Because calculators only work with decimals, you can only key-enter decimals or whole numbers. If a percent rate is given as a fraction, you must first convert the fraction to a decimal. For example, you have to convert $5\frac{1}{2}$% to 5.5% before using a calculator. As mentioned earlier, the conversion of a percent to a decimal is performed by the calculator. Therefore, you do not need to actually convert the percent to a decimal.

To find a percentage with a calculator, take the following steps:

1. Key-enter the base.
2. Key-enter the multiplication key, $\boxed{\times}$.
3. Key-enter the percent rate (in its percent form).
4. Key-enter the percent key, $\boxed{\%}$.

EXAMPLE:

Find 6% of $45.00 by using a calculator.

Key-enter	Display	
45	45.	
$\boxed{\times}$	45.	
6	6.	
$\boxed{\%}$	2.7	*percentage*

Key-enter the base, 45, followed by the multiplication key. Then key-enter the percent rate, 6, followed by the percent key to obtain the percentage, 2.7. Note that you do not need to key-enter the equals key, nor do you need to convert the percent to its decimal equivalent. These two operations are performed automatically by the calculator.

The rest of this section explains how to use the percentage formula with various bases and rates.

PART A Multiplying by Percents Between 1% and 100%

When multiplying to find a percentage (P), change the rate (R) to its decimal or fractional equivalent, depending on which is easier to use with the given base. Then multiply the base (B) by the decimal or fraction.

The following examples are worked with both decimals and fractions. You will find that some problems are easier to solve with decimals, while others are easier to solve with fractions.

EXAMPLES:

a. Find 4% of $275.

$$P = B \times R, \text{ where } R = 4\% = 0.04 = \frac{1}{25}$$

$$P = \$275 \times 0.04 = \$11.00$$

or

$$P = \$275 \times \frac{1}{25} = \frac{\overset{11}{\cancel{\$275}}}{1} \times \frac{1}{\underset{1}{\cancel{25}}} = \$11$$

b. 15% of 2,020 VCRs are defective. How many are defective?

$$P = B \times R, \text{ where } R = 15\% = 0.15 = \frac{3}{20}$$

$$P = 2{,}020 \times 0.15 = 303 \text{ VCRs}$$

or

$$P = \frac{\overset{101}{\cancel{2{,}020}}}{1} \times \frac{3}{\underset{1}{\cancel{20}}} = 303 \text{ VCRs}$$

c. What is $2\frac{1}{4}\%$ of $1,500?

$$P = B \times R, \text{ where } R = 2\tfrac{1}{4}\% = 0.0225 = \frac{9}{400}$$

$$P = \$1{,}500 \times 0.0225 = \$33.75$$

or

$$P = \$1{,}500 \times \frac{9}{400}$$

$$= \frac{\overset{15}{\cancel{\$1{,}500}}}{1} \times \frac{9}{\underset{4}{\cancel{400}}} = \frac{\$135}{4} = \$33.75$$

The next three examples illustrate three special cases that can be solved very quickly by using shortcuts. Since 1% = 0.01, 1% of a base (B) is found by moving the decimal point of the base *two* places to the left. Since 10% = 0.10 = 0.1, 10% of a base (B) is found by moving the decimal point of the base *one* place to the left. Since 100% = 1 and multiplying any number by 1 equals that same number, 100% of a base (B) is always equal to the base.

EXAMPLES:

a. Find 1% of $48.79.

$$P = B \times R$$

$$P = \$48.79 \times 0.01 = \$48.79 = \$0.49$$

b. How much is 10% of $365?

$$P = B \times R$$

$$P = \$365 \times 0.1 = \$365. = \$36.50$$

c. What is 100% of $318.00?

$$P = B \times R$$

$$P = \$318.00 \times 1 = \$318.00$$

PART B Using Fractional Equivalents of Percents

Table 2-1 in Chapter 2 gives the fractional equivalents of percents that frequently occur in percentage problems. Using the fractional equivalent of a percent is often shorter and simpler than using the decimal equivalent, especially without a calculator. Therefore, the fractional equivalent should be used when convenient, particularly when the denominator is a factor of the base. If there are common divisors, you can usually work faster by multiplying mentally than by using a calculator.

EXAMPLES:

a. Find 75% of $420.

$$P = B \times R, \text{ where } R = 75\% = \frac{3}{4}$$

$$P = \frac{\overset{105}{\cancel{\$420}}}{1} \times \frac{3}{\underset{1}{\cancel{4}}} = \$315$$

b. What is $33\frac{1}{3}\%$ of 480 meters?

$$P = B \times R, \text{ where } R = 33\frac{1}{3}\% = \frac{1}{3}$$

$$P = \frac{\cancel{480}^{160}}{1} \times \frac{1}{\cancel{3}_{1}} = 160 \text{ meters}$$

c. How much is $66\frac{2}{3}\%$ of \$2,496?

$$P = B \times R, \text{ where } R = 66\frac{2}{3}\% = \frac{2}{3}$$

$$P = \frac{\cancel{\$2{,}496}^{832}}{1} \times \frac{2}{\cancel{3}_{1}} = \$1{,}664$$

Note that if this example were done in decimals, to get an accurate answer you would have to multiply \$2,496 by 66.666667%. On a calculator, $\$2{,}496 \times 66.666667\% = \$1{,}664$.

PART C Multiplying by Percents Less Than 1%

When a percent is less than 1%, be careful converting the percent to a decimal or a fraction. Remember that if you have converted correctly, the decimal equivalent must have at least two zeros to the immediate right of the decimal point, and the fractional equivalent must have at least two zeros in the denominator (before being reduced).

EXAMPLES:

a. What is $\frac{1}{2}\%$ of \$200.00?

$$P = B \times R, \text{ where } R = \frac{1}{2}\% = 0.005 = \frac{1}{200}$$

$$P = \$200 \times 0.005 = \$1.00$$

or

$$P = \frac{\cancel{\$200}^{1}}{1} \times \frac{1}{\cancel{200}_{1}} = \$1.00$$

After the percentage has been found, mentally find 1% of the base by moving its decimal point two places to the left. The percentage found when multiplying by a percent less than 1% must be smaller than 1% of the base.

Check: $\$200 \times 1\% = \$2.00$. The percentage found in this example (\$1.00) is in fact half of \$2.00.

b. Find $0.625\% \times \$560$.

$$P = B \times R, \text{ where } R = 0.625\% = 0.00625$$
$$= \frac{1}{160}$$

$$P = \$560 \times 0.00625 = \$3.50$$

or

$$P = \frac{\cancel{\$560}^{7}}{1} \times \frac{1}{\cancel{160}_{2}} = \frac{\$7}{2} = \$3\frac{1}{2} = \$3.50$$

Check: $\$560 \times 1\% = \$5.60$. The percentage found in this example (\$3.50) is less than \$5.60.

Not only can 1% of the base be used as a check, but it can also be used to find the percentage when the percent is less than 1%. A fractional percent, whether it is expressed as a decimal or a fraction, equals that same fractional part of 1%.

EXAMPLE:

$$\frac{1}{2}\% = 0.5\% = 0.005 = \frac{1}{200}$$

$$1\% \qquad = 0.01 = \frac{1}{100}$$

$$\frac{1}{2} \times 1\% = \frac{1}{2} \times 0.01 = \frac{0.01}{2} = 0.005$$

or

$$\frac{1}{2} \times 1\% = \frac{1}{2} \times \frac{1}{100} = \frac{1}{200}$$

Thus, another method of multiplying the base by a fractional percent is to first find 1% of the base, and then multiply the result by the decimal or fraction in front of the percent sign.

EXAMPLES:

a. What is $\frac{3}{8}$% of $72.32?

$P = B \times R$

$1\% \times \$72.32 = \$0.7232$

$P = \$0.7232 \times \frac{3}{8} = \$0.2712 = \$0.27$

b. Find $\frac{5}{6}$% of $212.40.

$P = B \times R$

$1\% \times \$212.40 = \$2.1240$

$P = \dfrac{\$2.1240}{1} \times \dfrac{5}{6} = \dfrac{\$10.6200}{6} = \$1.77$

PART D Multiplying by Percents Greater Than 100%

When the percent is greater than 100%, be careful to correctly convert the percent to a decimal or a mixed number. Remember to move the decimal point only two places to the left to find the decimal equivalent.

EXAMPLES:

a. What is 200% of $550?

$P = B \times R$, where $R = 200\% = 2$

$P = \$550 \times 2 = \$1,100$

b. Find 125% of 788.

$P = B \times R$, where $R = 125\% = 1.25 = 1\frac{1}{4}$

$P = 788 \times 1.25 = 985$

or

$P = 788 \times 1\frac{1}{4} = \dfrac{\overset{197}{\cancel{788}}}{1} \times \dfrac{5}{\cancel{4}_1} = 985$

c. What is 150% of $1,780?

$P = B \times R$, where $R = 150\% = 1.5 = 1\frac{1}{2}$

$P = \$1,780 \times 1.5 = \$2,670$

or

$P = \$1,780 \times 1\frac{1}{2} = \dfrac{\overset{890}{\cancel{\$1,780}}}{1} \times \dfrac{3}{\cancel{2}_1} = \$2,670$

Remember that when the base is multiplied by a rate that is greater than 100%, the percentage will always be larger than the base. If the rate is 200%, the percentage will be two times the base; if the rate is 300%, the percentage will be 3 times the base, etc.

PART E Applications

An application of percents is the distribution of overhead expenses among the various departments of a company. Overhead expenses are those expenses, such as heat, light, rent, janitorial service, and insurance, which are necessary to run a business. One way to distribute overhead expenses is to charge each department a percent of the overhead expenses based on the percent of floor space occupied by each department.

EXAMPLE:

A company's overhead expenses for one month total $18,400. Distribute this amount among the five departments, based on the percent of floor space occupied.

Department	Percent of Floor Space
A	14.3
B	33.3
C	16.7
D	25.2
E	10.5
Total	= 100.0

Note that the percents must add up to exactly 100% in order to distribute the full amount of overhead.

Chapter 3   Working with Percents

Department A: $18,400 × 14.3%
$18,400 × 0.143 = $2,631.20
Department B: $18,400 × 33.3%
$18,400 × 0.333 = $6,127.20
Department C: $18,400 × 16.7%
$18,400 × 0.167 = $3,072.80
Department D: $18,400 × 25.2%
$18,400 × 0.252 = $4,636.80
Department E: $18,400 × 10.5%
$18,400 × 0.105 = $1,932.00

The tabulated answer to the problem is as follows:

Department	Percent of Floor Space	Amount
A	14.3	$ 2,631.20
B	33.3	6,127.20
C	16.7	3,072.80
D	25.2	4,636.80
E	10.5	1,932.00
Total	= 100.0	$18,400.00

COMPLETE ASSIGNMENT 13

Name _____

Date _____

ASSIGNMENT 13
SECTION 3.2

Finding the Percentage

	Perfect Score	Student's Score
PART A	20	
PART B	20	
PART C	20	
PART D	20	
PART E	20	
TOTAL	100	

PART A Multiplying by Percents Between 1% and 100%

DIRECTIONS: Find the percentage. Round off your answers to two decimal places (2 points for each correct answer).

1. 25% of 24.4 = _____

2. 6% of 0.927 = _____

3. 3.4% of $375 = _____

4. 40% of 0.275 = _____

5. $8\frac{2}{3}$% of $2.50 = _____

6. 20% of 37.25 = _____

7. 22% of $3.65 = _____

8. 66% of 0.638 = _____

9. $14\frac{1}{4}$% of $860 = _____

10. 5.15% of $7,500 = _____

STUDENT'S SCORE _____

PART B Using Fractional Equivalents of Percents

DIRECTIONS: Find the percentage by changing the percent to its fractional equivalent. Round off your answers to two decimal places (2 points for each correct answer).

11. 25% of 148 = _____

12. 59% of 450 = _____

131

13. $6\frac{2}{3}\%$ of $90.45 = $ _____

14. $8\frac{1}{3}\%$ of $60.48 = $ _____

15. $6\frac{1}{4}\%$ of $6,400 = $ _____

16. $37\frac{1}{2}\%$ of 7,200 = _____

17. 80% of 0.35 = _____

18. $33\frac{1}{3}\%$ of $320.40 = $ _____

19. $12\frac{1}{2}\%$ of 0.56 = _____

20. 75% of 0.248 = _____

STUDENT'S SCORE _____

PART C *Multiplying by Percents Less Than 1%*

DIRECTIONS: Find the percentage. Round off your answers to two decimal places (2 points for each correct answer).

21. $\frac{1}{4}\%$ of 148 = _____

22. $\frac{1}{2}\%$ of 2,000 = _____

23. $\frac{2}{3}\%$ of 2,400 = _____

24. 0.39% of $200 = _____

Chapter 3   Working with Percents

25. 0.15% of $4,000 = _____

26. 0.01% of $827.20 = _____

27. $\frac{3}{4}$% of $4,000 = _____

28. $\frac{1}{6}$% of $64.20 = _____

29. $\frac{1}{8}$% of $8.24 = _____

30. 0.0125% of 2,000 = _____

STUDENT'S SCORE _____

PART D   *Multiplying by Percents Greater Than 100%*

DIRECTIONS: Find the percentage. Round off your answers to two decimal places (2 points for each correct answer).

31. 150% of $7.50 = _____

32. 110% of $6.48 = _____

33. $120\frac{1}{2}$% of $56 = _____

34. 200% of 150 = _____

35. 240% of 1.35 = _____

36. 500% of $90 = _____

37. 280% of $3,600 = _____

38. 115% of $3,300 = _____

39. 110% of 8,000 = _____

40. 214% of $6.45 = _____

STUDENT'S SCORE _____

PART E Applications

DIRECTIONS: Use a calculator to find the percentage as indicated (6 points for Problem 41 and 7 points each for Problems 42 and 43).

41. Fielding's Department Store had $12,000,000 in sales last year. The percentage of sales volume for each of six departments is listed below. Calculate the sales in dollars of each department and write your answers on the lines provided.

Department	Percent of Sales Volume	Amount
Women's Apparel	23.9	$ _____
Men's Apparel	20.2	_____
Children's Apparel	11.0	_____
Cosmetics	13.38	_____
Shoes	19.02	_____
Linens	12.5	_____
Total	100.0%	$12,000,000

42. The Mason Supply Company had a total of $30,360 for overhead expenses for the month of May. Distribute this amount among the eight departments, according to the percent of floor space occupied by each. Write your answers on the lines provided.

Department	Percent of Floor Space	Amount
A	14.8	$ _____
B	17.2	_____
C	$16\frac{2}{3}$	_____
D	6.9	_____
E	20.4	_____
F	$8\frac{1}{3}$	_____
G	15.7	_____
Total	100.0%	$30,360.00

Chapter 3   Working with Percents

43. The cost for Bill Daniels to have a new home built was estimated at $71,800. Of this amount, 15% was for electrical work; $\frac{1}{2}$% for insurance; 32% for labor; 30% for materials and supplies; $12\frac{1}{2}$% for plumbing; and 2% for miscellaneous expenses. The rest was the building contractor's profit. Find the amount estimated for each item. Write your answers on the lines provided. Check your work by adding the amounts.

	Percent	Amount
Electrical work	_____	$ _____
Insurance	_____	_____
Labor	_____	_____
Materials and supplies	_____	_____
Plumbing	_____	_____
Miscellaneous expenses	_____	_____
Profit	_____	_____
Total	100%	$71,800.00

STUDENT'S SCORE _____

Chapter 3 Working with Percents

SECTION 3.3 Finding the Rate

Percents are used to compare one item to another. The percent of rainy days to total days, the percent of cost to selling price, the percent of rent to total expenses, and the percent of wins to total games are a few examples.

When you are asked to find what percent one number is of another, you are expected to find what part one number is of another. As mentioned earlier in this chapter, a percentage refers to a part of, or portion of, a whole. A percentage problem is usually expressed in one of two ways.

1. 36 is what percent of 90? (36 is what part of 90?)
2. What percent of 90 is 36? (What part of 90 is 36?)

Remember that the numbers given in the above problem are the percentage (P = 36) and the base (B = 90). You are asked to find the rate (R). The formula for finding the rate can be found easily with the basic formula, P = B × R. Divide both sides of the equation by B so that R stands alone.

$$\frac{P}{B} = \frac{\cancel{B} \times R}{\cancel{B}} \quad \text{or} \quad \frac{P}{B} = R \quad \text{or} \quad R = \frac{P}{B}$$

Since the fraction bar in $\frac{P}{B}$ indicates division, the formula for finding R can be written as:

$$R = P \div B$$
$$\text{Rate} = \text{Percentage} \div \text{Base}$$

Before using the formula to find the rate, it is important to determine which of the given numbers in a problem is the percentage (P) and which is the base (B). The base is easily recognized because it always follows the word *of*. Be aware that the base is *not* always the larger of the two numbers, as you will see in Part C of this section. The percentage is the other given number.

To find what percent one number is of another, follow these two steps:

1. Divide the percentage by the base.
2. Change the quotient to a percent by multiplying it by 100 and adding a percent sign.

EXAMPLE:

42 is what percent of 168?

B = 168 and P = 42

Step 1. R = P ÷ B

R = 42 ÷ 168 = 0.25

Step 2. 0.25 = 0.25 = 25%

In Step 2, remember to use the shortcut: *To multiply any number by 100, move the decimal point two places to the right*. To check, multiply the given base by the rate. The product should equal the given percentage.

Check: $168 \times 25\% = \frac{\cancel{168}^{42}}{1} \times \frac{1}{\cancel{4}_1} = 42$

A standard calculator does *not* provide a conversion from the quotient, 0.25, to the percent rate, 25%. You must know that 0.25 is the decimal equivalent of 25%. You must mentally convert this decimal equivalent to the percent rate by moving the decimal point two places to the right. Once again, you must understand the problem's solution to be able to use the calculator successfully.

PART A Finding Percents Between 1% and 100%

The most common percents are between 1% and 100%. Generally, a percent is expressed to one decimal place. Therefore, division to find a percent is often carried out to four decimal places, so that the quotient can be rounded off to three places. Two of these three places are for moving the decimal point to the right when converting to a percent.

In the following examples, the steps are numbered to show both the division and the conversion to a percent.

EXAMPLES:

a. 25 is what percent of 120?

B = 120 and P = 25

Step 1. R = P ÷ B

R = 25 ÷ 120 = 0.2083

= 0.208

Step 2. 0.208 = 0.208 = 20.8%

Check: 120 × 20.8% = 120 × 0.208

= 24.96 (a little less than 25)

Note that when the quotient is rounded off, the check will not be exactly equal to the given percentage. The check will be a little *less* than the given percentage if the third decimal place in the quotient is rounded *down*.

b. $25.96 is what percent of $685?

B = $685 and P = $25.96

Step 1. R = P ÷ B

R = $25.96 ÷ $685 = 0.0378

= 0.038

Step 2. 0.038 = 0.038 = 3.8%

Check: $685 × 3.8% = $685 × 0.038

= $26.03 (a little more than $25.96)

Note that when the third decimal place in the quotient is rounded *up,* the check will be a little *more* than the given percentage.

c. What percent of $450.50 is $302.25?

B = $450.50 and P = $302.25

Step 1. R = P ÷ B

R = $302.25 ÷ $450.50 = 0.6709

= 0.671

Step 2. 0.671 = 0.671 = 67.1%

Check: $450.50 × 67.1% = $450.50 × 0.671

= $302.2855
(a little more than $302.25)

These examples are more easily solved with a calculator, because tedious division is not required and the checks are almost exact.

EXAMPLES:

a. 25 is what percent of 120?

	Key-enter	Display	
	25	25.	
	÷	25.	
	120	120.	
	=	0.2083333	*percent*
Check:	×	0.2083333	
	120	120.	
	=	24.999996	= 25

b. $25.96 is what percent of $685?

	Key-enter	Display	
	25.96	25.96	
	÷	25.96	
	685	685.	
	=	0.0378978	*percent*
Check:	×	0.0378978	
	685	685.	
	=	25.959993	= $25.96

c. What percent of $450.50 is $302.25?

	Key-enter	Display	
	302.25	302.25	
	÷	302.25	
	450.50	450.50	
	=	0.6709211	*percent*
Check:	×	0.6709211	
	450.50	450.50	
	=	302.24995	= $302.25

As in Part A of Section 3.2, there are three special cases which should always be performed mentally. The next examples are worked in detail to show you why you can find the rates mentally in these special cases.

EXAMPLES:

a. When P × 100 = B, then R = 1%.

R = ?   R = P ÷ B
P = 5.76   R = 5.76 ÷ 576
B = 576   R = 0.01 = 1%

Chapter 3 Working with Percents

b. When $P \times 10 = B$, then $R = 10\%$.

$R = ?$ $R = P \div B$
$P = 57.6$ $R = 57.6 \div 576$
$B = 576$ $R = 0.1 = 0.10 = 10\%$

c. When $P = B$, then $R = 100\%$.

$R = ?$ $R = P \div B$
$P = 576$ $R = 576 \div 576$
$B = 576$ $R = 1 = 100\%$

Note that any number is 100% of itself.

Occasionally, a percent must be expressed to two decimal places. In this case, the division must be carried out to five places, so that the quotient can be rounded off to four places.

EXAMPLE:

What percent of 460 is 250?

$B = 460$ and $P = 250$

Step 1. $R = P \div B$

$R = 250 \div 460 = 0.54347$
$= 0.5435$

Step 2. $0.5435 = 0.5435 = 54.35\%$

Check: $460 \times 54.35\% = 460 \times 0.5435$
$= 250.01$ (a little more than 250)

PART B Finding Percents Less Than 1%

When the base is more than 100 times greater than the percentage, the required percent is less than 1%. Therefore, the number in front of the percent sign must be a fraction or a decimal less than 1. As usual, write a zero to the left of the decimal point in the final answer when there is no digit there.

EXAMPLES:

a. 2 is what percent of 400?

$B = 400$ and $P = 2$

Step 1. $R = P \div B$

$R = 2 \div 400 = 0.005$

Step 2. $0.005 = 0.005 = 0.5\% = \frac{1}{2}\%$

Check: $400 \times \frac{1}{2}\% = 400 \times 0.005$
$= 2$

b. What percent of $50.63 is 48¢?

$B = \$50.63$ and $P = \$0.48$

Step 1. $R = P \div B$

$R = \$0.48 \div \$50.63 = 0.0094$
$= 0.009$

Step 2. $0.009 = 0.009 = 0.9\%$

Check: $\$50.63 \times 0.9\% = \$50.63 \times 0.009$
$= \$0.45567$ (a little less than $0.48)

c. What percent of 2,400 is 8?

$B = 2,400$ and $P = 8$

Step 1. $R = P \div B$

$R = 8 \div 2,400 = 0.00333$

Step 2. $0.00333 = 0.00333 = 0.333\%$
$= \frac{1}{3}\%$

Check: $2,400 \times \frac{1}{3}\% = \dfrac{\overset{8}{\cancel{2{,}400}}}{1} \times \dfrac{1}{\underset{1}{\cancel{300}}} = 8$

In these examples, note that in Step 1 the quotient in its decimal form has at least two zeros to the immediate right of the decimal point. This is true for decimal equivalents of all percents less than 1%.

PART C Finding Percents Greater Than 100%

When the base is less than the percentage, the required percent is greater than 100%. Therefore, the number in front of the percent sign must be a whole number, a mixed number, or a decimal greater than 100. Remember to correctly identify the base and the percentage. The number following the word *of* is always the base, even if it is smaller than the percentage.

EXAMPLES:

a. 693 is what percent of 231?

$B = 231$ and $P = 693$

Step 1. $R = P \div B$

$R = 693 \div 231 = 3$

Step 2. $3 = 3.00 = 3.00 = 300\%$

Check: $231 \times 300\% = 231 \times 3$
$= 693$

b. $110.10 is what percent of $101.00?

$B = \$101.00$ and $P = \$110.10$

Step 1. $R = P \div B$

$R = \$110.10 \div \$101.00 = 1.09009$
$= 1.0901$

Step 2. $1.0901 = 1.0901 = 109.01\%$

Check: $\$101.00 \times 109.01\% = \$101 \times 1.0901$
$= \$110.1001$
(a little more than $110.00)

c. $1\frac{7}{8}$ is what percent of $\frac{5}{6}$?

$B = \frac{5}{6}$ and $P = 1\frac{7}{8}$

Step 1. $R = P \div B$

$R = 1\frac{7}{8} \div \frac{5}{6} = \frac{15}{8} \times \frac{6}{5} = \frac{9}{4}$

$= 2\frac{1}{4} = 2.25$

Step 2. $2.25 = 2.25 = 225\%$

Check: $\frac{5}{6} \times 225\% = \frac{5}{6} \times 2.25$

$= \frac{5}{6} \times 2\frac{1}{4}$

$= \frac{5}{6} \times \frac{9}{4} = \frac{15}{8}$

$= 1\frac{7}{8}$

Omitting Step 2 is a frequent error in solving these problems. The quotient must be multiplied by 100 before the percent sign is added.

PART D Applications

In analyzing its business operations, a company is often interested in the percent of increase or decrease in its sales for comparable periods. Generally, this is expressed as a rate of change. (See Section 1.3, Part F.) This is the percent that the increase or decrease in sales is of the earlier period's total sales. A plus sign is placed before a rate of increase, and a minus sign before a rate of decrease.

EXAMPLE:

Given the following data, what is the rate of change in each month's sales?

Month	Sales Last Year	Sales This Year
Sept.	$26,000	$30,000
Oct.	15,000	18,000
Nov.	14,000	17,000
Dec.	21,000	18,000

The first step in solving this problem is to find the *amount* of change. Finding the difference between the sales for each month this year and last year results in the following:

Sept.: $30,000 - \$26,000 = \$4,000$ (inc.)
Oct.: $18,000 - \$15,000 = \$3,000$ (inc.)
Nov.: $17,000 - \$14,000 = \$3,000$ (inc.)
Dec.: $21,000 - \$18,000 = \$3,000$ (dec.)

The amount of change for each month is the increase or decrease, or P in the formula $R = P \div B$.

The second step is to identify which of the given numbers represent the base (B). Would the base be "Last Year" figures or "This Year" figures? In any comparison that involves periods of time, the earlier period always represents the base. This is true whether the amount in the later period is larger or smaller. Thus, each number in the "Last Year" column becomes the base.

Chapter 3 Working with Percents

The final step is to find the rate of change (R) for each month. Using the formula, $R = P \div B$:

Sept.: $R = \$4{,}000 \div \$26{,}000$

$ = 0.154 = +15.4\%$

Oct.: $R = \$3{,}000 \div \$15{,}000$

$ = 0.2 = +20.0\%$

Nov.: $R = \$3{,}000 \div \$14{,}000$

$ = 0.214 = +21.4\%$

Dec.: $R = \$3{,}000 \div \$21{,}000$

$ = 0.143 = -14.3\%$

To check, multiply the rate of change (R) by the amount in the earlier period (B); the product should be the increase or decrease (P).

The complete business report for the above problem is shown in Table 3-1. In most business reports arranged in columns and rows, the dollar sign and percent sign are used only with the numbers in the first row and the totals. (Note that the report in Table 3-1 does not have totals.) Also, in most reports, the rate is correctly expressed to one decimal place. Therefore, when the rate is an exact whole number, as for October, a zero is placed in the *tenths* position, making it 20.0%. This indicates that the tenths digit was not accidentally omitted.

Month	Last Year	This Year	Amount of Change	Rate of Change
September	$26,000	$30,000	$ 4,000	+ 15.4%
October	15,000	18,000	3,000	+ 20.0
November	14,000	17,000	3,000	+ 21.4
December	21,000	18,000	− 3,000	− 14.3

TABLE 3-1 Report of Monthly Sales

COMPLETE ASSIGNMENT 14

Name _____

Date _____

ASSIGNMENT 14
SECTION 3.3

Finding the Rate

	Perfect Score	Student's Score
PART A	30	
PART B	30	
PART C	30	
PART D	10	
TOTAL	100	

PART A Finding Percents Between 1% and 100%

DIRECTIONS: Find the percent as indicated and check. Round off your answers to one decimal place (3 points for each correct answer).

1. What percent of 25 is 3?

2. What percent of 60 is 15?

3. $2.64 is what percent of $50?

4. $19.50 is what percent of $260?

5. What percent is $13\frac{3}{4}$ of $187\frac{1}{2}$?

6. What percent is 15.9 of $238\frac{1}{2}$?

7. What percent of 0.56 is 0.36?

8. What percent of $5.00 is $3.25?

9. $350.50 is what percent of $525.75?

10. $5\frac{1}{2}$ is what percent of $60\frac{1}{2}$?

STUDENT'S SCORE _____

PART B Finding Percents Less Than 1%

DIRECTIONS: Find the percent as indicated and check. Express the percent either as a decimal rounded off to two decimal places or as a fraction reduced to lowest terms (3 points for each correct answer).

11.   What percent of 800 is 5?   17. $5.50 is what percent of $2,200?

                        _____   _____

12.   What percent of 500 is 0.5?

                        _____   18. $1.80 is what percent of $450?

                                                                   _____

13. $5.10 is what percent of $1,893?

                                                 19.   What percent of 90 is $\frac{9}{20}$?
                        _____
                                                                   _____

14.   What percent is $1.00 of $145?

                        _____

15.   What percent is 0.48 of 60?   20. $\frac{3}{8}$ is what percent of $41\frac{2}{3}$?

                        _____   _____

16.   200 is what percent of 52,000?

                        _____

STUDENT'S SCORE _____

Chapter 3   Working with Percents   145

PART C Finding Percents Greater Than 100%

DIRECTIONS: Find the percent as indicated and check. Round off your answers to one decimal place (3 points for each correct answer).

21.   What percent of 500 is 1,000?   26.   What percent is $373.06 of $162.20?

   _____   _____

22.   What percent of $2.75 is $2.75?   27.   What percent of 3.10 is 12.48?

   _____   _____

23. $6\frac{3}{4}$ is what percent of $1\frac{1}{8}$?   28.   What percent of 8.21 is 54.81?

   _____   _____

24. $\frac{3}{4}$ is what percent of $\frac{2}{3}$?   29.   614.68 is what percent of 270.4?

   _____   _____

25.   What percent is 468.3 of 267.6?   30. $10 is what percent of $4.30?

   _____   _____

STUDENT'S SCORE _____

PART D Applications

DIRECTIONS: Complete the following report. For each percent of change, place a plus sign before an increase and a minus sign before a decrease. Round off your answers to one decimal place (1 point for each correct answer).

MONTHLY SALES

	Month	Last Year	This Year	Amount of Change	Rate of Change
31.	July	$12,000	$15,000	$ _____	_____
32.	August	22,000	25,300	_____	_____
33.	September	24,000	20,000	_____	_____
34.	October	27,000	36,000	_____	_____
35.	November	28,000	24,500	_____	_____

STUDENT'S SCORE _____

Chapter 3   Working with Percents

SECTION 3.4 Finding the Base of a Percentage

The first type of percentage problem, finding the percentage, was covered in Section 3.2 of this chapter. The second type, finding the rate when the percentage and the base are given, was covered in Section 3.3. In this section, you will learn how to solve the third type of percentage problem, finding the base (B) when the rate (R) and the percentage (P) are given. These problems are usually expressed in one of two ways. For example:

1. 5% of what number is 25?
2. 25 is 5% of what number?

Remember that you have two known factors in the above problem: the percentage (P = 25) and the rate (R = 5%). The unknown number is the base (B). The formula to find the base can be found with the basic formula, P = B × R. Divide both sides of the equation by R so that B stands alone:

$$\frac{P}{R} = \frac{B \times \cancel{R}}{\cancel{R}} \quad \text{or} \quad \frac{P}{R} = B \quad \text{or} \quad B = \frac{P}{R}$$

Since the fraction bar in $\frac{P}{R}$ indicates division, the formula also can be stated as:

$$B = P \div R$$
Base = Percentage ÷ Rate

Before using the formula, convert the rate (R) to either a decimal or a fraction.

EXAMPLE:

5% of what number is 25?

$B = P \div R$

$B = 25 \div 5\% = 25 \div 0.05$

$ = 500$

or

$$B = 25 \div \frac{5}{100} = \frac{\cancel{25}^5}{1} \times \frac{100}{\cancel{5}_1}$$

$ = 500$

It is especially important to check this type of percentage problem, since it is the most difficult of the three types. To check, multiply the base (the answer) by the given rate (R); the product should equal the given percentage (P).

Check: $500 \times 5\% = 500 \times 0.05$
$\phantom{\text{Check: } 500 \times 5\%} = 25$

PART A Finding the Base When the Given Percent is Between 1% and 100%

Most applications that find the base use percents between 1% and 100%. When the percent is between 1% and 100%, the base is always larger than the percentage.

As stated earlier, the percent can be converted to either a decimal or a fraction. Without a calculator, it is usually easier to use the fractional equivalent of the percent.

EXAMPLES:

a. 20% of what number is 50?

$B = P \div R$

$B = 50 \div 20\% = 50 \div \frac{1}{5}$

$B = \frac{50}{1} \times \frac{5}{1} = 250$

Check: $250 \times 20\% = \dfrac{\cancel{250}^{50}}{1} \times \dfrac{1}{\cancel{5}_1} = 50$

b. 60¢ is $33\frac{1}{3}\%$ of what number?

$B = P \div R$

$B = \$0.60 \div 33\frac{1}{3}\% = \$0.60 \div \frac{1}{3}$

$B = \frac{\$0.60}{1} \times \frac{3}{1} = \$1.80$

Check: $\$1.80 \times 33\frac{1}{3}\% = \dfrac{\cancel{\$1.80}^{0.60}}{1} \times \dfrac{1}{\cancel{3}_1} = \$0.60$

c. $66\frac{2}{3}\%$ of what number is 150?

$B = P \div R$

$B = 150 \div 66\frac{2}{3}\% = 150 \div \frac{2}{3}$

$B = \frac{\cancel{150}^{75}}{1} \times \frac{3}{\cancel{2}_1} = 225$

Check: $225 \times 66\frac{2}{3}\% = \frac{\cancel{225}^{75}}{1} \times \frac{2}{\cancel{3}_1} = 150$

With a calculator, the decimal equivalent of the percent is easier to use. When converting the percent to a decimal, remember to drop the percent sign and move the decimal point two places to the left.

EXAMPLES:

a. 570 is 30% of what number?

$B = P \div R$

$B = 570 \div 30\% = 570 \div 0.3$

$ = 1,900$

Check: $1,900 \times 30\% = 1,900 \times 0.3$

$\phantom{\text{Check: }1,900 \times 30\%} = 570$

b. $120.23 is 15% of what number?

$B = P \div R$

$B = \$120.23 \div 15\% = \$120.23 \div 0.15$

$ = \$801.53$

Check: $\$801.53 \times 15\% = \$801.53 \times 0.15$

$\phantom{\text{Check: }\$801.53 \times 15\%} = \$120.2295$

$\phantom{\text{Check: }\$801.53 \times 15\%} = \$120.23$

With a calculator, two methods of finding a percentage can be used. The preceding *Example a* is used to illustrate these two methods in the following example.

EXAMPLE:

570 is 30% of what number?

a.

Key-enter	Display
570	570.
÷	570.
.3	0.3
=	1900

b.

Key-enter	Display
570	570.
÷	570.
30	30.
%	1900

With the first method, 30% must be converted to its decimal equivalent before the problem can be solved. With the second method, the percent key automatically converts 30% to its decimal equivalent.

As in Sections 3.2 and 3.3, there are three simple cases that should always be performed mentally.

EXAMPLES:

a. When R = 1%, then B = P × 100.

$B = ?$ $\quad\quad B = P \div R$

$P = 6.47$

$R = 1\% = 0.01 \quad B = 6.47 \div 0.01$

$ = 647$

or

$B = 6.47 \times 100 = 6.47 = 647$

b. When R = 10%, then B = P × 10.

$B = ?$ $\quad\quad B = P \div R$

$P = 64.7$

$R = 10\% = 0.1 \quad B = 64.7 \div 0.1$

$ = 647$

or

$B = 64.7 \times 10 = 64.7 = 647$

c. When R = 100%, then B = P.

$B = ?$ $\quad\quad B = P \div R$

$P = 647$

$R = 100\% = 1 \quad B = 647 \div 1$

$ = 647$

Again, note that 100% of a number is the same number.

Chapter 3 Working with Percents

PART B Finding the Base When the Given Percent is Less Than 1%

To find a number when the given percent is less than 1%, the same formula is used: $B = P \div R$. Before applying the formula, be careful when converting the percent. When converting to a decimal equivalent, be sure that there are at least two zeros to the immediate right of the decimal point. When converting to a fractional equivalent, remember that there must be at least two ending zeros in the denominator before the fraction is reduced. Use the shortcut to convert a percent in which the number is a fraction: *Drop the percent sign and write two zeros to the right of the number in the denominator;* for example, $\frac{3}{8}\% = \frac{3}{800}$; $\frac{4}{5}\% = \frac{4}{500}$; and $\frac{2}{3}\% = \frac{2}{300}$.

EXAMPLES:

a. $\frac{2}{5}\%$ of what number is 60?

$B = P \div R$

$B = 60 \div \frac{2}{5}\% = 60 \div \frac{2}{500}$

$B = \frac{\cancel{60}^{30}}{1} \times \frac{500}{\cancel{2}_{1}} = 15{,}000$

Check: $15{,}000 \times \frac{2}{5}\% = \frac{\cancel{15{,}000}^{30}}{1} \times \frac{2}{\cancel{500}_{1}} = 60$

b. 0.65% of what number is $22.68?

$B = P \div R$

$B = \$22.68 \div 0.65\% = \$22.68 \div 0.0065$

$\quad = \$3{,}489.23$

Check:

$\$3{,}489.23 \times 0.65\% = \$3{,}489.23 \times 0.0065$

$\quad = \$22.679995$

$\quad = \$22.68$

c. $15 is $\frac{3}{8}\%$ of what number?

$B = P \div R$

$B = \$15 \div \frac{3}{8}\% = \$15 \div \frac{3}{800}$

$B = \frac{\cancel{\$15}^{5}}{1} \times \frac{800}{\cancel{3}_{1}} = \$4{,}000$

Check:

$\$4{,}000 \times \frac{3}{8}\% = \frac{\cancel{\$4{,}000}^{5}}{1} \times \frac{3}{\cancel{800}_{1}} = \$15$

PART C Finding the Base When the Given Percent is Greater Than 100%

To find the base when the given percent is greater than 100%, the same formula is used:

$B = P \div R$

Before applying the formula, take particular care in converting the percent. Remember that the converted percent must be a whole number, a mixed number, or a decimal greater than 1.

When the given percent is greater than 100%, the required base must be less than the given percentage.

EXAMPLES:

a. 125% of what number is $7.25?

$B = P \div R$

$B = \$7.25 \div 125\% = \$7.25 \div 1.25$

$\quad = \$5.80$

or

$B = \$7.25 \div 125\% = \$725 \div 1\frac{1}{4}$

$B = \frac{\$7.25}{1} \times \frac{4}{5} = \frac{\$29.00}{5} = \$5.80$

Check: $\$5.80 \times 125\% = \$5.80 \times 1.25$

$\quad = \$7.25$

b. $80.45 is 210% of what number?

$B = P \div R$

$B = \$80.45 \div 210\% = \$80.45 \div 2.1$
$ = \$38.31$

Check: $\$38.31 \times 210\% = \$38.31 \times 2.1$
$\phantom{\text{Check: } \$38.31 \times 210\%} = \$80.451$
$\phantom{\text{Check: } \$38.31 \times 210\%} = \$80.45$

c. 30.82 is 335% of what number?

$B = P \div R$

$B = 30.82 \div 335\% = 30.82 \div 3.35$
$ = 9.2$

Check: $9.2 \times 335\% = 9.2 \times 3.35$
$\phantom{\text{Check: } 9.2 \times 335\%} = 30.82$

PART D Applications

One application of finding the base when the percentage and rate are given is finding the selling price of an article when the cost of the article and the percent that the cost is of the selling price are known. In terms of base, rate, and percentage:

Known cost of the article	= P
Known percent that the cost is of the selling price	= R
Unknown selling price	= B

Thus, the applicable formula is $B = P \div R$. This formula can be restated as:

Selling Price = Given Cost ÷ Given Percent

In other words, divide the given cost by the given percent to find the selling price. To check, multiply the answer (selling price) by the given percent; the product should equal the cost.

EXAMPLE:

Find the selling price of a telephone which costs $88 if the cost is 80% of the selling price.

$B = P \div R$

Selling Price = Cost ÷ Percent
$\phantom{\text{Selling Price}} = \$88 \div 80\%$
$\phantom{\text{Selling Price}} = \$88 \div 0.8$
$\phantom{\text{Selling Price}} = \$110$

Check: $\$110 \times 80\% = \$110 \times 0.8$
$\phantom{\text{Check: } \$110 \times 80\%} = \$88$

Generally, sales are advertised as a percent off the original price. For example, a "30–50% off" sale means 30–50% off the original price. It is possible to calculate the original price of an article by knowing the sale price and the discounted percent, or the "percent off."

EXAMPLE:

Find the original price of a radio which is on sale for $60 if the sale price is 20% off the original price.

$60 is 80% (100% − 20%) of the original price.

$B = P \div R$

Original Price = Cost ÷ Percent
$\phantom{\text{Original Price}} = \$60 \div 80\%$
$\phantom{\text{Original Price}} = \$60 \div 0.8$
$\phantom{\text{Original Price}} = \$75$

Check: $\$75 \times 80\% = \$75 \times 0.8$
$\phantom{\text{Check: } \$75 \times 80\%} = \$60$

or

$\$75 \times 20\% = \$75 \times 0.2$
$ = \$15 \text{ (discount)}$

$\$75 - \$15 = \$60$

COMPLETE ASSIGNMENT 15

Name _____

Date _____

ASSIGNMENT 15
SECTION 3.4

Finding the Base of a Percentage

	Perfect Score	Student's Score
PART A	45	
PART B	15	
PART C	30	
PART D	10	
TOTAL	100	

PART A Finding the Base When the Given Percent is Between 1% and 100%

DIRECTIONS: Find the number as indicated and check. Round off your answers to two decimal places (3 points for each correct answer).

1. 116 is 12% of what number?

2. 518 is 80% of what number?

3. 66 is 40% of what number?

4. 130 is $16\frac{1}{4}$% of what number?

5. 65% of what number is $62.92?

6. 45% of what number is $15.75?

7. $26.48 is $82\frac{1}{4}$% of what number?

8. $4.16 is $32\frac{3}{4}$% of what number?

9. 95% of what number is $31.35?

10. 85% of what number is $10.40?

11. 25% of what number is $\frac{1}{2}$?

12. 75% of what number is $\frac{9}{16}$?

13. $0.87\frac{1}{2}$ is $62\frac{1}{2}$% of what number?

14. 0.5 is $12\frac{1}{2}$% of what number?

15. 65% of what number is 0.13?

STUDENT'S SCORE _____

Unit 1   Fundamentals

PART B Finding the Base When the Given Percent is Less Than 1%

DIRECTIONS: Find the number as indicated and check. Round off your answers to two decimal places (3 points for each correct answer).

16. $9.00 is $\frac{1}{3}$% of what number?

19.   0.125% of what number is $18.75?

20.   42¢ is 0.07% of what number?

17. $1.50 is $\frac{1}{2}$% of what number?

18.   0.375% of what number is $2.25?

STUDENT'S SCORE _____

PART C Finding the Base When the Given Percent is Greater Than 100%

DIRECTIONS: Find the number as indicated and check. Round off your answers to two decimal places (3 points for each correct answer).

21.   5 is 125% of what number?

25.   65¢ is 172% of what number?

22. $3\frac{3}{4}$ is 150% of what number?

26. $1.95 is 130% of what number?

23.   150% of what number is 2,700?

27. $162\frac{1}{2}$% of what number is 75¢?

24.   160% of what number is 4.6?

28.   375% of what number is $7.53?

Chapter 3   Working with Percents   153

29. $5,200 is 325% of what number?   **30.** $900 is 225% of what number?

                                                _____   _____

STUDENT'S SCORE _____

PART D Applications

DIRECTIONS: Find the selling price (2 points for each correct answer).

31.   Cost = $464; percent of selling price = $62\frac{1}{2}$%; selling price =   _____

32.   Cost = $6.45; percent of selling price = 60%; selling price =   _____

33.   Cost = $54.99; percent of selling price = 65%; selling price =   _____

DIRECTIONS: Find the original price (2 points for each correct answer).

34.   Sale price = $9.00; discount rate = 30%; original price =   _____

35.   Sale price = $29.99; discount rate = 50%; original price =   _____

STUDENT'S SCORE _____

CHAPTER 4

Working with Weights and Measures

A system of weights and measures is vital to business and industry. Standardized units of measurement are needed in the manufacture and sale of goods so that sizes and weights have the same meaning for all concerned.

Most of the world uses the **International Metric System** of measurement, often called only **SI** (for *Le Système International d'Unités*). It is a modernized metric system adopted in 1960. The United States is in the process of converting to the metric system from the **customary or English system**. Some of this conversion has already been completed, and much more will take place in the future.

Metric measurements that are not exact are normally written as decimal numbers. For your information, most European countries use a comma in place of a decimal point in a decimal number. Thus, 5.329 is written as 5,329. This method of writing decimals will *not* be used in this text, since it is inconsistent with calculators and computers and might cause confusion. Also, instead of using commas to separate number groups, Europeans leave spaces between them. For example, 4,236,078 is written as 4 236 078. This method of writing large numbers will also *not* be used in this text.

In this chapter, you will learn how to work in each system and how to use the measures in applications.

SECTION 4.1 Working with the Customary and the Metric Systems

A particular weight or measure, whether customary or metric, consists of a number followed by the name of the measurement unit (sometimes called the **denomination**), such as 5 feet, 2 meters, 3 pounds, or 4 kilograms. To save time and space, *symbols* representing the various units are used. No period is placed after the symbol, and no "s" is used for plurals. Thus, the measures previously mentioned would appear as:

    5 ft   2 m   3 lb   4 kg

Symbols that are commonly used in both systems are shown in Tables 4-1 and 4-2.

While the customary system has been in existence for a long time, it can be difficult to learn. Originally, an inch was equal to three barleycorns laid end to end; a yard was the distance from the tip of a king's nose to the end of his thumb when his arm was stretched out. Eventually, of course, all units were defined more precisely. However, when you examine Table 4-1, you can see that you must know how many smaller units there are in the next larger unit to convert a measurement from one unit to another. There is no set pattern. For example, 12 inches equal one foot, three feet equal one yard, and $5\frac{1}{2}$ yards equal one rod. To add to the difficulty, there are two measures for capacity: a dry measure and a liquid measure. Finally, the names of the measuring units in the customary system show no particular pattern.

Table 4-2 gives the most common metric weights and measures. Note that the measures in the metric system are based on multiples of 10. Also note that there is a pattern to the smaller and larger parts of a base unit of measure. For example, consider the following three base units: the meter (a measure of length), the liter (a measure of capacity), and the gram (a measure of weight). Larger and smaller units of measure are created by combining a base unit with a prefix. (A *prefix* is one or more letters

	Unit	Symbol	Equivalents in Other Units
Linear (Length or Distance)	inch*	in.	0.083 feet or 0.028 yards
	foot	ft	12 inches or 0.333 yards
	yard	yd	3 feet or 36 inches
	rod	rd	$5\frac{1}{2}$ yards or $16\frac{1}{2}$ feet
	mile	mi	320 rods or 5,280 feet
Area	square inch	sq in.	0.007 square feet
	square foot	sq ft	144 square inches
	square yard	sq yd	9 square feet
	square rod	sq rd	$30\frac{1}{4}$ square yards
	acre	A	160 square rods
	square mile	sq mi	640 acres or 1 section (sec)
	township	twp	36 square miles
Volume	cubic inch	cu in.	0.00058 cubic feet
	cubic foot	cu ft	1,728 cubic inches
	cubic yard	cu yd	27 cubic feet
Weight	grain	gr	0.037 drams
	dram	dr	$27\frac{11}{32}$ grains
	ounce	oz	16 drams
	pound	lb	16 ounces
	hundredweight	cwt	100 pounds
	ton	T	2,000 pounds or 20 hundredweight
Capacity (Dry Measure)	pint	pt	$\frac{1}{2}$ quart
	quart	qt	2 pints
	peck	pk	8 quarts
	bushel	bu	4 pecks
Capacity (Liquid Measure)	pint	pt	16 ounces
	quart	qt	2 pints
	gallon	gal	4 quarts
	barrel	bbl	$31\frac{1}{2}$ gallons
	cubic foot	cu ft	$7\frac{1}{2}$ gallons
Household Measure	teaspoon	tsp or t	0.333 tablespoons
	tablespoon	tbsp or T	3 teaspoons
	cup	c	16 tablespoons
	pint	pt	2 cups
Time	second	s	0.017 minutes
	minute	min	60 seconds
	hour	h	60 minutes
	day	da	24 hours
	week	wk	7 days
	month	mo	$4\frac{1}{3}$ weeks (generally)
	year	yr	365 days (366 in a leap year)

*The symbol for ''inch'' ends with a period to prevent confusion between the word ''in'' and this symbol.

TABLE 4-1 *Customary System of Weights and Measures*

	Unit	Symbol	Equivalent to	Equivalents in Meters	Number of Units in One Meter
Linear	millimeter	mm		0.001 m	1 m = 1,000 mm
	centimeter	cm	10 millimeters	0.01 m	1 m = 100 cm
	decimeter	dm	10 centimeters	0.1 m	1 m = 10 dm
	meter	m	10 decimeters	base unit	base unit
	dekameter	dam	10 meters	10 m	1 m = 0.1 dam
	hectometer	hm	10 dekameters	100 m	1 m = 0.01 hm
	kilometer	km	10 hectometers	1000 m	1 m = 0.001 km
Area (Dry land)	centare	ca	1 square meter		
	hectare	ha	10,000 square meters		
Area	square millimeter	mm^2			
	square centimeter	cm^2	100 square millimeters		
	square decimeter	dm^2	100 square centimeters		
	square meter	m^2	100 square decimeters		
Volume	cubic millimeter	mm^3			
	cubic centimeter*	cm^3	1,000 cubic millimeters		
	cubic decimeter**	dm^3	1,000 cubic centimeters		
	cubic meter	m^3	1,000 cubic decimeters		

	Unit	Symbol	Equivalent to	Equivalents in Grams	Number of Units in One Gram
Weight	milligram	mg		0.001 g	1 g = 1,000 mg
	centigram	cg	10 milligrams	0.01 g	1 g = 100 cg
	decigram	dg	10 centigrams	0.1 g	1 g = 10 dg
	gram	g	10 decigrams	base unit	base unit
	dekagram	dag	10 grams	10 g	1 g = 0.1 dag
	hectogram	hg	10 dekagrams	100 g	1 g = 0.01 hg
	kilogram***	kg	10 hectograms	1,000 g	1 g = 0.001 kg
	metric ton	t	1,000 kilograms		

	Unit	Symbol	Equivalent to	Equivalents in Liters	Number of Units in One Liter
Capacity (Liquid and Dry Measure)	milliliter	mL		0.001 L	1 L = 1,000 mL
	centiliter	cL	10 milliliters	0.01 L	1 L = 100 cL
	deciliter	dL	10 centiliters	0.1 L	1 L = 10 dL
	liter	L	10 deciliters	base unit	base unit
	dekaliter	daL	10 liters	10 L	1 L = 0.1 daL
	hectoliter	hL	10 dekaliters	100 L	1 L = 0.01 hL
	kiloliter	kL	10 hectoliters	1,000 L	1 L = 0.001 kL

*Used for measuring very small quantities of liquid such as medicine: 1 cubic centimeter (cc) = 1 milliliter.
**Used for measuring most quantities of liquid such as gasoline and milk: 1 cubic decimeter (dm^3) = 1 liter.
***The kilogram is the basic unit which is in common everyday use: 1 kg = weight of 1 liter of water.

TABLE 4-2 *Metric System of Weights and Measures*

that precede a word and change its meaning.) *Kilo* is a prefix meaning 1,000, and *centi* is a prefix meaning 0.01. When these prefixes are combined with the base unit *meter,* two new units of measure are formed: *kilometer* and *centimeter.*

There are two types of prefixes. Those used for measures *larger* than the base unit are of Greek origin, such as kilo, hecto, and deka. Those used for measures *smaller* than the base unit are of Latin origin, such as deci, centi, and milli. The meanings of these prefixes are illustrated in the following table.

Larger-than-Base Prefixes

Prefix	Meaning
kilo	1,000 × base unit
hecto	100 × base unit
deka	10 × base unit

Smaller-than-Base Prefixes

Prefix	Meaning
deci	0.1 × base unit
centi	0.01 × base unit
milli	0.001 × base unit

Note that when the prefixes are written on a straight line, they correspond to the place values of our decimal number system:

kilo	hecto	deka	**base unit**	deci	centi	milli
↑	↑	↑	↑	↑	↑	↑
1,000	100	10	1	0.1	0.01	0.001

According to the above illustrations, some metric measurements can be defined as follows:

1. A milliliter is a thousandth of a liter or 0.001 liter.
2. A centigram is a hundredth of a gram or 0.01 gram.
3. A kilometer is one thousand times a meter or 1,000 meters.

There are other prefixes, but they are used mostly for scientific work. Of those given, hecto, deka, and deci are not used very much in daily life. For example, the 400 meter dash is never called the 4 hectometer dash, nor are 40 liters of gasoline called 4 dekaliters. Nevertheless, Table 4-2 includes these prefixes to show how all the measures in the system are related.

Note that the gram in Table 4-2 is listed as a base unit. A gram, about the weight of a paper clip, is simply too small to be commonly used. More commonly used is the kilogram (1,000 g), so some persons refer to it as the *basic* unit for measuring weight. Grams are used extensively in laboratories and drugstores, where small weights are common.

PART A Changing to the Next Smaller or the Next Larger Unit

In the customary system, a measure can consist of two or more parts, such as 9 feet 3 inches. In the metric system, measures are rarely expressed in more than one unit. Decimal parts of a unit are used, such as 4.5 centimeters instead of 4 centimeters 5 millimeters.

Changing to the Next Smaller or the Next Larger Unit in the Customary System. To change a measure from a given unit to the next smaller one, multiply the quantity in the given measure by the number of the smaller units contained in one unit of the given measure. Continue this process if even smaller units are desired.

EXAMPLES:

a. Change $4\frac{1}{3}$ yards to feet.

1 yd = 3 ft

$$4\frac{1}{3} \times 3 = \frac{13}{3} \times 3 = 13 \text{ ft}$$

b. Change $\frac{1}{2}$ ton to pounds.

1 T = 2,000 lb

$$\frac{1}{2} \times 2,000 = 1,000 \text{ lb}$$

c. Change 2 square yards to square inches.

1 sq yd = 9 sq ft
2 × 9 = 18 sq ft

1 sq ft = 144 sq in.
18 × 144 = 2,592 sq in.

Chapter 4 Working with Weights and Measures

Using a calculator to change a measure from a given unit to the next smaller unit is a simple task. Follow the same logic presented in the previous examples.

EXAMPLES:

a. Change $4\frac{1}{3}$ yards to feet.

Key-enter	Display	
4.3333	4.3333	
$\times$	4.3333	
3	3.	
=	12.9999	= 13

The decimal equivalent of $4\frac{1}{3}$ is 4.3333. The answer is 12.9999 feet, or 13 feet.

b. Change $\frac{1}{2}$ ton to pounds.

Key-enter	Display
0.5	0.5
$\times$	0.5
2000	2000.
=	1000.

The decimal equivalent of $\frac{1}{2}$ is 0.5. The answer is 1,000 pounds.

c. Change 2 square yards to square inches.

Key-enter	Display	
2	2.	
$\times$	2.	
9	9.	
=	18.	*sq ft*
$\times$	18.	
144	144.	
=	2592.	*sq in.*

The answer is 2,592 square inches.

If the given measure has two or more parts, such as 5 yards 3 feet or 6 yards 4 feet 2 inches, follow these steps:

1. Start with the largest unit (the first one) and change it to the next smaller unit.
2. Add the converted units from Step 1 to the same units of the given measure.
3. Continue this process until the desired unit is reached.

EXAMPLES:

a. Change 3 feet 8 inches to inches.

1 ft = 12 in.
3 $\times$ 12 = 36 in.
36 in. + 8 in. = 44 in.

b. Change 5 gallons 3 quarts 1 pint to pints.

1 gal = 4 qt
5 $\times$ 4 = 20 qt
20 qt + 3 qt = 23 qt

1 qt = 2 pt
23 $\times$ 2 = 46 pt
46 pt + 1 pt = 47 pt

To change a measure from a given unit to the next larger one, divide the quantity in the given measure by the number of the given units contained in one unit of the larger measure. Continue this process if larger units are desired.

EXAMPLES:

a. Change 40 ounces to pounds and ounces.

1 lb = 16 oz
40 $\div$ 16 = 2.5 lb
2.5 lb = 2 lb 8 oz

Note that if you want the answer expressed only in pounds, it would be $2\frac{8}{16}$ lb, $2\frac{1}{2}$ lb, or 2.5 lb.

b. Change 45 cubic feet to cubic yards.

1 cu yd = 27 cu ft
45 $\div$ 27 = 1 cu yd 18 cu ft
$= 1\frac{18}{27}$ cu yd
$= 1\frac{2}{3}$ cu yd

c. Change 35 quarts to gallons and quarts.

1 gal = 4 qt
35 $\div$ 4 = 8.75 gal
8.75 gal = 8 gal 3 qt

Using a calculator to change a measure to the next larger unit is not a straightforward task. Follow the same logic presented in the preceding examples, but convert the decimal part of the answer to the smaller unit.

EXAMPLES:

a. Change 40 ounces to pounds and ounces.

1 lb = 16 oz

Key-enter	Display	
40	40.	
÷	40.	
16	16.	
=	2.5	*lb*
−	2.5	
2	2.	
=	0.5	*decimal part*
×	0.5	
16	16.	
=	8.	*oz*

The answer is 2 lb 8 oz. The decimal part of the answer is 0.5. Multiply 0.5 by 16 to obtain 8 ounces. Then add the whole-number part of the answer, 2, to obtain the final answer of 2 lb 8 oz.

b. Change 45 cubic feet to cubic yards.

1 cu yd = 27 cu ft

Key-enter	Display	
45	45.	
÷	45.	
27	27.	
=	1.6666667	*cu yd*

The answer is 1.6666667 cu yd, or $1\frac{2}{3}$ cu yd.

c. Change 35 quarts to gallons and quarts.

1 gal = 4 qt

Key-enter	Display	
35	35.	
÷	35.	
4	4.	
=	8.75	*gal*
−	8.75	
8	8.	
=	0.75	*decimal part*
×	0.75	
4	4.	
=	3.	*qt*

The answer is 8 gal 3 qt. The decimal part of the answer is 0.75. Multiply 0.75 by 4 to obtain 3 quarts. Then add the whole-number part of the answer, 8, to obtain the final answer of 8 gal 3 qt.

If the given measure has two or more parts, follow these steps:

1. Start with the *smallest* unit (the last one) and change it to the next larger unit.
2. Add the converted units from Step 1 to the same units of the given measure.
3. Continue this process until the desired unit is reached.

EXAMPLES:

a. Change 3 hours 48 minutes 15 seconds to hours.

1 min = 60 s
15 ÷ 60 = 0.25 min
48 min + 0.25 min = 48.25 min

1 h = 60 min
48.25 ÷ 60 = 0.8042 h
3 h + 0.8042 h = 3.8042 h

b. Change 3 quarts 1 pint to part of a gallon.

1 qt = 2 pt
$1 \div 2 = \frac{1}{2}$ qt

$3 \text{ qt} + \frac{1}{2} \text{ qt} = 3\frac{1}{2}$ qt

1 gal = 4 qt

$3\frac{1}{2} \div 4 = \frac{7}{2} \times \frac{1}{4}$

$= \frac{7}{8}$ gal

Changing to the Next Smaller or the Next Larger Unit in the Metric System. Changing a metric measure from a given unit to another unit, either smaller or larger, is very easy. Each unit is 10 times the next smaller unit in linear, weight, and capacity measurements; 100 times the next smaller unit in area measurements; and 1,000 times the next smaller unit in volume measurements.

To change a measure from a given unit to the next smaller one, multiply the quantity in the given measure by the number of the smaller units contained in one unit of the given measure. The number of smaller units contained in one unit of the given measure is always 10, 100, or 1,000, depending on the kind of measure. Therefore, use the shortcut for multiplying. If the quantity is a whole number, write one zero to the right of the number for each zero in the multiplier. If the quantity is a decimal number, move the decimal point one place to the right for each zero in the multiplier.

EXAMPLES:

a. Change 7 centigrams to milligrams.

1 cg = 10 mg
7 × 10 = 70 mg

b. Change 4.5 square decimeters to square centimeters.

1 dm^2 = 100 cm^2
4.5 × 100 = 450 cm^2

c. Change 85.321 cubic meters to cubic decimeters.

1 m^3 = 1,000 dm^3
85.321 × 1,000 = 85,321 dm^3

If a still smaller unit is desired, keep multiplying by 10, 100, or 1,000 until the required unit is reached. The work can be shortened if a base unit is reached in the process. Multiply the quantity of the base unit by the number of the required smaller units contained in one base unit, as given in Table 4-2.

EXAMPLES:

a. Change 1.5 kilograms to centigrams.

1 kg = 1,000 g
1.5 × 1,000 = 1,500 g

1 g = 100 cg
1,500 × 100 = 150,000 cg

b. Change 1.3 square meters to square centimeters.

1 m^2 = 100 dm^2
1.3 × 100 = 130 dm^2

1 dm^2 = 100 cm^2
130 × 100 = 13,000 cm^2

Note that this calculation can be done in one step, as follows:

1.3 × 100 × 100 = 13,000 cm^2

c. Change 0.5 cubic decimeters to cubic millimeters.

1 dm^3 = 1,000 cm^3
0.5 × 1,000 = 500 cm^3

1 cm^3 = 1,000 mm^3
500 × 1,000 = 500,000 mm^3

or

0.5 × 1,000 × 1,000 = 500,000 mm^3

To change a measure from a given unit to the next larger one, divide the quantity in the given measure by the number of the given units contained in one unit of the larger measure. Use the shortcut for division by moving the decimal point one place to the left for each zero in the divisor.

EXAMPLES:

a. Change 3,250 milliliters to centiliters.

1 cL = 10 mL
3,250 ÷ 10 = 325 cL

b. Change 47.25 square decimeters to square meters.

1 m^2 = 100 dm^2
47.25 ÷ 100 = 0.4725 m^2

c. Change 8,325 cubic centimeters to cubic decimeters.

1 dm^3 = 1,000 cm^3
8,325 ÷ 1,000 = 8.325 dm^3

If a still larger unit is desired, keep dividing by 10, 100, or 1,000 until the required unit is reached. The work can be shortened if a base unit is reached in the process. Divide the quantity of the base unit by the number of base units contained in one unit of the required larger measure.

EXAMPLES:

a. Change 9,000 milliliters to dekaliters.

1 L = 1,000 mL
9,000 ÷ 1,000 = 9L

1 daL = 10L
9 ÷ 10 = 0.9 daL

b. Change 5,505 square centimeters to square meters.

$1 \text{ dm}^2 = 100 \text{ cm}^2$
$5,505 \div 100 = 55.05 \text{ dm}^2$
$1 \text{ m}^2 = 100 \text{ dm}^2$
$55.05 \div 100 = 0.5505 \text{ m}^2$

c. Change 100,000 cubic millimeters to cubic decimeters.

$1 \text{ cm}^3 = 1,000 \text{ mm}^3$
$100,000 \div 1,000 = 100 \text{ cm}^3$
$1 \text{ dm}^3 = 1,000 \text{ cm}^3$
$100 \div 1,000 = 0.1 \text{ dm}^3$

After you have learned to change one metric unit to another by using these methods, you may wish to use another conversion method. This does not require a table, but you must know the order of the prefixes. Since it is easier to work with the prefixes when they are listed horizontally, that arrangement is used here.

                 base
kilo   hecto   deka   **unit**   deci   centi   milli

Start at the prefix of the given measurement and count the number of places up to the prefix to which the given measure is to be changed. Then follow these three rules:

1. For linear measures (length or distance), when changing to a *smaller* unit, move the decimal point in the quantity the same number of places to the right. Write as many zeros to the right as necessary. This is the same as multiplication by a multiple of 10. When changing to a *larger* unit, move the decimal point in the quantity the same number of places to the left. Write as many zeros to the left as necessary. This is the same as division by a multiple of 10.

EXAMPLES:

a. Change 12.75 kiloliters to deciliters. Starting at *kilo*, count four places to the right, to *deci*.

12.75 kL = 12.7500 kL = 127,500 dL

b. Change 1,670 millimeters to meters. Starting at *milli*, count three places to the left, to the base unit (meters).

1,670 mm = 1,670 mm = 1.67 m

2. For square measures (area), count the places to the right or left as shown. However, when you move the decimal point to the right or left, move it *twice* the number of places counted.

EXAMPLES:

a. Change 9.125 square meters to square millimeters. Starting at the *base unit*, count three places to the right, to *milli*. Because of the square units, move the decimal point *six* places to the right.

$9.125 \text{ m}^2 = 9.125000 \text{ m}^2$
           = 9,125,000 \text{ mm}^2$

b. Change 25,000 square centimeters to square meters. Starting at *centi*, count two places to the left, to the *base unit*. Because of the square units, move the decimal point *four* places to the left.

$25,000 \text{ cm}^2 = 25,000 \text{ cm}^2$
             = 2.5000 \text{ m}^2 \text{ or } 2.5 \text{ m}^2$

3. For cubic measures (volume), count the places to the right or left as shown. However, when you move the decimal point to the right or left, move it *three times* the number of places counted.

EXAMPLES:

a. Change 4.625 cubic meters to cubic millimeters. Starting at the *base unit*, count three places to the right, to *milli*. Because of the cubic units, move the decimal point *nine* places to the right.

$4.625 \text{ m}^3 = 4.625000000 \text{ m}^3$
          = 4,625,000,000 \text{ mm}^3$

b. Change 325,650 cubic centimeters to cubic meters. Starting at *centi*, count two places to the left, to the *base unit*. Because of the cubic units, move the decimal point *six* places to the left.

$325,650 \text{ cm}^3 = 325,650. \text{ cm}^3$
            = 0.325650 \text{ m}^3 \text{ or } 0.32565 \text{ m}^3$

You should be aware that many of these examples of converting measurements to other units do not occur in ordinary usage. They are given here only to illustrate conversions within the metric system.

Chapter 4 Working with Weights and Measures

PART B Converting from One System to the Other

Because the metric system is not used in all countries, you should know how to convert a measure in one system to its equivalent in the other system. Table 4-3 shows some commonly used metric and customary equivalents.

Converting Measures and Weights. To change a measure expressed in one system to its equivalent in the other system, multiply the quantity in the given measure by the number of equivalent units in the other system.

EXAMPLES:

a. Convert 10 centimeters to inches.

   1 cm = 0.394 in.
   10 × 0.394 = 3.94 in.

	Customary to Metric	**Metric to Customary**
Linear	1 in. = 2.54 cm or 25.4 mm 1 ft = 0.305 m or 30.5 cm 1 yd = 0.914 m or 9.14 dm 1 rd = 5.029 m 1 mi = 1.609 km	1 mm = 0.0394 in. 1 cm = 0.394 in. 1 dm = 3.937 in. 1 m = 39.37 in. or 3.281 ft or 1.094 yd 1 km = 0.621 mi
Area	1 sq in. = 6.452 cm^2 1 sq ft = 0.093 m^2 1 sq yd = 0.836 m^2 1 A = 0.405 ha	1 cm^2 = 0.155 sq in. 1 m^2 = 10.764 sq ft 1 km^2 = 0.386 sq mi 1 ha = 2.471 A
Volume	1 cu in. = 16.387 cm^3 1 cu ft = 0.0283 m^3 1 cu yd = 0.765 m^3	1 cm^3 = 0.061 cu in. 1 dm^3 = 0.0353 cu ft 1 m^3 = 1.308 cu yd
Weight	1 gr = 0.0648 g 1 dr = 1.772 g 1 oz = 28.350 g 1 lb = 453.592 g or 0.454 kg 1 T = 0.907 t	1 mg = 0.0154 gr 1 cg = 0.154 gr 1 dag = 1.543 gr 1 g = 0.0353 oz 1 kg = 2.205 lb 1 t = 1.102 T
Capacity (Dry Measure)	1 pt = 0.551 L 1 qt = 1.101 L	1 L = 1.816 pt 1 L = 0.908 qt
Capacity (Liquid Measure)	1 pt = 0.473 L 1 qt = 0.946 L 1 gal = 3.785 L	1 L = 2.114 pt 1 L = 1.057 qt 1 L = 0.264 gal
Household Measure	1 t = 5 mL 1 T = 15 mL 1 c (dry)* = 275 mL 1 c (liquid) = 237 mL 1 pt = 0.473 L	1 mL = 0.2 t 1 dL = 6.8 T 1 L = 3.6 c (dry) 1 L = 4.2 c (liquid)

Most dry ingredients are weighed: 1 c sugar = 190 g; 1 c flour = 140 g

TABLE 4-3 *Equivalents of Metric and Customary Measures*

b. Convert 350 feet to meters.

1 ft = 0.305 m
350 × 0.305 = 106.75 m

c. Convert 80 miles to kilometers.

1 mi = 1.609 km
80 × 1.609 = 128.72 km

d. Convert 250 kilometers to miles.

1 km = 0.621 mi
250 × 0.621 = 155.25 mi

e. Convert 3 pounds to kilograms.

1 lb = 453.592 g = 0.454 kg
3 × 0.454 = 1.362 kg

f. Convert 75 kilograms to pounds.

1 kg = 2.205 lb
75 × 2.205 = 165.375 lb

g. Convert 5 liters to gallons.

1 L = 0.264 gal
5 × 0.264 = 1.32 gal

h. Convert 18 quarts (dry) to liters.

1 qt = 1.101 L
18 × 1.101 = 19.818 L

i. Convert 10 cups (liquid) to liters.

1 c = 237 mL = 0.24 L
10 × 0.24 = 2.4 L

j. Convert 90 hectares to acres.

1 ha = 2.471 A
90 × 2.471 = 222.39 A

k. Convert 30 acres to hectares.

1 A = 0.405 ha
30 × 0.405 = 12.15 ha

Converting Temperatures. Two common units for measuring temperature are **Celsius** and **Fahrenheit**. The following examples of temperatures are expressed in both Celsius and Fahrenheit.

	Celsius	Fahrenheit
Water boils at	100°C	212°F
Water freezes at	0°C	32°F
Normal human body temperature	37°C	98.6°F
Pleasant summer day	24°C	75°F

In either system, a temperature of four degrees below zero, for example, is written as −4°F or −4°C. We would say *minus 4 degrees* or *4 degrees below 0*. However, −4°F is *not* equivalent to −4°C. To convert a Fahrenheit temperature to a Celsius temperature, use the following equation:

$$C = \frac{5}{9}(F - 32°)$$

Subtract 32° from the given Fahrenheit temperature and multiply the result by $\frac{5}{9}$. The 32° in the formula is a Fahrenheit temperature.

EXAMPLES:

a. Change 212°F to Celsius.

212° − 32° = 180°
180° × $\frac{5}{9}$ = 100°C

b. Change 14°F to Celsius.

14° − 32° = 18° below 0° *or* −18°
−18° × $\frac{5}{9}$ = −10°C

To convert a Celsius temperature to a Fahrenheit temperature, use the following equation:

$$F = \frac{9}{5}C + 32°$$

Multiply the given Celsius temperature by $\frac{9}{5}$ and add 32° to the result.

EXAMPLES:

a. Change 30°C to Fahrenheit.

30° × $\frac{9}{5}$ = 54°
54° + 32° = 86°F

b. Change −40°C to Fahrenheit.

−40° × $\frac{9}{5}$ = −72°
−72° + 32° = −40°F

Note that −40° is the only temperature that is the same on both scales.

Chapter 4 Working with Weights and Measures

PART C Adding Measures

Measures can be added when they are expressed in the same unit and even when they are expressed in two or more units. Each operation is explained and illustrated in the customary system first and then in the metric system. As you will see, the work is much simpler in the metric system.

Adding Measures Expressed in the Same Unit. The procedure to add measures expressed in the same unit is the same in both systems. Add the quantities and attach the same unit to the sum.

EXAMPLE:

13 oz + 20 oz + 7 oz = 40 oz
= 2 lb 8 oz
= 2.5 lb

In the customary system, the answer is usually changed to larger units when possible, unless the original unit is specified.

EXAMPLE:

15 mm + 34 mm + 34 mm = 83 mm

In the metric system, the answer is usually expressed in the original unit, unless another unit is specified.

Adding Measures Consisting of Two or More Units. To add measures which consist of two or more units in the customary system, first arrange them so that like units are vertically aligned (inches under inches, feet under feet, etc.). Then add each column separately. Starting with the sum at the extreme right, convert units to the next larger unit wherever possible, but keep any remainder in the smaller unit. Then, add this converted result to any like units in the column immediately to the left. Change this new sum in the same way. Continue with this process, one at a time, through the last column on the left.

EXAMPLES:

a. Add 2 feet 11 inches, 1 foot 8 inches, and 2 feet 9 inches.

2 ft 11 in.
1 ft 8 in.
2 ft 9 in.
─────────
5 ft 28 in.

28 in. = 2 ft 4 in.

5 ft
2 ft 4 in.
─────────
7 ft 4 in.

7 ft = 2 yd 1 ft

2 yd 1 ft
     4 in.
─────────
2 yd 1 ft 4 in.

The final answer is 2 yd 1 ft 4 in.

b. Add 3 cups 8 tablespoons 2 teaspoons, 4 cups 1 teaspoon, and 11 tablespoons 2 teaspoons.

3 c 8 T 2 t
4 c   1 t
    11 T 2 t
─────────────
7 c 19 T 5 t

5 t = 1 T 2 t

7 c 19 T
     1 T 2 t
─────────────
7 c 20 T 2 t

20 T = 1 c 4 T

7 c   2 t
1 c 4 T
─────────────
8 c 4 T 2 t

The final answer is 8 c 4 T 2 t.

In the metric system, a measure is generally expressed in only one unit, but it may be necessary to add measures given in various different units. Before adding, change all the measures to a common unit—usually the one specified for the answer.

EXAMPLE:

Add 3.21 liters, 5 centiliters, 380 milliliters, 9.8 liters, 500 centiliters, and 309 milliliters.

3.21 L =   3.21 L
5 cL   =   0.05 L
380 mL =   0.38 L
9.8 L   =   9.8   L
500 cL =   5.0   L
309 mL =   0.309 L
            ────────
            18.749 L

The final answer is 18.749 L.

PART D Subtracting Measures

Measures expressed in the same unit or in different units may also be subtracted.

Subtracting Measures Expressed in the Same Unit. In either system, when both the minuend and the subtrahend are expressed in the same unit, the subtraction is done as usual and the unit is attached to the difference.

EXAMPLES:

a. Take 39 gal from 60 gal.

$$\begin{array}{r} 60 \text{ gal} \\ -39 \text{ gal} \\ \hline 21 \text{ gal} \end{array}$$

b. Take 48 m from 129.8 m.

$$\begin{array}{r} 129.8 \text{ m} \\ -48.0 \text{ m} \\ \hline 81.8 \text{ m} \end{array}$$

Subtracting Measures Consisting of Two or More Units. When the measures consist of two or more units in the customary system, arrange like units under each other. If the minuend does not have some of the units contained in the subtrahend, allow for them with 0 quantities. Then subtract. If it is necessary to borrow from the next larger unit, convert the borrowed 1 to the number of smaller units contained in one of the larger units and add this quantity to the given number of smaller units.

EXAMPLE:

Take 4 yards 2 feet 9 inches from 6 yards 2 feet 2 inches.

$$\begin{array}{r} 6 \text{ yd } 2 \text{ ft } 2 \text{ in.} \\ -4 \text{ yd } 2 \text{ ft } 9 \text{ in.} \end{array}$$

Since 9 in. is greater than 2 in., borrow 1 ft (12 in.) from the 2 ft. 2 ft 2 in. becomes 1 ft 14 in.

$$\begin{array}{r} \phantom{6 \text{ yd }} 1 \phantom{\text{ ft }} 14 \phantom{\text{ in.}} \\ 6 \text{ yd } \cancel{2} \text{ ft } \cancel{2} \text{ in.} \\ -4 \text{ yd } 2 \text{ ft } 9 \text{ in.} \\ \hline \phantom{6 \text{ yd } 2 \text{ ft } 1}5 \text{ in.} \end{array}$$

Since 2 ft is greater than 1 ft, borrow 1 yd (3 ft) from the 6 yd. 6 yd 1 ft becomes 5 yd 4 ft.

$$\begin{array}{r} \phantom{6 \text{ yd }}5 \phantom{\text{ ft}} 4 \phantom{\text{ in.}} \\ \cancel{6} \text{ yd } \cancel{2} \text{ ft } \cancel{2} \text{ in.} \\ -4 \text{ yd } 2 \text{ ft } 9 \text{ in.} \\ \hline 1 \text{ yd } 2 \text{ ft } 5 \text{ in.} \end{array}$$

The final answer is 1 yd 2 ft 5 in.

Occasionally, it is necessary to go two or more places to the left to borrow 1 from a larger unit. Convert the borrowed 1 to the next smaller unit. Then keep borrowing and converting until the desired smaller unit is reached.

EXAMPLE:

Take 14 gallons 2 quarts 1 pint from 18 gallons.

$$\begin{array}{r} 18 \text{ gal } 0 \text{ qt } 0 \text{ pt} \\ -14 \text{ gal } 2 \text{ qt } 1 \text{ pt} \end{array}$$

Since the minuend has no quarts or pints, borrow 1 gal (4 qt) from the 18 gal. 18 gal becomes 17 gal 4 qt. Then borrow 1 qt (2 pt) from the 4 qt. 17 gal 4 qt becomes 17 gal 3 qt 2 pt. Then subtract.

$$\begin{array}{r} \phantom{\text{gal }}3\phantom{\text{ qt }} \phantom{\text{ pt}} \\ 17\phantom{\text{ gal }} \cancel{4}\phantom{\text{ qt }} 2\phantom{\text{ pt}} \\ \cancel{18} \text{ gal } \cancel{0} \text{ qt } \cancel{0} \text{ pt} \\ -14 \text{ gal } 2 \text{ qt } 1 \text{ pt} \\ \hline 3 \text{ gal } 1 \text{ qt } 1 \text{ pt} \end{array}$$

The final answer is 3 gal 1 qt 1 pt.

When the minuend and subtrahend are given in different units in the metric system, both must be expressed in a common unit—usually the one in which the answer is desired.

EXAMPLE:

Take 50 milliliters from 2 liters. Express the answer in liters.

50 mL = 0.05 L

$$\begin{array}{r} 2.00 \text{ L} \\ -0.05 \text{ L} \\ \hline 1.95 \text{ L} \end{array}$$

The final answer is 1.95 L.

PART E Multiplying Measures

A measure expressed in one or more units may be multiplied by a number or by another measure.

Chapter 4   Working with Weights and Measures

Multiplying One Unit by a Number. In both systems, when a measure in only one unit is multiplied by a number, the product is expressed in the same unit. It may be changed to other units if desired.

EXAMPLES:

a. Multiply 10 inches by 7. Express the answer in feet and inches.

10 in. × 7 = 70 in.

1 ft = 12 in.
70 ÷ 12 = 5 ft 10 in.

b. Multiply 200 milligrams by 25. Express the answer in grams.

200 mg × 25 = 5,000 mg

1 g = 1,000 mg
5,000 ÷ 1,000 = 5 g

Multiplying Measures Consisting of Two or More Units by a Number. To multiply a measure having two or more units in the customary system by a number, first multiply each part separately. Then, starting at the right, change each part to larger units wherever possible, as shown in Part C of this section.

EXAMPLE:

Multiply 6 yards 1 foot 10 inches by 2.

6 yd 1 ft 10 in.
× 2
---
12 yd 2 ft 20 in.

20 in. = 1 ft 8 in.

12 yd 2 ft
    1 ft 8 in.
---
12 yd 3 ft 8 in.

3 ft = 1 yd

12 yd   8 in.
 1 yd
---
13 yd   8 in.

The final answer is 13 yd 8 in.

Because a measure is expressed as one unit in the metric system, multiplication is straightforward.

EXAMPLE:

Multiply 6.043 meters by 2. Express the answer in meters.

6.043 m
× 2

12.086 m

You may check and find that 6 yd 1 ft 10 in. is equal to 6.043 meters and that 13 yd 8 in. is equal to 12.086 m. Thus, the two preceding examples are identical. However, the solution in the metric system is much less complicated.

Multiplying Measures Expressed in Different Units. Measures of length or distance are called **linear** (line) **measures.** The product of two linear measures is a **square measure** (a measure used for area). The product of three linear measures is a **cubic measure** (a measure used for volume).

Measures must be in the same unit to be multiplied together. For example, feet can be multiplied only by feet to get square feet, and meters must be multiplied by meters and meters to get cubic meters.

If the measures to be multiplied are in different units or if they consist of two or more parts (as they may in the customary system), express them in the same unit before multiplying them. Then express the answer as specified.

EXAMPLES:

a. Multiply 3 feet 2 inches by 5 feet 7 inches. Express the answer in square yards, square feet, and square inches.

3 ft 2 in. = 38 in.
5 ft 7 in. = 67 in.

38 in. × 67 in. = 2,546 sq in.

1 sq ft = 144 sq in.
2,546 ÷ 144 = 17 sq ft 98 sq in.

1 sq yd = 9 sq ft
17 ÷ 9 = 1 sq yd 8 sq ft

2,546 sq in. = 1 sq yd 8 sq ft 98 sq in.

b. Find the product of 1 meter, 0.8 meter, and 10 centimeters. Express the answer in cubic centimeters.

1 m = 100 cm; 0.8 m = 80 cm
100 cm × 80 cm × 10 cm = 80,000 cm^3

There are some exceptions to the rule that measures must be in the same unit for multiplication. Most of the exceptions are in scientific fields, but a common one occurs with lumber in the customary system.

A **board foot,** the unit by which lumber is sold in the United States, is 1 foot long, 1 foot wide, and 1 inch thick. The standard way of describing a piece of lumber, however, is in terms of thickness (in inches) and width (in inches). Thus, a *2 by 4* is a board 2 inches thick and 4 inches wide. The length is expressed in feet. Therefore, when board feet are calculated, the width must be converted to a fraction of a foot.

To find the number of board feet in a piece of lumber, multiply the thickness in inches by the width in feet and the length in feet.

EXAMPLE:

Find the number of board feet in a 2 by 4 by 12 piece of lumber.

$4 \text{ in.} = \frac{1}{3} \text{ ft}$

$2 \text{ in.} \times \frac{1}{3} \text{ ft} \times 12 \text{ ft} = 8 \text{ board ft}$

You should be aware that the thickness and width of an unfinished piece of lumber may be a little more or a little less than the measurements specified. All finished lumber, however, is standardized at a smaller thickness and width than the stated measurements. Thus, a 2 by 4 is standardized at $1\frac{5}{8}$ inches by $3\frac{5}{8}$ inches, but its cost is based on 2 inches by 4 inches.

The calculation of board feet can become quite complicated. Thus, the lumber industry has made elaborate tables to find the total number of board feet when the number of pieces and their thickness, width, and length are given.

PART F Dividing Measures

As in the multiplication of measures, a measure expressed in one or more units may be divided by a number or by another measure.

Dividing One Unit by a Number. In both systems, when a measure in only one unit is divided by a number, the quotient is expressed in the same unit. It may be changed to other units if desired.

EXAMPLES:

a. Divide 23 yards by 4.

$23 \text{ yd} \div 4 = 5\frac{3}{4} \text{ yd } or \text{ } 5 \text{ yd } 2 \text{ ft } 3 \text{ in.}$

Note: $\frac{3}{4} \times 3 \text{ ft} = \frac{9}{4} \text{ ft} = 2\frac{1}{4} \text{ ft,}$
and $\frac{1}{4} \times 12 \text{ in.} = \frac{12}{4} \text{ in.} = 3 \text{ in.}$
so $5\frac{3}{4} \text{ yd} = 5 \text{ yd } 2\frac{1}{4} \text{ ft} = 5 \text{ yd } 2 \text{ ft } 3 \text{ in.}$

b. Divide 42 meters by 4. Express the answer in meters, centimeters, and millimeters.

$42 \text{ m} \div 4 = 10\frac{1}{2} \text{ m} = 10.5 \text{ m}$
$= 1,050 \text{ cm}$
$= 10,500 \text{ mm}$

Dividing Measures Consisting of Two or More Units by a Number. To divide a measure consisting of two or more units in the customary system by a number, start with the largest unit at the left. If the division is not exact, change the remainder to the next smaller unit and add any like units in the measure. Then, continue dividing in the same way for the remaining units.

EXAMPLE:

Divide 20 gallons 3 quarts 1 pint by 12.

$$\begin{array}{r} 1 \text{ gal} \quad\quad 2 \text{ qt} \quad\quad 1\frac{11}{12} \text{ pt} \\ \hline 12 \overline{)20 \text{ gal} \quad\quad 3 \text{ qt} \quad\quad 1 \text{ pt}} \\ \underline{12} \quad\quad + 32 \text{ qt} + \,\, 22 \text{ pt} \\ 8 \text{ gal} \,\nearrow\, 35 \text{ qt} \quad\quad 23 \text{ pt} \\ \\ 12\overline{)35 \text{ qt}} \,\, 12\overline{)23 \text{ pt}} \\ \underline{24 \text{ qt}} \quad\quad \underline{12} \\ 11 \text{ qt} \quad\quad 11 \end{array}$$

The final answer is 1 gal 2 qt $1\frac{11}{12}$ pt.

Because a measure is expressed in one unit in the metric system, division is straightforward.

Chapter 4 Working with Weights and Measures

EXAMPLE:

Divide 85.8 liters by 12. Express the answer in liters.

85.8 L ÷ 12 = 7.15 L

Dividing Measures Expressed in Different Units. Generally one measure can be divided by another measure only when both are expressed in the same unit. This is true in both the customary and the metric systems. The quotient in dividing measures is a number expressed in pieces, packages, etc. As in multiplication, there are some exceptions that are not discussed here.

EXAMPLES:

a. A bolt of goods contains 48 yards. How many pieces 4 feet long can be cut from this bolt?

1 yd = 3 ft
48 × 3 = 144 ft

144 ft ÷ 4 ft = 36 pieces

b. The liquid in a 4.5 liter container is to be poured into small bottles containing 75 milliliters each. How many bottles are needed?

4.5 L = 4,500 mL
4,500 mL ÷ 75 mL = 60 bottles

COMPLETE ASSIGNMENT 16

Name _____

Date _____

ASSIGNMENT 16
SECTION 4.1

Working with the Customary and the Metric Systems

	Perfect Score	Student's Score
PART A	40	
PART B	20	
PART C	10	
PART D	10	
PART E	10	
PART F	10	
TOTAL	100	

PART A Changing to the Next Smaller or the Next Larger Unit

DIRECTIONS: Change the measures and express your answers in the units indicated on the answer lines (2 points for each correct answer).

1. 33 tablespoons = _____ t

2. $\frac{3}{4}$ yard = _____ in.

3. $3\frac{1}{4}$ pounds = _____ oz

4. 250 pounds = _____ cwt

5. 0.25 hour = _____ min

6. 8.4 square feet = _____ sq in.

7. $\frac{1}{2}$ ton = _____ lb

8. 3 gallons = _____ qt

9. 45 inches = _____ ft _____ in.

10. 45 square feet = _____ sq yd

11. 20 ounces = _____ lb

12. 15 quarts = _____ gal

13. 55 minutes = _____ h

14. $5\frac{1}{4}$ feet = _____ yd

15. 90 cubic feet = _____ cu yd   18. 5,000 grams = _____ kg

16. 84 meters = _____ mm   19. ¾ liter = _____ mL

17. 382 milligrams = _____ g   20. 5.4 cubic meters = _____ cm³

STUDENT'S SCORE _____

PART B Converting from One System to the Other

DIRECTIONS: Change the measures and express your answers in the units indicated on the answer lines (2 points for each correct answer).

21. 12 feet = _____ m   26. 1,500 grams = _____ lb

22. 12 quarts (liquid) = _____ L   27. 20 liters = _____ qt (liquid)

23. 10 liters = _____ qt (dry)   28. 200 grams = _____ oz

24. 250 meters = ___ yd ___ ft ___ in.   29. 25 cubic feet = _____ m³

25. 10 pounds = _____ kg   30. 80 square meters = _____ sq yd

STUDENT'S SCORE _____

Chapter 4   Working with Weights and Measures

PART C   Adding Measures

DIRECTIONS: Add the measures and express the sums in the units indicated on the answer lines (2 points for each correct answer).

31. 48 min   _____ h _____ min   34.   85 cL + 200 mL + 8 L = _____ L
     18 min
     24 min

32. 5 lb 21 oz   _____ lb _____ oz
     4 lb 10 oz   35.   78 cm + 3.5 m + 562 cm = _____ cm
     3 lb 5 oz

33. 2 sq yd 8 sq ft 95 sq in.
     5 sq yd   118 sq in.
              5 sq ft   26 sq in.

     _____ sq yd _____ sq ft _____ sq in.   STUDENT'S SCORE _____

PART D   Subtracting Measures

DIRECTIONS: Subtract the measures and express your answers in the units indicated on the answer lines (2 points for each correct answer).

36. 2 yd 1 ft − 1 ft 7 in. =   38.   5 gal 1 qt − 2 gal 3 qt 1 pt =

        _____ yd _____ ft _____ in.   _____ gal _____ qt _____ pt

37.   12 cu yd 8 cu ft   39.   50.1 cm − 17.3 mm = _____ mm
      − 3 cu yd 16 cu ft 560 cu in.

        _____ cu yd _____ cu ft _____ cu in.

                                         40.   0.75 kg − 288 g = _____ kg

STUDENT'S SCORE _____

PART E Multiplying Measures

DIRECTIONS: Multiply the measures and express the products in the units indicated on the answer lines (1 point for each correct answer).

41. 5 gal 3 pt   ____ gal ____ qt ____ pt
     ×   5

42. 2 lb 10 oz   ____ lb ____ oz
     ×   7

43. 5 ft 7 in. × 9 ft 8 in. =

     ____ sq yd ____ sq ft ____ sq in.

44. 6 ft 8 in. × 9 ft 4 in. =

     ____ sq yd ____ sq ft ____ sq in.

45. 3 ft × 5 ft × 5 ft =

     ____ cu yd ____ cu ft

46. 2 yd 2 ft × 5 yd 1 ft × 8 yd 2 ft =

     ____ cu yd ____ cu ft

47. 375 g × 25 =   ____ kg

48. 115 mL × 20 =   ____ L

49. 3.2 m × 50 cm =   ____ m^2

50. 0.85 m × 30 mm × 12 cm =   ____ cm^3

STUDENT'S SCORE _____

PART F Dividing Measures

DIRECTIONS: Divide the measures and express your answers in the units indicated on the answer lines (2 points for each correct answer).

51. 12 lb 9 oz ÷ 3 =   ____ lb ____ oz

52. 20 yd 2 ft 3 in. ÷ 9 = ____ yd ____ in.

53. 3,822 mL ÷ 3 =   ____ L

54. 34 gal 2 qt ÷ 12 =

     ____ gal ____ qt ____ pt

55. 21,000 g ÷ 30 =   ____ kg

STUDENT'S SCORE _____

Chapter 4 Working with Weights and Measures

SECTION 4.2 Applications Using Weights and Measures

Measures have widespread use in business and industry. Linear, square, and cubic measures are used in construction, manufacturing, engineering, etc.

In addition, many of the other measures are used in buying and selling goods. Prices are often quoted for a specific number of pieces, such as a dozen, a hundred, or a thousand; for a specific weight, such as a pound or a kilogram; or for a specific volume, such as a gallon or a liter.

In both the metric and the customary systems, the cost of an item may be quoted in terms of a basic unit, and orders for that item are generally stated in the same unit. Thus, if the price is quoted per pound, the quantity ordered is in pounds (or parts of a pound). If the price is quoted per kilogram, the quantity ordered is in kilograms (or parts of a kilogram). Of course, if the partial pound or partial kilogram is very small, the weight may be stated in ounces or grams.

However, when the quantity ordered is expressed in a unit other than the unit in the quoted price, the problems of conversion differ between the two systems. In the metric system, it is very easy to change from one unit to another, because the system is based on 10. This simplicity is a major reason for using the metric system.

EXAMPLE:

Find the cost of 240 cm of copper wire priced at $1.05 per meter.

240 cm = 2.4 m
2.4 × $1.05 = $2.52

When the quantity ordered is in a different unit in the customary system, changing to the unit in the quoted price requires more work. Because of the greater difficulty with the customary system and the need for more practice with it, Parts B through D of this section will discuss and illustrate only examples in the customary system.

PART A Perimeter, Area, and Volume

Perimeter, area, and volume are measurable properties that are as important in everyday experience as they are in business and industry. A simple sketch showing the measurements will often help in solving perimeter, area, and volume problems.

Finding the Perimeter. The **perimeter** of a surface is the distance around the surface. It is always expressed in linear units. Consider a **rectangle,** a surface with four right angles (square corners), like a football field. The perimeter (P) of a rectangle is equal to the sum of twice its length (l) and twice its width (w).

Perimeter of rectangle = (2 × l) + (2 × w)

Generally, the same unit of measurement is used for the length and the width, but this is not absolutely necessary. The length could be expressed in yards and the width in feet; the total would then be expressed in either yards or feet. In metric measurements, the length could be expressed in meters and the width in centimeters; the total would then be expressed in either meters or centimeters. If the length and the width are expressed in different units of measurement, convert them to the same unit before finding the perimeter.

EXAMPLES:

a. How many feet of link fence are needed to enclose a rectangular yard that is 70 feet long and 50 feet wide? The number of feet needed is the perimeter.

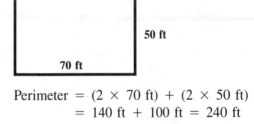

Perimeter = (2 × 70 ft) + (2 × 50 ft)
= 140 ft + 100 ft = 240 ft

b. Base molding is to be installed around a large room that measures 8.5 meters long and 650 centimeters wide. There is one door which is one meter wide. How many meters of molding are needed?

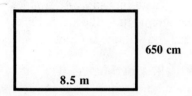

650 cm = 6.5 m
Perimeter = (2 × 8.5 m) + (2 × 6.5 m)
= 17 m + 13 m = 30 m
Molding needed = 30 m − 1 m = 29 m

The **diameter** of a circle is the length of a straight line across the circle which passes through the center of the circle. The **radius** of a circle is one-half of its diameter, or the distance from the center of the circle to its edge. The perimeter, or **circumference,** of a circle is the distance around the circle and is equal to pi times the diameter, or 2 times pi times the radius. Pi is represented by the Greek symbol π and has a numerical value of $\frac{22}{7}$ or 3.1416.

Circumference of circle = $\pi \times d$
= $2 \times \pi \times r$

EXAMPLE:

How much rope is needed to secure a circular pool cover if the distance across the pool is 9 feet? The rope will be purchased by the foot.

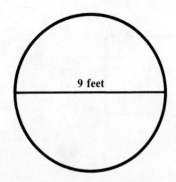

Perimeter = 3.1416 × 9
= 28.2744 ft *or* 29 ft

Finding the Area. An **area** refers to the amount of space within a set of lines and is always expressed in square units. For example, to find the area of a rectangular space, such as a wall, the side of a box, the land on which a building stands, etc., multiply its length (l) by its width (w).

Area of rectangle = l × w

Remember that the length and the width must be expressed in the same unit before multiplying. That is, both length and width must be expressed in feet, in yards, in meters, etc.

Carpeting and linoleum are generally sold by the square yard or square meter. Some widths of carpeting are sold by the *running yard* or *running meter,* but the width is taken into consideration in determining the price. With regard to paint, the area that can be covered by the contents of a can of paint is usually stated in square feet or square meters.

EXAMPLES:

a. A hallway rug is 3 meters long and 125 centimeters wide. The rug costs $33 per running meter. What is the total cost of the rug? What is the cost per square meter?

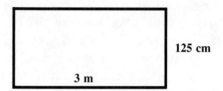

Total cost = 3 m × $33 = $99

125 cm = 1.25 m
Area = 3 m × 1.25 m = 3.75 m^2

Cost per m^2 = $99 ÷ 3.75
= $26.40

b. The label on a can of paint states that the paint will cover 650 square feet. Is one can enough to paint the walls and ceiling of a room that is 15 feet 6 inches long, 12 feet 3 inches wide, and 8 feet high?

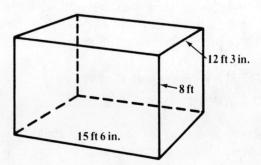

Area of front
and back walls = 2 × 15.5 ft × 8 ft
= 248 sq ft

Area of end walls = 2 × 12.25 ft × 8 ft
= 196 sq ft

Area of ceiling = 12.25 ft × 15.5 ft
= 189.875 sq ft

Add all areas: 248   sq ft
                 196   sq ft
               + 189.875 sq ft
                 633.875 sq ft

Thus a single can of paint, covering 650 square feet, should be enough to paint this room.

The area of a circle refers to the closed space within the circle. The area of a circle is equal to π times the radius times itself (the radius squared). Remember that the radius is one-half of the diameter.

$$\text{Area of circle} = \pi \times r \times r$$
$$= \pi \times r^2$$

The notation r^2 is read as *r squared* or *r to the second power*. This is called exponential notation, or *exponentiation*.

EXAMPLE:

Find the total cost of the material needed to cover a circular pool that is 7 feet across if the material costs $22 per square foot.

Radius of pool = $\frac{1}{2}$ × diameter

= $\frac{1}{2}$ × 7 ft

= $\frac{7}{2}$ ft

Area of pool = $\pi \times r^2$
= $\pi \times r \times r$

= $\frac{22}{7} \times \frac{7}{2} \times \frac{7}{2}$

= $\frac{\overset{11}{\cancel{22}}}{\underset{1}{\cancel{7}}} \times \frac{\cancel{7}}{2} \times \frac{7}{2}$

= $\frac{77}{2}$ sq ft

= 38.5 sq ft

Total cost = 38.5 × $22
= $847

Finding the Volume. The term **volume** refers to the inside space or capacity of a container, a room, a hole, etc., and is always expressed in cubic units. To find the volume of a rectangular solid, multiply its length (l) by its width (w) by its height (h).

Volume of rectangular solid = l × w × h

These three measures, or *dimensions,* must all be expressed in the same unit to have a product in cubic units.

EXAMPLES:

a. Find the cost of the concrete in a patio that is 15 feet long, 12 feet wide, and 6 inches thick if the concrete costs $60 per cubic yard.

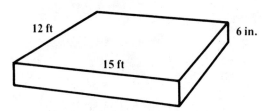

6 in. = 0.5 ft

Volume = 15 ft × 12 ft × 0.5 ft
= 90 cu ft

1 cu yd = 27 cu ft

90 ÷ 27 = $3\frac{1}{3}$ cu yd

Cost = $3\frac{1}{3}$ × $60 = $200

b. Find the number of cubic centimeters in a box that is 1.1 meters long, 0.6 meter wide, and 25 centimeters deep.

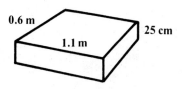

1.1 m = 110 cm; 0.6 m = 60 cm

Volume = 110 cm × 60 cm × 25 cm
= 165,000 cm³

PART B Ounces, Pounds, and Tons

With respect to customary weights, if both the price per unit and the quantity are expressed in ounces, pounds, or tons, the solution is straightforward. However, if the quantity is given in a unit other than the unit used in pricing, or if the quantity is given in a combination of units, then some conversions are necessary for a correct solution.

Finding the Cost When the Unit Price is per Ounce. Multiply the given number of pounds by 16 and add any ounces given in the quantity to find the total number of ounces. Remember that you are changing to a smaller unit; therefore, the number of ounces must be greater than the number of pounds. Multiply the total number of ounces by the unit price to find the cost.

EXAMPLES:

a. Find the cost of $1\frac{1}{2}$ pounds at $1.20 per ounce.

$$\text{Number of oz} = 1\frac{1}{2} \times 16 = 24 \text{ oz}$$

$$\text{Cost} = 24 \times \$1.20 = \$28.80$$

b. What is the cost of 2 pounds 10 ounces at $2.10 per ounce?

$$\begin{aligned}\text{Number of oz} &= (2 \times 16) + 10 \\ &= 32 + 10 \\ &= 42 \text{ oz}\end{aligned}$$

$$\text{Cost} = 42 \times \$2.10 = \$88.20$$

Finding the Cost When the Unit Price is per Pound. To change the given number of ounces in the quantity to a fraction of a pound, divide the number of ounces by 16. Then, add any pounds that are given in the quantity. Generally, a fraction of a pound is expressed as a decimal. Multiply the total number of pounds by the unit price to find the cost.

EXAMPLES:

a. Find the cost of 10 ounces at $5 per pound.

$$\begin{aligned}\text{Number of lb} &= 10 \div 16 \\ &= 0.625 \text{ lb}\end{aligned}$$

$$\begin{aligned}\text{Cost} &= 0.625 \times \$5 \\ &= \$3.125 \text{ or } \$3.13\end{aligned}$$

b. What is the cost of 3 pounds 5 ounces at 79¢ per pound?

$$\begin{aligned}\text{Number of lb} &= (5 \div 16) + 3 \\ &= 0.3125 + 3 \\ &= 3.3125 \text{ lb}\end{aligned}$$

$$\begin{aligned}\text{Cost} &= 3.3125 \times \$0.79 \\ &= \$2.616875 \text{ or } \$2.62\end{aligned}$$

Finding the Cost When the Unit Price is per Ton. Materials such as coal, steel, iron, grain, and fertilizer are often sold with the weight stated in pounds in order to figure shipping charges. The unit price, however, is per ton. To find the number of tons, divide the quantity in pounds by 2,000. Use the shortcut for dividing by 2,000. First, divide the number of pounds by 2. Then, move the decimal point of the result three places to the left. Multiply the total number of tons by the price per ton.

EXAMPLES:

a. Find the cost of 6,600 pounds at $275 per ton.

$$\begin{aligned}\text{Number of T} &= 6,600 \div 2,000 \\ &= (6,600 \div 2) \div 1,000 \\ &= 3,300 \div 1,000 \\ &= 3,300. \\ &= 3.3 \text{ T}\end{aligned}$$

$$\begin{aligned}\text{Cost} &= 3.3 \times \$275 \\ &= \$907.50\end{aligned}$$

b. What is the cost of 530 pounds at $75 per ton?

$$\begin{aligned}\text{Number of T} &= 530 \div 2,000 \\ &= (530 \div 2) \div 1,000 \\ &= 265 \div 1,000 \\ &= 265. \\ &= 0.265 \text{ T}\end{aligned}$$

$$\begin{aligned}\text{Cost} &= 0.265 \times \$75 \\ &= \$19.875 \text{ or } \$19.88\end{aligned}$$

PART C Units, Dozens, and Gross

There are 12 units in one **dozen** and 12 dozen in one **gross**. There are 144 (12 × 12) units in one gross. When the quantity ordered and the unit price are expressed in different units, convert the units of the quantity to the units of the unit price. Then, multiply to find the total cost.

Chapter 4   Working with Weights and Measures

Finding the Cost When the Unit Price is per Unit. When the quantity ordered is stated in dozens or gross but the price is per unit, find the total number of units by multiplying the number of dozens by 12 and the gross by 144.

EXAMPLES:

a. What is the cost of 50 dozen at $2\frac{1}{4}$¢ per unit?

Number of units = 50 × 12
= 600 units

$2\frac{1}{4}$¢ = 2.25¢ = $0.0225

Cost = 600 × $0.0225
= $13.50

b. Find the cost of $2\frac{1}{3}$ gross at 9¢ per unit.

Number of units = $2\frac{1}{3}$ × 144
= 336 units

9¢ = $0.09

Cost = 336 × $0.09
= $30.24

Finding the Cost When the Unit Price is per Dozen. When the quantity ordered is stated in units but the price is per dozen, divide the number of units by 12 to find the number of dozens. Then, multiply the number of dozens by the price per dozen to find the total cost. Any portion of one dozen can be expressed as a decimal part.

EXAMPLE:

What is the cost of 30 units at $1.10 per dozen?

Number of dozens = 30 ÷ 12
= 2.5 doz

Cost = 2.5 × $1.10
= $2.75

Finding the Cost When the Unit Price is per Gross. When the quantity ordered is stated in units but the price is per gross, divide the number of units by 144 to find the number of gross. Then, multiply the number of gross by the price per gross to find the total cost. Any portion of a gross can be expressed as a decimal part.

EXAMPLE:

Find the cost of 486 units at $96 per gross.

Number of gross = 486 ÷ 144
= 3.375 gross

Cost = 3.375 × $96
= $324

When the quantity ordered is stated in dozens but the price is per gross, divide the number of dozens by 12 to find the number of gross. Then multiply the number of gross by the price per gross to find the total cost.

EXAMPLE:

What is the cost of 15 dozen at $65 per gross?

Number of gross = 15 ÷ 12
= 1.25 gross

Cost = 1.25 × $65
= $81.25

PART D   C, Cwt, and M

Building materials, hardware, food, office supplies, etc., are often priced by the hundred (per **C**), by the hundredweight (per **cwt**), or by the thousand (per **M**). The unit *cwt* is usually stated on a 100-pound bag or box of something like flour or sugar.

Finding the Number of Cs or cwts. When the quantity ordered is stated in units, linear or square feet, or pounds, but the unit price is per C or per cwt, divide the quantity by 100 to find the number of Cs or cwts. Then, multiply the number of Cs or cwts by the unit price to find the total cost. If the quantity ordered is less than 100, express the number of Cs or cwts as a decimal.

EXAMPLES:

a. Find the cost of 1,000 linear feet at $62.50 per C.

Number of Cs = 1,000 ÷ 100
= 10 C

Cost = 10 × $62.50
= $625

b. What is the cost of 75 units at $124 per C?

Number of Cs = 75 ÷ 100
= 0.75 C

Cost = 0.75 × $124 = $93

c. Find the cost of 520 pounds at $8.90 per cwt.

Number of cwts = 520 ÷ 100
= 5.2 cwt

Cost = 5.2 × $8.90
= $46.28

Finding the Number of Ms. When the quantity ordered is stated in units, board feet, or running yards, but the unit price is per M, divide the quantity by 1,000 to find the number of Ms. Then, multiply the number of Ms by the unit price to find the total cost. If the quantity is less then 1,000, express the number of Ms as a decimal.

EXAMPLES:

a. What is the cost of 7,420 units at $120 per M?

Number of Ms = 7,420 ÷ 1,000
= 7.42 M

Cost = 7.42 × $120
= $890.40

b. Find the cost of 500 board feet at $795 per M.

Number of Ms = 500 ÷ 1,000
= 0.5 M

Cost = 0.5 × $795
= $397.50

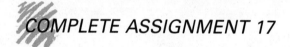
COMPLETE ASSIGNMENT 17

Name _____

Date _____

ASSIGNMENT 17
SECTION 4.2

Applications Using Weights and Measures

	Perfect Score	Student's Score
PART A	20	
PART B	30	
PART C	30	
PART D	20	
TOTAL	100	

PART A Perimeter, Area, and Volume

DIRECTIONS: Solve the problems and show all your work. Write your answers on the lines provided (2 points for each correct answer).

1. A pool area is 30 feet long and 25 feet wide. How much fencing is needed to enclose the pool area?

   _____

2. A square city lot is 27.3 meters per side. If fencing material costs $1.75 per meter, how much will it cost to fence the land?

   _____

3. A lake borders the long side of a rectangular lot that is 100 m long and 60 m wide. How much fencing is needed to enclose the other three sides?

   _____

4. A new rim is needed for a circular silo that measures 50 m across. If the material costs $30 per running meter, how much will it cost to replace the old rim?

   _____

5. How many gallons of paint are needed to cover the four walls and ceiling of a room that is 24 ft long by 30 ft wide by 10 ft high? A gallon of paint covers 400 sq ft.

   _____

Unit 1   Fundamentals

6. A floor is 7.2 meters by 5.4 meters. How much does it cost to cover it with linoleum that costs $25 per square meter?

_____

7. A strip of carpet is 30 inches wide by 90 inches long. At $6 per running foot, what is the cost per square foot for the carpet?

_____

8. At $72 per cubic meter, what is the cost of concrete in a sidewalk that is 16 meters long 120 centimeters wide and 15 centimeters thick?

_____

9. Good topsoil costs $19 per cubic yard. How much will it cost to put 6 inches of topsoil on a garden that is 12 feet by 15 feet?

_____

10. How many liters of water are needed to fill an aquarium 50 centimeters long, 20 centimeters wide, and 25 centimeters deep?

_____

STUDENT'S SCORE _____

PART B   Ounces, Pounds, and Tons

DIRECTIONS: *Find the cost (2 points for each correct answer).*

11.   3 lb at 50¢ per oz =   _____

12.   4 lb at 49¢ per oz =   _____

Chapter 4 Working with Weights and Measures

13. 2 lb 12 oz at 40¢ per oz = _____

14. 7 lb 12 oz at $37\frac{1}{2}$¢ per oz = _____

15. 5 lb at $1 per oz = _____

16. 19 lb 2 oz at $12\frac{1}{2}$¢ per oz = _____

17. 8 oz at $1.50 per lb = _____

18. 20 oz at $15 per lb = _____

19. 15 lb 12 oz at 20¢ per lb = _____

20. 15 lb 4 oz at $1.50 per lb = _____

21. 3,600 lb at $30 per T = _____

22. 875 lb at $120 per T = _____

23. 175 lb at $400 per T = _____

24. 16,840 lb at $42 per T = _____

25. 20,000 lb at $50 per T = _____

STUDENT'S SCORE _____

PART C Units, Dozens, and Gross
DIRECTIONS: Find the cost (2 points for each correct answer).

26. 16 articles at $10.98 per doz. = _____

27. 84 pieces at 30¢ per doz. = _____

28. 16 units at $9.60 per doz. = _____

29. 240 pieces at $1.75 per doz. = _____

30. 10 articles at $5.25 per doz. = _____

31. 20 pieces at $18 per doz. = _____

32. 680 units at $2 per gross = _____

33. 198 articles at $1.98 per gross = _____

34. 25 pieces at $20 per gross = _____

35. 90 units at $134 per gross = _____

36. 36 doz. at $12 per gross = _____

37. 30 doz. at $15 per gross = _____

38. 7.8 doz. at $2 per unit = _____

39. 14 gross at 12¢ per unit = _____

40. $5\tfrac{1}{2}$ gross at 2¢ per unit = _____

STUDENT'S SCORE _____

PART D C, Cwt, and M

DIRECTIONS: Find the cost (2 points for each correct answer).

41. 600 units at $4.25 per C = _____

42. 800 pieces at $18 per C = _____

43. 26 units at $6.10 per C = _____

44. 65 linear feet at $70 per C = _____

45. 318 lb at $6 per cwt = _____

46. 95 lb at $16.80 per cwt = _____

47. 15,100 lb at $0.50 per lb = _____

48. 16,000 board feet at $40 per M = _____

49. 2,940 pieces at $3.10 per M = _____

50. 2,000 board feet at $618 per M = _____

STUDENT'S SCORE _____

CHAPTER 5

Estimations, Graphs, and Shortcuts

Most computation today is done with calculators or computers. Estimating with rounded-off numbers can be used to get approximations of sizes, costs, or values. You estimate to determine if the result of a calculation is about right. Thus, estimation provides an additional check of your work, particularly placement of the decimal point. If you use a hand-held calculator in your work, it is important to have a good idea of what is a reasonable answer. It is very easy to make significant errors by pressing the wrong key, pressing a key too lightly, or pressing a key too many times.

Section 5.2 of this chapter shows you how to prepare and use the three most common graphs: the line graph, the bar graph, and the circle graph. These graphs are used by businesses and their shareholders, bankers, vendors, and customers to interpret mathematical information representing the business environment.

Shortcuts are very helpful in estimation, because shortcuts increase speed, improve accuracy, and provide procedures that can be done mentally. Many shortcuts are actually faster than using a calculator. Therefore, they are efficient methods of checking your work. Whether you use shortcuts in the original calculations or to check previous work, you may discover that they are quite convenient.

SECTION 5.1 Estimations

An **estimate** is almost never the correct answer to a problem; it is *approximately* the correct answer. An estimated calculation is done with rounded-off numbers to get a rough idea of what the actual results should be.

Estimates are given in many fields of business, such as construction, remodeling, repair work, etc. An estimate gives the customer some idea of costs. Experience in a particular business is needed to prepare a good estimate for that business. For this reason, business estimates are not presented here. You will learn how to estimate an answer to mathematics problems.

Up to this point, you have only rounded off answers. Further, you rounded only to a specific number of decimal places. The same general procedure is used to round off a number to the nearest unit, ten, hundred, thousand, etc. For *all* rounding off, the first two steps are the same.

1. Look at the digit immediately to the right of the required place; remaining digits further to the right are not used. You may want to draw a vertical line after the required place and underline the digit immediately to its right.
2. If the digit to the immediate right of the required place is less than 5, make no change in the digit in the required place. If the digit to the immediate right of the required place is 5 or more, add 1 to the digit in the required place.

In rounding off to the nearest unit, ten, hundred, thousand, etc., there are two more steps.

3. Drop any decimal places as well as the decimal point.
4. When rounding off to the nearest ten, hundred, thousand, etc., change all the digits to the right of the required place to zeros.

When you have completed rounding off, always check the result. It should be *about* the same as the original number; that is, it should be either a little smaller or a little larger.

PART A Rounding Off to the Nearest Unit

In business, an amount of money given in dollars and cents is often rounded off to the nearest dollar. Rounding off to the nearest dollar is the same as rounding off to the nearest unit.

By following Step 1 for all rounding off, you can tell that the digit to the immediate right of the units place is in the *tenths* position. The application of Steps 2 and 3 is explained in detail in the following examples.

When the Digit in the Tenths Position Is Less Than 5. Make no change in the units digit if the tenths place digit is less than 5. Drop the decimal point and all the decimal places. The rounded-off result is the same as the original whole-number part.

EXAMPLES:

a. $289.4776 = 289|.4776 = 289|.4776 = 289$

The tenths place digit is 4. Since this is less than 5, make no change in the 9. Drop the decimal point and all decimal places.

b. $\$4,653.49 = \$4,653|.49 = \$4,653|.49$
$= \$4,653$

When the Digit in the Tenths Position Is 5 or More. Add 1 to the units digit if the tenths place digit is 5 or more. Then, drop the decimal point and all decimal places. The rounded-off result is *one* more than the original whole-number part.

EXAMPLES:

a. $\$371.53 = \$371|.53 = \$371|.53 = \$372$

The tenths place digit is 5. Therefore, add 1 to the 1 in the units place, making it a 2. Drop the decimal point and all decimal places.

b. $\$513.82 = \$513|.82 = \$513|.82 = \$514$

The tenths place digit is 8. Since this is more than 5, add 1 to the 3 to make it 4. Drop the decimal point and all decimal places.

PART B Rounding Off to the Nearest Ten, Hundred, Thousand, etc.

To round off to the nearest ten, you should look only at the digit in the *units* position. The units place digit is the first digit to the left of the decimal point. It is also the last digit in a whole number. To round off to the nearest hundred, you should look only at the digit in the *tens* position. To round off to the nearest thousand, you should look only at the digit in the *hundreds* position. Disregard all digits more than one place to the right of the required place.

Step 2 for all rounding off still holds true. If the digit to the immediate right of the required place is less than 5, make no change in the digit in the required place; if it is 5 or more, add 1 to the digit in the required place.

Apply Step 3 only if the number being rounded off is a decimal. Drop the decimal point and all decimal places. If the number being rounded off is a whole number, omit Step 3.

Lastly, apply Step 4 as given. Change all digits to the right of the required place to zeros.

The following examples illustrate, in detail, Steps 1 through 4 for all rounding off.

EXAMPLES:

a. Round off 1,043.81 to the nearest ten.

$1,043.81 = 1,04|3.81 = 1,04|3.81$
$= 1,040$

The digit in the units position is 3. Since this is less than 5, make no change in the 4. Drop the .81 and change the 3 to 0.

b. Round off $928.30 to the nearest ten dollars.

$\$928.30 = \$92|8.30 = \$92|8.30 = \$930$

The digit in the units position is 8. Since this is more than 5, add 1 to the 2, to make it 3. Drop the .30 and change the 8 to 0.

c. Round off 6,321 to the nearest hundred.

$6,321 = 6,3|21 = 6,3|21 = 6,300$

The digit in the tens position is 2. Since this is less than 5, make no change in the 3. Change the 21 to 00.

Chapter 5 Estimations, Graphs, and Shortcuts

d. Round off 682 to the nearest hundred.

$$682 = 6|82 = 6|\overline{8}2 = 700$$

The digit in the tens position is 8. Since this is more than 5, add 1 to the 6 to make it 7. Change the 82 to 00.

e. Round off $726,399.56 to the nearest thousand dollars.

$$\$726,399.56 = \$726,|399.56$$
$$= \$726,|\overline{3}99.\cancel{56}$$
$$= \$726,000$$

The digit in the hundreds position is 3. Since this is less than 5, make no change in the 6. Drop the .56 and change the 399 to 000.

f. Round off $19,500 to the nearest thousand dollars.

$$\$19,500 = \$19,|500 = \$19,|\overline{5}\cancel{00}$$
$$= \$20,000$$

The digit in the hundreds position is 5. Since this is 5, add 1 to the 19 to make it 20. Change the 500 to 000.

To round off numbers to larger place values, follow the same procedure. For example, 22,998 to the nearest ten thousand is 20,000; 26,298 to the nearest ten thousand is 30,000; 649,999 to the nearest hundred thousand is 600,000; 750,000 to the nearest hundred thousand is 800,000; 7,443,210 to the nearest million is 7,000,000; and 7,543,210 to the nearest million is 8,000,000.

PART C Estimating Sums and Differences

An estimate shows you *approximately* how large the answer should be. The main reason for estimating is to find out if an answer is about right.

Estimates by different people may vary slightly, depending on how the numbers were rounded off. Differing estimates may be satisfactory, however, since an estimate is only a rough answer. An estimate will show whether or not a decimal point is in the right place or if you have pressed the correct keys on a calculator. However, it will not show minor errors in the calculation.

Estimating Sums. To estimate the sum of a column of figures, you should generally round off all the addends to the *same* place value—ten, hundred, thousand, etc., whichever is the largest. Then, only a single digit from each number needs to be added, since the rest of the number consists of zeros. In the rounding off, some of the numbers may be so small that they are rounded off to zeros. This does not affect the estimate to any great extent, since other numbers may be rounded *up* as much or more.

To estimate the sum of a column of figures, round off the addends and add the non-zero digits in the rounded-off addends. In the examples that follow, the rounded-off numbers are given only to show the procedure.

EXAMPLES:

a.
```
 $ 77.35 ---    $ 80   ( 8)
   44.67 ---      40   ( 4)
    8.27 ---      10   ( 1)
   94.49 ---      90   ( 9)
    3.47 ---       0   ( 0)
  +65.01 ---      70   ( 7)
  $293.26        $290   (29)
```

b.
```
 $   68.33 ---    100   ( 1)
    595.24 ---    600   ( 6)
     36.42 ---    000   ( 0)
    129.87 ---    100   ( 1)
  1,459.95 ---  1,500   (15)
    307.06 ---    300   ( 3)
  $2,596.87     2,600   (26)
```

Note that the addends in this example are rounded off to the nearest hundred, since there is only one number in the thousands.

c.
```
 $10,025.93 --- 10,000   (10)
      68.24 ---  0,000   ( 0)
       1.37 ---  0,000   ( 0)
     511.55 ---  1,000   ( 1)
     432.50 ---  0,000   ( 0)
      50.78 ---  0,000   ( 0)
  $11,090.37     11,000   (11)
```

Note that the four numbers that are rounded off to 0,000 in this example are not included in the estimate. Their sum is a little more than 500, which is made up when the 511.55 is rounded up to 1,000. Thus, the estimate is close to the actual total of $11,090.37. Since the first addend is so large, the addends can be rounded off to the nearest thousand.

Estimating Differences. In estimating the difference between two numbers, the minuend and the subtrahend may be rounded off to *different* place values, depending on the problem. The only guideline is that you must be able to mentally subtract the rounded-off numbers.

EXAMPLES:

a. $\begin{array}{r} 578.67 \\ -219.89 \\ \hline 358.78 \end{array}$ --- $\begin{array}{r} 600 \\ 200 \\ \hline 400 \end{array}$

b. $\begin{array}{r} 430.26 \\ -49.75 \\ \hline 380.51 \end{array}$ --- $\begin{array}{r} 400 \\ 50 \\ \hline 350 \end{array}$ or $\begin{array}{r} 430 \\ -50 \\ \hline 380 \end{array}$

c. $\begin{array}{r} 7{,}147.88 \\ -67.93 \\ \hline 7{,}079.95 \end{array}$ --- $\begin{array}{r} 7{,}000 \\ 100 \\ \hline 6{,}900 \end{array}$ or $\begin{array}{r} 7{,}000 \\ 70 \\ \hline 6{,}930 \end{array}$

When the minuend is much larger than the subtrahend, only that part of the minuend directly involved in the subtraction needs to be considered when rounding off. The digits to its left are brought down to the estimate.

EXAMPLE:

$\begin{array}{r} 27{,}492.36 \\ -38.67 \\ \hline 27{,}453.69 \end{array}$ --- $\begin{array}{r} (274)90 \\ -40 \\ \hline 27{,}450 \end{array}$

PART D Estimating Products

Estimating a product is a check on the size of the product. Here, you will learn how to estimate products when the factors are rounded off to numbers *greater* than 1 (such as to the nearest ten, hundred, thousand, etc.) and when the factors are rounded off to numbers *less* than 1 (such as to the nearest tenth, hundredth, or thousandth). The multiplicand and the multiplier, generally referred to as the *factors,* can be rounded off to *different* place values.

Rounding Off Factors to the Nearest Ten, Hundred, Thousand, etc. To estimate the product of two factors rounded off to the nearest ten, multiply the non-zero digits of the factors mentally. After that product, write one zero for each zero in the rounded-off factors. This number is the estimated product.

EXAMPLES:

a. $\begin{array}{r} 88 \\ \times\ 36 \\ \hline 3{,}168 \end{array}$ --- $\begin{array}{r} 90 \\ \times\ 40 \\ \hline 3{,}600 \end{array}$

Round *up* both factors to the nearest ten. Multiply the non-zero digits, 9 and 4, to get 36. The total number of zeros in the rounded-off factors is two; write two zeros after the 36 to get the estimated product, 3,600. Note that the actual product is less than the estimated product. This is because both factors were rounded *up* to get the estimate.

b. $\begin{array}{r} 21 \\ \times\ 28 \\ \hline 588 \end{array}$ --- $\begin{array}{r} 20 \\ \times\ 30 \\ \hline 600 \end{array}$

To estimate the product of two factors rounded off to the nearest hundred, multiply the non-zero digits of the factors mentally. After that product, write one zero for each zero in the rounded-off factors. This number is the estimated product.

EXAMPLES:

a. $\begin{array}{r} 487 \\ \times\ 369 \\ \hline 179{,}703 \end{array}$ --- $\begin{array}{r} 500 \\ \times\ 400 \\ \hline 200{,}000 \end{array}$

b. $\begin{array}{r} 321 \\ \times\ 289 \\ \hline 92{,}769 \end{array}$ --- $\begin{array}{r} 300 \\ \times\ 300 \\ \hline 90{,}000 \end{array}$

To estimate the product of two factors rounded off to the nearest thousand, multiply the non-zero digits of the factors mentally. After that product, write one zero for each zero in the rounded-off factors. This number is the estimated product.

EXAMPLES:

a. $\begin{array}{r} 4{,}870 \\ \times\ 3{,}691 \\ \hline 17{,}975{,}170 \end{array}$ --- $\begin{array}{r} 5{,}000 \\ \times\ 4{,}000 \\ \hline 20{,}000{,}000 \end{array}$

b. $\begin{array}{r} 4{,}782 \\ \times\ 3{,}010 \\ \hline 14{,}393{,}820 \end{array}$ --- $\begin{array}{r} 5{,}000 \\ \times\ 3{,}000 \\ \hline 15{,}000{,}000 \end{array}$

To estimate the product of two factors, you can round off the factors to *different* place values, depending on the problem. You may also choose to only round off one of the factors. As in addition and subtraction, the only guideline is that you must be able to mentally multiply the rounded-off numbers.

Chapter 5   Estimations, Graphs, and Shortcuts

An estimate should never be more difficult than the actual problem.

EXAMPLES:

a. 368.1 ---- 400
    $\times$69.8 ---- $\times$70
    25,693.38 28,000

b. 1,102 ---- 1,000
    $\times$9.7 ---- $\times$10
    10,689.4 10,000

c. 7,198 ---- 7,000
    $\times$2 ---- $\times$2
    14,396 14,000

The closer the rounded-off factors are to the actual factors, the closer the estimated product will be to the actual product.

EXAMPLES:

a. 368.1 ---- 400 370
    $\times$69.8 ---- $\times$70 or $\times$70
    25,693.38 28,000 25,900

b. 1,102 ---- 1,000 1,100
    $\times$9.7 ---- $\times$10 or $\times$10
    10,689.4 10,000 11,000

c. 7,198 ---- 7,000 7,200
    $\times$2 ---- $\times$2 or $\times$2
    14,396 14,000 14,400

Rounding Off Factors to the Nearest Tenth, Hundredth, Thousandth, etc. To estimate a product when at least one of the factors is a decimal less than 1, round off the decimal(s) to the nearest tenth, hundredth, thousandth, etc. Also, round off any factors greater than 1. Then, multiply the non-zero digits of the factors mentally and correctly place the decimal point in the estimated product. The correct number of decimal places is the total number of decimal places in the rounded-off factors.

EXAMPLES:

a. 0.03125 ------- 0.03
    $\times$49.5 ------- $\times$50
    1.546875 1.50

Round off the decimal 0.03125 to the nearest hundredth, 0.03. Round off the decimal 49.5 to the nearest ten, 50. Multiply the rounded-off factors and correctly position the decimal point to get the estimated product, 1.50.

b. 18,890 --------- 20,000 19,000
    $\times$0.018 --------- $\times$0.02 or $\times$0.02
    340.02 400.00 380.00

Note that the actual product is less than the estimated product. This is because both factors were rounded *up* to get the estimate.

PART E Estimating Quotients

Like estimating a product, estimating a quotient is a check on the size of the quotient. Estimating is also a check on the position of the decimal point in the actual quotient. Here, you will learn how to estimate quotients when the divisor is rounded off to a number *greater* than 1 (such as to the nearest ten, hundred, thousand, etc.) and when the divisor is rounded off to a number *less* than 1 (such as to the nearest tenth, hundredth, thousandth, etc.).

When the Divisor is a Number Greater Than 1. To estimate the quotient of two numbers, each of which is a whole number or a decimal greater than 1, round off both numbers to values that enable you to divide mentally. This will happen when the rounded-off dividend and divisor are factors of one another.

EXAMPLES:

a. 416 ÷ 48

400 ÷ 50 = 8

416 ÷ 48 = 8.667

Round off the divisor, 48, to 50. Round off the dividend, 416, to 400 rather than 420, because the division can then be done mentally. The estimate should always be less difficult than the actual problem. The decimal point in the actual quotient appears to be in the correct position, since the actual quotient is close to the estimate.

b. 1,176.1 ÷ 6.273

1,200 ÷ 6 = 200

1,176.1 ÷ 6.273 = 187.49

c. $9.05 \div 18.32$

$9 \div 18 = 0.5$

$9.05 \div 18.32 = 0.494$

To estimate a quotient when the dividend is a decimal and the divisor is a whole number or a decimal greater than 1, round off the dividend to the nearest tenth, hundredth, thousandth, etc. Round off the divisor to the nearest ten, hundred, thousand, etc. Then, estimate the quotient. Remember to round off both numbers to values that enable you to divide mentally.

EXAMPLE:

$0.883 \div 315$

$0.9 \div 300 = 0.003$

$0.883 \div 315 = 0.0028$

When the Divisor is a Decimal Less Than 1. To estimate a quotient when the divisor is a decimal less than 1, round off the divisor to the nearest tenth, hundredth, thousandth, etc. Round off both numbers to values that enable you to divide mentally. This will happen when the rounded-off dividend and divisor are factors of one another. Finally, estimate the quotient. Be sure to put the decimal point in the correct position in the quotient.

EXAMPLE:

$174.61 \div 0.572$

$180 \div 0.6 = 300$

$174.61 \div 0.572 = 305.2622$

Round off the divisor, 0.572, to 0.6. Round off the dividend, 174.61, to 180 rather than 200, because the division can then be done mentally. Remember that the estimate should never be as difficult as the actual problem. Then divide the rounded-off dividend by the rounded-off divisor and correctly position the decimal point in the quotient. The estimated quotient is 300.

To estimate a quotient when both the dividend and the divisor are decimals less than 1, round off each number separately to the nearest tenth, hundredth, thousandth, etc.

EXAMPLE:

$0.0627 \div 0.00173$

$0.06 \div 0.002 = 30$

$0.0627 \div 0.00173 = 36.243$

The decimal point in the actual quotient appears to be in the correct place, since the actual quotient is close to the estimate.

PART F Checking Calculator Answers by Estimating

Estimating is the easiest method of checking answers found with a calculator. As stated earlier, the estimating should be easier than the original problem; you should be able to do it mentally. Remember that an estimate will not reveal minor errors in your calculations, but it should reveal a misplaced decimal point or a wrong key pressed on the calculator. Both of these mistakes would result in a *size* error. The following examples demonstrate how to use estimating to check work performed on a calculator.

EXAMPLES:

a. $32.758 + 367.61 + 1{,}560$

Key-enter	Display
32.758	32.758
$+$	32.758
367.61	367.61
$+$	400.368
1560	1560.
$=$	1960.368

Estimate: $30 + 370 + 1{,}600 = 2{,}000$

The total on the display and the estimate are approximately the same.

b. $203{,}200 - 4{,}985.4$

Key-enter	Display
203200	203200.
$-$	203200.
4985.4	4985.4
$=$	198214.6

Estimate: $200{,}000 - 5{,}000 = 195{,}000$

The difference on the display and the estimate are approximately the same.

Chapter 5 Estimations, Graphs, and Shortcuts

c. 4.33×0.0575

Key-enter	Display
4.33	4.33
×	4.33
.0575	0.0575
=	0.248975

Estimate: $4 \times 0.06 = 0.24$

The product on the display and the estimate are approximately the same.

d. $29{,}748 \div 485.4$

Key-enter	Display
29748	29748.
÷	29748.
485.4	485.4
=	61.285537

Estimate: $30{,}000 \div 500 = 60$

The quotient on the display and the estimate are approximately the same.

COMPLETE ASSIGNMENT 18

ASSIGNMENT 18
SECTION 5.1

Name

Date

Estimations

	Perfect Score	Student's Score
PART A	10	
PART B	30	
PART C	10	
PART D	10	
PART E	20	
PART F	20	
TOTAL	100	

PART A Rounding Off to the Nearest Unit

DIRECTIONS: Round off to the nearest unit or dollar as indicated (1 point for each correct answer).

Nearest Unit

1. 8.187 _____
2. 3.891 _____
3. 9.011 _____
4. 74.5 _____
5. 500.72 _____

Nearest Dollar

6. $9.49 _____
7. $97.50 _____
8. $378.02 _____
9. $3,999.99 _____
10. $4,800.47 _____

STUDENT'S SCORE _____

PART B Rounding Off to the Nearest Ten, Hundred, Thousand, etc.

DIRECTIONS: Round off to the nearest ten, hundred, thousand, etc., as indicated (1 point for each correct answer).

Nearest Ten

11. 25.99 _____
12. 16.58 _____
13. 53.99 _____
14. 15,328 _____
15. 9,825.55 _____

Nearest $10

16. $829.00 _____
17. $9.00 _____
18. $699.99 _____
19. $111.11 _____
20. $1,111.00 _____

		Nearest Hundred			**Nearest Thousand**
21.	765.21	_____	31.	750.00	_____
22.	56.63	_____	32.	5,133.99	_____
23.	49.99	_____	33.	5,999.99	_____
24.	1,142.32	_____	34.	551.11	_____
25.	950.00	_____	35.	10,321.89	_____

		Nearest $100			**Nearest $1,000**
26.	$3,649.99	_____	36.	$500.01	_____
27.	$94.95	_____	37.	$4,912.54	_____
28.	$48.95	_____	38.	$491.36	_____
29.	$5,048.95	_____	39.	$15,428.12	_____
30.	$2,950.00	_____	40.	$105,500.00	_____

STUDENT'S SCORE _____

PART C Estimating Sums and Differences

DIRECTIONS: Estimate the sums or differences as indicated. Write the rounded-off numbers on the lines at the right. Write the estimated answers on the double-ruled lines (1 point for each correct answer).

41.	32.71		42.	247.98	
	81.74			9.02	
	357.16			63.27	
	512.81			717.00	
	9.83			820.20	
	+932.11			+29.63	

Chapter 5   Estimations, Graphs, and Shortcuts

43.
```
    371.21
  1,183.71
  5,472.88
     37.42
    489.19
 + 852.74
```

44.
```
  7,000.43
     69.08
    400.00
    196.78
      5.66
 + 651.95
```

45.
```
   4,217.14
     953.27
  84,912.40
   6,228.78
   3,101.24
     832.47
     215.98
   8,188.45
   1,744.21
 +   947.82
```

46.
```
    87.50
 − 18.50
```

47.
```
   352.74
 − 28.96
```

48.
```
   958.65
 − 86.80
```

49.
```
  4,728.04
 −   79.49
```

50.
```
  2,108.35
 −  267.44
```

STUDENT'S SCORE _____

PART D Estimating Products

DIRECTIONS: Estimate the products. Show the rounded-off factors used to find the estimated product (1 point for each correct answer).

	Rounded-Off Factors				**Estimated Product**
51. 75 × 38	_____	×	_____	=	_____
52. 29 × 8	_____	×	_____	=	_____
53. 2,942 × 0.94	_____	×	_____	=	_____
54. 405 × 1.15	_____	×	_____	=	_____
55. 32,412 × 0.002	_____	×	_____	=	_____
56. 410.9 × 0.06	_____	×	_____	=	_____
57. 0.007 × 61.8	_____	×	_____	=	_____
58. 10.81 × 0.28	_____	×	_____	=	_____
59. 30,989.9 × 0.08	_____	×	_____	=	_____
60. 582.3 × 382.79	_____	×	_____	=	_____

STUDENT'S SCORE _____

Chapter 5 Estimations, Graphs, and Shortcuts   197

PART E Estimating Quotients

DIRECTIONS: Estimate the quotients. Show the rounded-off dividends and divisors used. Use a calculator to find the actual quotients. Round the quotients to two places (2 points for each correct answer).

		Rounded-Off Dividend		Rounded-Off Divisor		Estimated Quotient	Actual Quotient
61.	623.24 ÷ 292.1	_____	÷	_____	=	_____	_____
62.	876.05 ÷ 315	_____	÷	_____	=	_____	_____
63.	87.01 ÷ 27.94	_____	÷	_____	=	_____	_____
64.	80.98 ÷ 4.126	_____	÷	_____	=	_____	_____
65.	62.09 ÷ 0.018	_____	÷	_____	=	_____	_____
66.	88.1 ÷ 0.34	_____	÷	_____	=	_____	_____
67.	0.0171 ÷ 0.201	_____	÷	_____	=	_____	_____
68.	0.414 ÷ 0.825	_____	÷	_____	=	_____	_____
69.	8 ÷ 215	_____	÷	_____	=	_____	_____
70.	47 ÷ 250	_____	÷	_____	=	_____	_____

STUDENT'S SCORE _____

Unit 1   Fundamentals

PART F Checking Calculator Answers by Estimating

DIRECTIONS: *The given answers to the following problems were found with a calculator. Estimate an answer for each problem, and write your estimates on the lines provided. For each problem, use your estimated result to determine if the calculator result is correct. If the calculator result is correct, write a check mark in the "Calculator Result" column; if the calculator result is wrong, write an X in the "Calculator Result" column (2 points for each correct answer).*

		Estimated Result	**Calculator Result**
71.	81.74 + 3,279.81 = 11,453.81	_____	_____
72.	7,000 + 812.83 + 1,320 = 9,132.83	_____	_____
73.	4,217.14 + 231 = 4,448.14	_____	_____
74.	1,081.29 − 714 = 367.29	_____	_____
75.	4,813 − 2,988 = 45,175	_____	_____
76.	309.07 × 12.5 = 4,883.75	_____	_____
77.	29.68 × 30.008 = 890.63744	_____	_____
78.	100,800 × 4.032 = 4,064,256	_____	_____
79.	275 ÷ 0.155 = 1.774	_____	_____
80.	194.96 ÷ 2.437 = 80	_____	_____

STUDENT'S SCORE _____

Chapter 5 Estimations, Graphs, and Shortcuts

SECTION 5.2 Graphs

Mostly numerical information is found on invoices, purchase orders, payroll checks, income statements, balance sheets, inventory cards, sales orders, etc. Mathematics is used to prepare all of these documents. In a modern business, there are also tools, such as graphs, that assist in the collection, preparation, dissemination, and use of the information from these documents.

It is common to summarize a group of numbers in a table, instead of listing the numbers in a sentence or a paragraph like the following example:

In 1985, Agee Enterprises had sales of $160,000,000; in 1986, $175,000,000; in 1987, $210,000,000; in 1988, $195,000,000; and in 1989, $240,000,000.

This data can be presented more clearly in a table, such as Table 5-1.

Year	Annual Sales
1989	$240,000,000
1988	195,000,000
1987	210,000,000
1986	175,000,000
1985	160,000,000

TABLE 5-1 *Five-Year Sales of Agee Enterprises, 1985 to 1989*

To work with the tabular data, you may have to do some calculations mentally or with a calculator. Therefore, it is often more helpful to display the data in a graph instead of a table. A **graph** is a diagram showing the relationships between different data. A graph summarizes the data, so you can quickly grasp its importance. The graph can readily answer some important questions. Was there a yearly increase in sales? If not, in what years was there a decrease? How do 1990 sales compare to 1989 sales?

The most common kinds of graphs are the line graph, the bar graph, and the circle graph. Each of these can be useful in different situations.

PART A Line Graphs

The **line graph** is especially useful for displaying time series data, such as that given in Table 5-1. In time series data, there is a number (such as a sales figure) for each time period (such as a year).

The line graph is the simplest graph to construct. The first step is to draw a horizontal line perpendicular to a vertical line, usually with the horizontal line at the bottom and the vertical line at the left. The two lines are called the **horizontal axis** and the **vertical axis.** For time series data, write the years (or months, weeks, etc.) at even intervals below the horizontal axis.

Usually, you start at the left, with the earliest time period. Place the other data (such as dollars) to the left of the vertical axis, also at even intervals. Normally, you start at the bottom, with the value zero (0), and work upward in equal increments. For example, an increment of 50 represents numbers from 0 to 50 to 100 to 150; an increment of 100 represents numbers from 0 to 100 to 200 to 300, and so on. Label each axis with an appropriate title.

If the numbers on the vertical axis are very large, such as thousands, millions, or billions, you can use fractional parts. For example, instead of placing the actual number *50,000,000* on the vertical axis, you can divide it by 1,000,000 and use the number 50. All other data to be placed on the vertical axis must also be divided by 1,000,000. Label the vertical axis to reflect this change, as shown in Figure 5-1.

The next step is plotting the data. To plot the data in Table 5-1, start with the data for 1985 by locating the appropriate sales amount on the vertical axis. Then make a dot corresponding to that amount on the vertical axis and 1985 on the horizontal axis. Graph paper can make this step easier and more accurate. When you have plotted the sales amounts for each of the five years, connect the dots with straight lines, as shown in Figure 5-1.

Although it is preferable to begin the vertical axis with the value zero (0) at the bottom, this can sometimes lead to wasted space. There is a way to

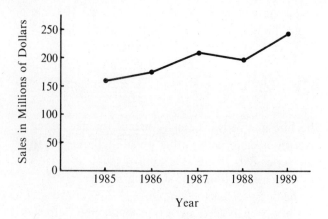

FIGURE 5-1 Line Graph Showing Five-Year Sales of Agee Enterprises, 1985 to 1989

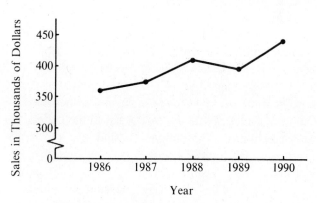

FIGURE 5-2 Line Graph Showing Five-Year Sales of On-Line, Inc., 1986 to 1990

avoid this. For example, the sales data for On-Line, Inc. are given in Table 5-2. Notice that the graph in Figure 5-2, which represents the data in Table 5-2, saves space by skipping directly from 0 to 300 on the vertical axis. A short, jagged line at the bottom of the vertical axis emphasizes the omission of values between zero (0) and the first convenient value on the vertical axis.

double the space between the marks on the vertical axis (decrease the increment by half) in Figure 5-2, you create the appearance of much greater change in annual sales, as shown in Figure 5-3.

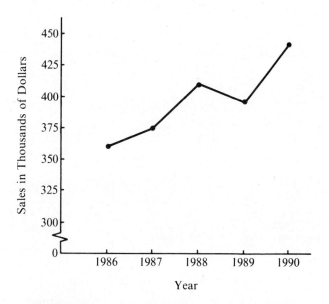

Year	Annual Sales
1990	$440,000
1989	395,000
1988	410,000
1987	375,000
1986	360,000

TABLE 5-2 Five-Year Sales of On-Line, Inc., 1986 to 1990

FIGURE 5-3 Line Graph Showing Five-Year Sales of On-Line, Inc., 1986 to 1990

The number used at the bottom of the vertical axis is arbitrary and depends on the desired appearance of the graph. The selection of a scale of numbers is also arbitrary. For example, if you

PART B Bar Graphs

The **bar graph** is also useful for displaying data. It is slightly more complicated than the line graph, but much more versatile, as you will see. The bar graph

Chapter 5 Estimations, Graphs, and Shortcuts

may be a simple bar graph, a composite bar graph, or a comparative bar graph.

Simple Bar Graphs. Follow these steps to construct a simple bar graph:

1. Draw the horizontal and vertical axes, just as for the line graph, and label each axis appropriately.
2. Locate the heights for the sales amounts in relation to the vertical axis.
3. Draw vertical bars down to the corresponding time periods.

A bar graph depicting the data for Agee Enterprises (given in Table 5-1) is shown in Figure 5-4. The vertical bars may be plain, or they may have diagonal lines, dots, or any pattern desired.

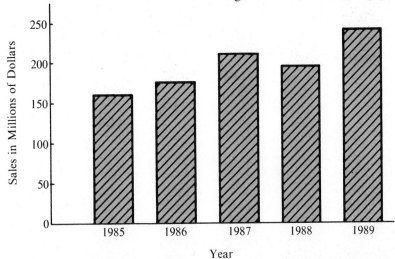

FIGURE 5-4 Vertical Bar Graph Showing Five-Year Sales of Agee Enterprises, 1985 to 1989

Sometimes, a bar graph may be more useful with the bars running horizontally instead of vertically. Every bar graph can be drawn either way. The decision depends on whether the person making the graph thinks it will be easier to read vertically or horizontally. The graph in Figure 5-4 is redrawn horizontally in Figure 5-5.

Composite Bar Graphs. Suppose that Agee Enterprises has three divisions: National, European, and Far East. The annual sales from 1985 to 1989 for each division are given in Table 5-3. To demonstrate the versatility of the bar graph, all the data in Table 5-3 is displayed in just one bar graph, shown in Figure 5-6.

To construct the composite bar graph shown in Figure 5-6, draw a vertical bar representing all three divisions for each year. Since the National division represents the largest portion of company sales, show it at the bottom of each bar. Now take the following steps:

1. In the vertical bar for the year 1985, first find the height of $145,000,000 (National division sales) and mark it.

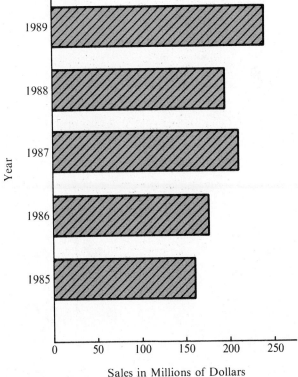

FIGURE 5-5 Horizontal Bar Graph Showing Five-Year Sales of Agee Enterprises, 1985 to 1989

Year	National	European	Far East	Total
1989	$160	$60	$20	$240
1988	140	40	15	195
1987	170	30	10	210
1986	152	15	8	175
1985	145	10	5	160

TABLE 5-3 *Five-Year Sales (in millions) of Agee Enterprises for Three Divisions, 1985 to 1989*

2. The European division had $10,000,000 in sales for 1985. Add this $10,000,000 to the National division's $145,000,000. Locate and mark the height of this total ($155,000,000) in the same vertical bar.
3. Add the Far East sales of $5,000,000 to $155,000,000. Locate and mark on the same bar the height of the total for all three divisions ($160,000,000).
4. Follow Steps 1 through 3 to draw the vertical bars for each of the remaining years.

To make the final composite bar graph easier to read, use different markings to distinguish the segments of each bar representing the three divisions. Then draw a legend, such as the one shown at the top of Figure 5-6, for these markings. You may place this legend elsewhere on the graph.

Comparative Bar Graphs. A third type of bar graph is the comparative bar graph. This graph is used to compare related data for two different time periods, companies, divisions, etc. If the data being compared are *not* time periods, then the horizontal axis should be labeled accordingly.

Suppose that Agee Enterprises wished to compare the annual sales in 1988 and 1989 for each of

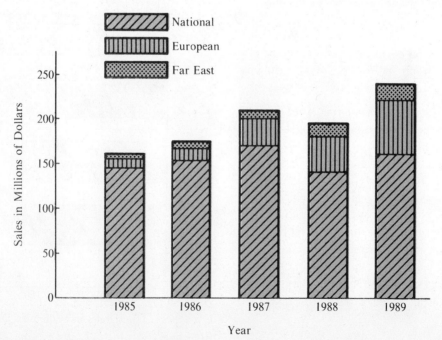

FIGURE 5-6 *Composite Bar Graph Showing Five-Year Sales of Agee Enterprises, 1985 to 1989*

its three divisions. You can still label the vertical axis in millions of dollars, but label the horizontal axis by division. Then, draw vertical bars at heights of $140,000,000 for National, $40,000,000 for European, and $15,000,000 for Far East to represent their sales for 1988. You now have a *partially completed* comparative bar graph, shown in Figure 5-7.

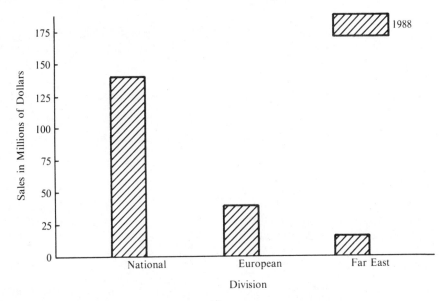

FIGURE 5-7 *Partially Completed Comparative Bar Graph Showing Divisional Sales of Agee Enterprises, 1988*

Since Figure 5-7 is just a simple bar graph, draw three more vertical bars, adjacent to those for 1988, to complete the comparative bar graph. Draw these bars at heights of $160,000,000 for National, $60,000,000 for European, and $20,000,000 for Far East to represent divisional sales for 1989. To avoid confusion, give the bars for 1989 different markings from those for 1988. The final comparative bar graph, with its legend, is shown in Figure 5-8.

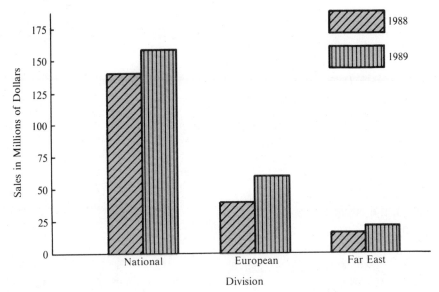

FIGURE 5-8 *Comparative Bar Graph Showing Divisional Sales of Agee Enterprises, 1988 and 1989*

PART C Circle Graphs

The circle graph, or pie chart, is somewhat more specialized than the line graph and the bar graph. The **circle graph** is used to illustrate percentages or parts that comprise a total.

To draw a circle graph, follow these steps:

1. Determine the percentages or the parts that comprise the total.
2. Draw a circle of any convenient size.
3. Separate the area of the circle into parts that visually represent the various proportions and label these parts.

EXAMPLE:

Draw a circle graph showing total sales by division of Agee Enterprises, based on the following data for 1989:

National	$160,000,000
European	60,000,000
Far East	20,000,000
Total Sales	$240,000,000

Step 1. National:

$$\$160{,}000{,}000 \div \$240{,}000{,}000 = 0.6667 = 66\tfrac{2}{3}\%$$

European:

$$\$60{,}000{,}000 \div \$240{,}000{,}000 = 0.25 = 25\%$$

Far East:

$$\$20{,}000{,}000 \div \$240{,}000{,}000 = 0.0833 = 8\tfrac{1}{3}\%$$

Step 2. Draw a circle.

Step 3. Represent the National division with two-thirds ($66\tfrac{2}{3}\%$) of the circle. Separate the remaining one-third into parts that are approximately $\tfrac{1}{4}$ of the whole circle (to represent the European division) and $\tfrac{1}{12}$ of the whole circle (to represent the Far East division).

The completed circle graph is shown in Figure 5-9.

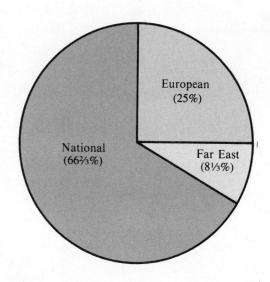

FIGURE 5-9 Circle Graph Showing Divisional Sales of Agee Enterprises, 1989

To draw a circle graph precisely, you may have to use a tool called a *protractor* to measure the angles. For practical purposes, however, you can make a close approximation of the angles. To approximate parts or percents in a circle graph, follow these steps:

1. Draw a circle.
2. Through the center of the circle, draw a line dividing the circle into two halves. Label one half as 50%. Then, draw a line, perpendicular to the first line, from the center of the circle to the edge. This line divides one of the halves into two parts, each of which is 25% of the whole circle. Label one of these parts as 25%. Finally, draw another line from the center of the circle to divide one of the fourths (25%) into two parts. Each of these parts is one-eighth of the whole circle. Label each part as 12.5%.

The circle should now look like the one in Figure 5-10.

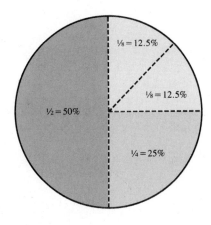

FIGURE 5-10 *Approximating Percents for a Circle Graph*

EXAMPLE:

Using the following data, draw a circle graph showing total sales of Agee Enterprises in 1989, broken down into Cost of Goods Sold (CGS), Operating Expenses (OE), and Net Profit (NP).

	CGS	$144,000,000
	OE	72,000,000
	NP	24,000,000
	Total Sales	$240,000,000

CGS: $144,000,000 ÷ $240,000,000 = 0.6
                                 = 60%

OE: $72,000,000 ÷ $240,000,000 = 0.3
                               = 30%

NP: $24,000,000 ÷ $240,000,000 = 0.1
                               = 10%

From Figure 5-10, you can estimate that the part of the circle for CGS (60%) would be larger than 50%. Also, the part of the circle for OE (30%) would be larger than 25%, and the part of the circle for NP (10%) would be slightly smaller than 12.5%. The resulting circle graph is shown in Figure 5-11.

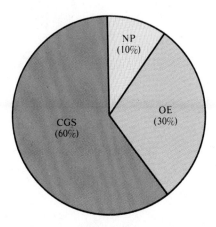

FIGURE 5-11 *Circle Graph Showing Cost of Goods Sold (CGS), Operating Expenses (OE), and Net Profit (NP) as Percentages of Total Sales of Agee Enterprises, 1989*

COMPLETE ASSIGNMENT 19

Name _____

Date _____

ASSIGNMENT 19
SECTION 5.2

Graphs

	Perfect Score	Student's Score
PART A	20	
PART B	60	
PART C	20	
TOTAL	100	

PART A *Line Graphs*

DIRECTIONS: *From the line graph for J.W. Brownies, shown below, answer the following questions to the nearest $5,000 (2 points for each correct answer).*

1. Estimate the sales in 1985. _____
2. Estimate the sales in 1988. _____
3. Which year had the largest gain? _____
4. Which year had the largest loss? _____
5. What was the total increase between 1985 and 1986? _____

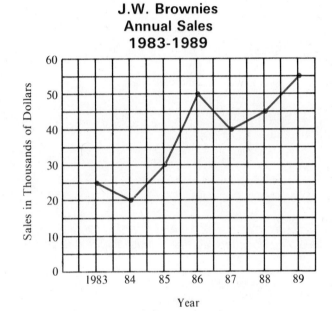

DIRECTIONS: *Draw a line graph from the data given in the table below for the sales figures, by quarter, of the Sea Jay Company during the last year (10 points).*

6.
Sea Jay Company
Quarterly Sales

March 31	$ 75,000
June 30	125,000
Sept. 30	100,000
Dec. 31	150,000

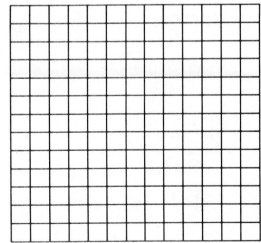

STUDENT'S SCORE _____

PART B Bar Graphs

DIRECTIONS: From the simple bar graph for J.W. Brownies, shown below, answer the following questions to the nearest $5,000 (2 points for each correct answer).

7. Estimate the sales in 1984. _____

8. Estimate the sales in 1989. _____

9. Which year had the smallest gain? _____

10. Which year had the smallest loss? _____

11. What was the increase from the worst year to the best year? _____

DIRECTIONS: From the composite bar graph for the Action Paint Company, shown below, answer the following questions to the nearest $50,000 (2 points for each correct answer).

12. What were the total company revenues in the worst year? _____

13. What were the revenues for Department B in its best year? _____

14. Which department had the most consistent growth in its revenues? _____

15. What were the total revenues for Department A over the three-year period? _____

16. In what year did the combined revenues of Departments B and C come closest to matching the revenue of Department A? _____

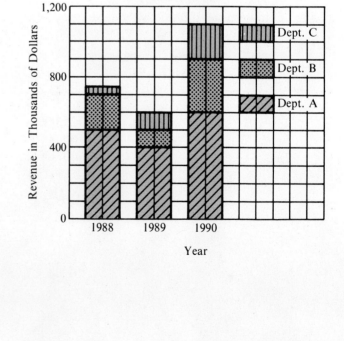

Chapter 5   Estimations, Graphs, and Shortcuts

DIRECTIONS: From the comparative bar graph for monthly sales, shown below, answer the following questions to the nearest $2,500 (2 points for each correct answer).

17. How much did Sarah sell in her best month? _____

18. How much did Bill sell in his worst month? _____

19. Who has the more consistent sales? _____

20. What was Sarah's total sales over the three-month time period? _____

21. What was the greatest difference between their sales in any single month? _____

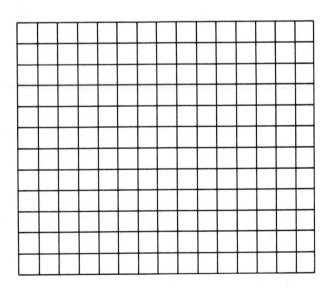

Monthly Sales by Sales Representative January, February, March

DIRECTIONS: Use the data given in Problem 6 on page 205 to construct a simple bar graph for the Sea Jay Company (10 points).

22.

DIRECTIONS: From the data given in the table below, construct a composite bar graph (10 points).

23.

Bayside Deli Monthly Sales (in pounds)

	June	July	August
Meats	400	600	500
Cheese	200	300	200
Salads	100	100	80

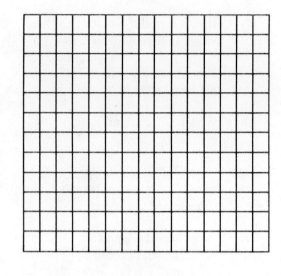

DIRECTIONS: From the data for the Bayside Deli, shown in the table below, construct a comparative bar graph comparing the sales of meats and cheese (10 points).

24.

Monthly Sales of Meats and Cheese (in pounds)

	June	July	August
Meats	400	600	500
Cheese	200	300	200

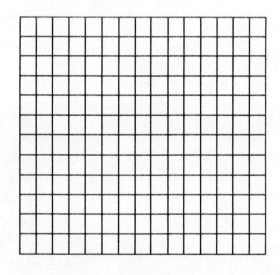

STUDENT'S SCORE _____

Chapter 5   Estimations, Graphs, and Shortcuts

PART C   Circle Graphs

DIRECTIONS: The Golden Acorn Restaurant has three special dinners (labeled A, B, and C), in addition to the other dinners on its regular menu. The circle graph below shows each dinner's percentage of the total number of meals served. Answer the following questions to the nearest 5% (2 points for each correct answer).

25. What percentage of the meals served are *not* the dinners labeled A, B, and C? _____

26. What percentage of the meals served are B and C dinners? _____

27. What is the total percentage of *all* dinners served? _____

28. Compare the proportion of Dinner C served with that of Dinners A and B combined. (C is greater than, less than, or equal to A plus B.) _____

29. From the circle graph alone, can you say anything about the relative profit earned on the various meals served at this restaurant? (Yes or No) _____

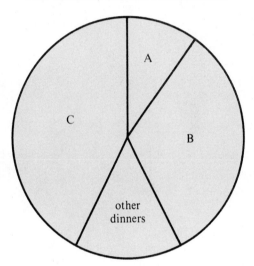

Dinners Served at The Golden Acorn Restaurant

DIRECTIONS: Use the data given in the table below to construct a circle graph showing each department's percentage of total annual receipts (10 points).

30.   **Dragon Leaf Restaurants
Annual Receipts
by Department**

Take-out	$1,250,000	=
Catering	1,875,000	=
Gift Shop	1,875,000	=
Dining Room	2,500,000	=

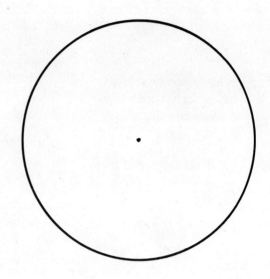

STUDENT'S SCORE _____

Chapter 5 Estimations, Graphs, and Shortcuts

SECTION 5.3 Shortcuts

There are many shortcuts for addition, subtraction, multiplication, and division, but most of them require learning new rules. Some shortcuts, however, are so useful that they are worth learning.

PART A Adding Digits from Left to Right

Adding two numbers (containing two or more digits) from left to right is much the same as adding them from right to left. Before you write the sum of the two digits being added, look at the digits to the right. If their sum is 10 or more, there will be a carry-over of 1 to the sum of the digits to the left. As you proceed in adding from left to right, always check for possible carry-overs.

The procedure is explained step by step in the following examples.

EXAMPLES:

a. 54
     +35

Look at the digits in each column. There is no carry-over. Therefore, adding from left to right is as straightforward as adding from right to left.

Step 1. Add $5 + 3$. Write the 8.

$$\begin{array}{r} 54 \\ +35 \\ \hline 8 \end{array}$$

Step 2. Add $4 + 5$. Write the 9.

$$\begin{array}{r} 54 \\ +35 \\ \hline 89 \end{array}$$

b. 38
     +27

Step 1. Start with $3 + 2$, but don't write the 5 yet. Look to the right, at $8 + 7$. There will be a carry-over of 1 to $3 + 2$, making it 6. Write the 6.

$$\begin{array}{r} 38 \\ +27 \\ \hline 6 \end{array}$$

Step 2. Go to $8 + 7$. Since the 1 (from 15) was already carried over, just write the 5.

$$\begin{array}{r} 38 \\ +27 \\ \hline 65 \end{array}$$

c. 539.41
     +321.38

Step 1. Start with $5 + 3$, but don't write the 8 yet. Look to the right, at $3 + 2$. Since there will be no carry-over, write the 8.

$$\begin{array}{r} 539.41 \\ +321.38 \\ \hline 8 \end{array}$$

Step 2. Go to $3 + 2$, but don't write the 5 yet. Look to the right, at $9 + 1$. There will be a carry-over of 1 to $3 + 2$, making it 6. Write the 6.

$$\begin{array}{r} 539.41 \\ +321.38 \\ \hline 86 \end{array}$$

Step 3. Go to $9 + 1$. The 1 (from 10) was already carried over, but don't write the 0 yet. Look to the right, at $4 + 3$. Since there will be no carry-over, write the 0.

$$\begin{array}{r} 539.41 \\ +321.38 \\ \hline 860. \end{array}$$

Step 4. Go to $4 + 3$, but don't write the 7 yet. Look to the right, at $8 + 1$. Since there will be no carry-over, write the 7.

$$\begin{array}{r} 539.41 \\ +321.38 \\ \hline 860.7 \end{array}$$

Step 5. Go to $1 + 8$. Write the 9.

$$\begin{array}{r} 539.41 \\ +321.38 \\ \hline 860.79 \end{array}$$

PART B Subtracting Digits from Left to Right

Subtracting digits from left to right is somewhat like adding from left to right, in that you always look at the combination to the right before writing down a result. Instead of looking for a carry-over, however, look to see if a 1 will need to be borrowed to subtract the combination to the right. If a 1 needs to be borrowed, decrease the digit in the minuend by 1, and then subtract. Remember, the minuend is the top number, and the subtrahend is the bottom number.

EXAMPLES:

a. 68
       −32

Look at the digits in each column. Borrowing is not necessary. Therefore, subtracting from left to right is as straightforward as subtracting from right to left.

Step 1. Subtract 6 − 3. Write the 3.

    68
   −32
   ———
     3

Step 2. Subtract 8 − 2. Write the 6.

    68
   −32
   ———
    36

b. 7,823
      −6,241

Step 1. Start with 7 − 6, but don't write the 1 yet. Look to the right, at 8 − 2. Since this can be subtracted without borrowing, write the 1.

    7,823
   −6,241
   ———————
    1,

Step 2. Go to 8 − 2, but don't write the 6 yet. Look to the right, at 2 − 4. To subtract 2 − 4, borrow 1 from the 8. 8 − 2 becomes 7 − 2. Write the 5.

    7,823
   −6,241
   ———————
    1,5

Step 3. Go to 2 − 4. Since 1 has already been borrowed, this is 12 − 4, but don't write the 8 yet. Look to the right, at 3 − 1. Since this can be subtracted without borrowing, write the 8.

    7,823
   −6,241
   ———————
    1,58

Step 4. Go to 3 − 1. Write the 2.

    7,823
   −6,241
   ———————
    1,582

c. 6,783.45
      −1,875.32

Step 1. Start with 6 − 1, but don't write the 5 yet. Look to the right, at 7 − 8. To subtract 7 − 8, borrow 1 from the 6. 6 − 1 becomes 5 − 1. Write the 4.

    6,783.45
   −1,875.32
   —————————
    4,

Step 2. Go to 7 − 8. Since 1 has already been borrowed, this is 17 − 8, but don't write the 9 yet. Look to the right, at 8 − 7. Since this can be subtracted without borrowing, write the 9.

    6,783.45
   −1,875.32
   —————————
    4,9

Step 3. Go to 8 − 7, but don't write the 1 yet. Look to the right, at 3 − 5. To subtract 3 − 5, borrow 1 from the 8. 8 − 7 becomes 7 − 7. Write the 0.

    6,783.45
   −1,875.32
   —————————
    4,90

Step 4. Go to 3 − 5. Since 1 has already been borrowed, this is 13 − 5, but don't write the 8 yet. Look to the right, at 4 − 3. Since this can be subtracted without borrowing, write the 8.

    6,783.45
   −1,875.32
   —————————
    4,908.

Chapter 5   Estimations, Graphs, and Shortcuts   215

Step 5. Go to 4 − 3, but don't write the 1 yet. Look to the right, at 5 − 2. Since this can be subtracted without borrowing, write the 1.

$$\begin{array}{r} 6{,}783.45 \\ -1{,}875.32 \\ \hline 4{,}908.1 \end{array}$$

Step 6. Go to 5 − 2. Write the 3.

$$\begin{array}{r} 6{,}783.45 \\ -1{,}875.32 \\ \hline 4{,}908.13 \end{array}$$

PART C Adding Two Fractions Whose Numerators Are 1

The sum of two fractions whose numerators are both 1 is a fraction whose numerator is the *sum* of the two denominators and whose denominator is the *product* of the two denominators.

EXAMPLES:

a. $\dfrac{1}{5} + \dfrac{1}{4} = \dfrac{5 + 4}{5 \times 4} = \dfrac{9}{20}$

b. $\dfrac{1}{4} + \dfrac{1}{3} = \dfrac{4 + 3}{4 \times 3} = \dfrac{7}{12}$

If the two original denominators have a common factor, the sum found with this shortcut will be a fraction that can be reduced to lower terms.

EXAMPLES:

a. $\dfrac{1}{6} + \dfrac{1}{9} = \dfrac{6 + 9}{6 \times 9} = \dfrac{15}{54} = \dfrac{5}{18}$

b. $\dfrac{1}{9} + \dfrac{1}{15} = \dfrac{9 + 15}{9 \times 15} = \dfrac{24}{135} = \dfrac{8}{45}$

PART D Subtracting Two Fractions Whose Numerators Are 1

The difference between two fractions whose numerators are both 1 is a fraction whose numerator equals the second denominator *minus* the first denominator and whose denominator is the *product* of the two denominators.

As in addition, if the two original denominators have a common factor, the difference found with this shortcut is a fraction that can be reduced to lower terms.

EXAMPLES:

a. $\dfrac{1}{5} - \dfrac{1}{9} = \dfrac{9 - 5}{9 \times 5} = \dfrac{4}{45}$

b. $\dfrac{1}{4} - \dfrac{1}{6} = \dfrac{6 - 4}{6 \times 4} = \dfrac{2}{24} = \dfrac{1}{12}$

c. $\dfrac{1}{6} - \dfrac{1}{9} = \dfrac{9 - 6}{9 \times 6} = \dfrac{3}{54} = \dfrac{1}{18}$

PART E Multiplying and Dividing by 0.1, 0.01, 0.001, etc.

The shortcut for multiplying or dividing by 0.1, 0.01, 0.001, etc., is to move the decimal point to either the left (multiplication) or the right (division). If the multiplicand or the dividend is a whole number, place a decimal point after the last digit, and then move the decimal point.

Multiplying by 0.1, 0.01, 0.001, etc. When the multiplier is 0.1, 0.01, 0.001, etc., all you need to do is move the decimal point of the multiplicand one place to the *left* for each decimal place in the multiplier. It may be necessary to add zeros to the left of the multiplicand in order to move the decimal point the correct number of places.

When the multiplier is 0.1, move the decimal point of the multiplicand one place to the left.

EXAMPLE:

$4.14 \times 0.1 = 4{.}14$
$= 0.414$

When the multiplier is 0.01, move the decimal point of the multiplicand two places to the left.

EXAMPLE:

$387 \times 0.01 = 387. \times 0.01$
$= 387.$
$= 3.87$

When the multiplier is 0.001, move the decimal point of the multiplicand three places to the left.

EXAMPLE:

32.8 × 0.001 = 032.8
= 0.0328

Since 0.1 equals $\frac{1}{10}$, 0.01 equals $\frac{1}{100}$, etc., multiplying by 0.1 (or $\frac{1}{10}$) is the same as dividing by 10; multiplying by 0.01 (or $\frac{1}{100}$) is the same as dividing by 100, etc.

Dividing by 0.1, 0.01, 0.001, etc. When the divisor is 0.1, 0.01, 0.001, etc., all you need to do is move the decimal point of the dividend one place to the right for each decimal place in the divisor. It may be necessary to add zeros to the right of the dividend in order to move the decimal point the correct number of places.

When the divisor is 0.1, move the decimal point of the dividend one place to the right.

EXAMPLE:

5.41 ÷ 0.1 = 5.41
= 54.1

When the divisor is 0.01, move the decimal point of the dividend two places to the right.

EXAMPLE:

8.3 ÷ 0.01 = 8.30
= 830

When the divisor is 0.001, move the decimal point of the dividend three places to the right.

EXAMPLE:

0.2732 ÷ 0.001 = 0.2732
= 273.2

Since 0.1 equals $\frac{1}{10}$, 0.01 equals $\frac{1}{100}$, etc., dividing by 0.1 (or $\frac{1}{10}$) is the same as multiplying by 10; dividing by 0.01 (or $\frac{1}{100}$) is the same as multiplying by 100, etc.

PART F Multiplying and Dividing by 50 and 25

A shortcut can be used to multiply and divide by 50 and 25. Express 50 as 100 ÷ 2 and 25 as 100 ÷ 4.

Multiplying by 50 and 25. To multiply any number by 50, first multiply it by 100, and then divide that product by 2. Use the shortcut for multiplying by 100; move the decimal point two places to the right.

EXAMPLES:

a. 48 × 50

48 × 100 = 4,800

4,800 ÷ 2 = 2,400

b. 9.012 × 50

9.012 × 100 = 901.2

901.2 ÷ 2 = 450.6

To multiply any number by 25, first multiply it by 100, and then divide that product by 4.

EXAMPLES:

a. 200 × 25

200 × 100 = 20,000

20,000 ÷ 4 = 5,000

b. 31.96 × 25

31.96 × 100 = 3,196

3,196 ÷ 4 = 799

Dividing by 50 and 25. When dividing any number by 50, remember that 50 equals 100 ÷ 2, or $\frac{100}{2}$. Also, remember that to divide any number by a fraction, you invert the fraction and then multiply. Therefore, dividing a number by 50 is the same as multiplying it by $\frac{2}{100}$. Multiply the number by 2, and then divide that product by 100. Use the shortcut for dividing by 100; move the decimal point two places to the left.

EXAMPLES:

a. 808 ÷ 50

$808 \div \frac{100}{2} = 808 \times \frac{2}{100}$

808 × 2 = 1,616

1,616 ÷ 100 = 16.16

Chapter 5 Estimations, Graphs, and Shortcuts

b. 4.75 ÷ 50

$4.75 \div \frac{100}{2} = 4.75 \times \frac{2}{100}$

$4.75 \times 2 = 9.50$

$9.50 \div 100 = 0.095$

When dividing any number by 25, remember that 25 equals $100 \div 4$, or $\frac{100}{4}$. Therefore, dividing a number by 25 is the same as multiplying it by $\frac{4}{100}$. Multiply the number by 4, and then divide that product by 100.

EXAMPLES:

a. 1,320 ÷ 25

$1,320 \div \frac{100}{4} = 1,320 \times \frac{4}{100}$

$1,320 \times 4 = 5,280$

$5,280 \div 100 = 52.8$

b. 6.23 ÷ 25

$6.23 \div \frac{100}{4} = 6.23 \times \frac{4}{100}$

$6.23 \times 4 = 24.92$

$24.92 \div 100 = 0.2492$

PART G Multiplying by 11

To multiply any two-digit number by 11 (11 is the multiplier), follow these steps:

1. Add the two digits of the multiplicand.
2. If the sum of these two digits is 9 or less, place this sum between the two digits of the multiplicand to form the product.

EXAMPLES:

a. 52 × 11

$5 + 2 = 7$ (less than 9)

$52 \times 11 = 572$

b. 36 × 11

$3 + 6 = 9$ (equal to 9)

$36 \times 11 = 396$

3. If the sum of the two digits of the multiplicand is greater than 9, add the 1 of this sum to the tens digit of the multiplicand. Next, place the units digit of this sum between the *new* tens digit and the units digit of the multiplicand.

EXAMPLES:

a. 75 × 11

$7 + 5 = 12$ (greater than 9)

$7 + 1 = 8$; 2 goes between 8 and 5

$75 \times 11 = 825$

b. 92 × 11

$9 + 2 = 11$ (greater than 9)

$9 + 1 = 10$; 1 goes between 10 and 2

$92 \times 11 = 1,012$

If either of the factors is a decimal, the number of decimal places in the product will equal the total number of decimal places in the factors.

EXAMPLES:

a. 2.6 × 11

$2 + 6 = 8$ (less than 9)

$26 \times 11 = 286$

$2.6 \times 11 = 28.6$ (one decimal place)

b. 51 × 0.11

$5 + 1 = 6$ (less than 9)

$51 \times 11 = 561$

$51 \times 0.11 = 5.61$ (two decimal places)

c. 0.79 × 0.011

$7 + 9 = 16$ (greater than 9)

$7 + 1 = 8$; 6 goes between 8 and 9

$79 \times 11 = 869$

$0.79 \times 0.011 = 0.00869$ (five decimal places)

PART H Some Miscellaneous Shortcuts

Adding and Subtracting 9. The fact that 9 is one less than 10 provides a shortcut for adding 9 to or subtracting 9 from a number.

To add 9 to any number, first add 10 to it, and then subtract 1 from that sum.

EXAMPLES:

a. $36 + 9$

$$36 + 9 = (36 + 10) - 1$$
$$= 46 - 1$$
$$= 45$$

b. $82 + 9$

$$82 + 9 = (82 + 10) - 1$$
$$= 92 - 1$$
$$= 91$$

To subtract 9 from any number, first subtract 10 from it, and then add 1 to that difference.

EXAMPLES:

a. $45 - 9$

$$45 - 9 = (45 - 10) + 1$$
$$= 35 + 1$$
$$= 36$$

b. $121 - 9$

$$121 - 9 = (121 - 10) + 1$$
$$= 111 + 1$$
$$= 112$$

You can also use this shortcut when adding or subtracting two-digit numbers ending in 9. 19 is one less than 20, 29 is one less than 30, 39 is one less than 40, etc. When working problems like these, do not recopy the problems and add or subtract in the usual way. Try to do all the work mentally.

EXAMPLES:

a. $75 + 39$

$$75 + 39 = (75 + 40) - 1$$
$$= 115 - 1$$
$$= 114$$

b. $67 + 79$

$$67 + 79 = (67 + 80) - 1$$
$$= 147 - 1$$
$$= 146$$

c. $96 - 49$

$$96 - 49 = (96 - 50) + 1$$
$$= 46 + 1$$
$$= 47$$

d. $134 - 69$

$$134 - 69 = (134 - 70) + 1$$
$$= 64 + 1$$
$$= 65$$

Multiplying a Three-Digit Number by a One-Digit Number When All the Digits are 4 or Less. When a three-digit number is multiplied by a one-digit number and all the digits are 4 or less, you can write the product immediately. Starting at the left, multiply mentally, and add as you go along: the hundreds digit times the multiplier, plus the tens digit times the multiplier, plus the units digit times the multiplier. It sounds more difficult than it really is.

EXAMPLES:

a. 212×4

$(200 \times 4) + (10 \times 4) + (2 \times 4)$

$800 + 40 + 8 = 848$

b. 341×3

$(300 \times 3) + (40 \times 3) + (1 \times 3)$

$900 + 120 + 3 = 1{,}023$

Multiplying a Two-Digit Number by Another Two-Digit Number. The shortcut for multiplying a two-digit number by another two-digit number is very useful, especially if the digits are small. However, it does require some practice. The steps in the multiplication are the same as in the long method, but the partial products are added mentally. Study the following examples.

EXAMPLES:

a.   32
      $\times$ 24

$$7\ \ 6 \quad 8 = 768$$

Step 1. Multiply 4×2. Write the 8 in the units position.

Step 2. Multiply 4×3 to get 12. Multiply 2×2 to get 4. Mentally add $12 + 4$ to get 16. Write the 6 in the tens position and carry the 1.

Chapter 5   Estimations, Graphs, and Shortcuts   **219**

Step 3. Multiply 2 × 3 to get 6. Add the 1 (carried over) to get 7. Write the 7 in the hundreds position.

Step 1. Multiply 8 × 4 to get 32. Write the 2 in the units position and carry the 3.

Step 2. Multiply 8 × 6 to get 48. Multiply 2 × 4 to get 8. Mentally add 48 + 8 to get 56. Add the 3 (carried over) to get 59. Write the 9 in the tens position and carry the 5.

b. 64
      ×28

Step 3. Multiply 2 × 6 to get 12. Add the 5 (carried over) to get 17. Write the 17 to the left.

$$\begin{array}{r} 64 \\ \times\ 28 \\ \hline 17\ \ 9\ \ \ 2 \end{array} = 1{,}792$$

COMPLETE ASSIGNMENT 20

Name _____

Date _____

ASSIGNMENT 20
SECTION 5.3

Shortcuts

	Perfect Score	Student's Score
PART A	15	
PART B	15	
PART C	15	
PART D	15	
PART E	10	
PART F	10	
PART G	10	
PART H	10	
TOTAL	100	

PART A Adding Digits from Left to Right

DIRECTIONS: Add from left to right only (1 point for each correct answer).

1. 54
   +23
   ____

2. 25
   +72
   ____

3. 46
   +23
   ____

4. 328
   +541
   ____

5. 721
   +137
   ____

6. 68
   +27
   ____

7. 68
   + 34
   ____

8. 29
   +31
   ____

9. 58
   +39
   ____

10. 385
    +473
    ____

11. 39.4
    +55.3
    ____

12. 10.1
    + 91.9
    ____

13. 35.07
    +36.91
    ____

14. 41.45
    +26.84
    ____

15. 33.22
    + 77.71
    ____

STUDENT'S SCORE _____

PART B Subtracting Digits from Left to Right

DIRECTIONS: Subtract from left to right only (1 point for each correct answer).

16. 74
    −42
    ____

17. 75
    −62
    ____

18. 59
    −17
    ____

19. 892
    −381
    ____

20. 531
    −421
    ____

21. 84
    −28
    ____

22. 54
    −18
    ____

23. 591
    −382
    ____

24. 312
    −291
    ____

25. 5.89
    −2.99
    ____

| 26. | 96.54
−41.29 | 27. | 10.64
− 8.38 | 28. | 129.86
− 38.95 | 29. | 41.82
−36.57 | 30. | 462.31
−271.09 |

STUDENT'S SCORE _____

PART C Adding Two Fractions Whose Numerators Are 1

DIRECTIONS: Add by using the shortcut. Reduce your answers to lowest terms (1 point for each correct answer).

31. $\frac{1}{3} + \frac{1}{2} =$ _____

32. $\frac{1}{2} + \frac{1}{4} =$ _____

33. $\frac{1}{5} + \frac{1}{6} =$ _____

34. $\frac{1}{3} + \frac{1}{6} =$ _____

35. $\frac{1}{5} + \frac{1}{2} =$ _____

36. $\frac{1}{5} + \frac{1}{8} =$ _____

37. $\frac{1}{8} + \frac{1}{18} =$ _____

38. $\frac{1}{6} + \frac{1}{7} =$ _____

39. $\frac{1}{5} + \frac{1}{12} =$ _____

40. $\frac{1}{20} + \frac{1}{12} =$ _____

41. $\frac{1}{8} + \frac{1}{7} =$ _____

42. $\frac{1}{9} + \frac{1}{12} =$ _____

43. $\frac{1}{10} + \frac{1}{15} =$ _____

44. $\frac{1}{6} + \frac{1}{15} =$ _____

45. $\frac{1}{20} + \frac{1}{10} =$ _____

STUDENT'S SCORE _____

Chapter 5   Estimations, Graphs, and Shortcuts

PART D Subtracting Two Fractions Whose Numerators Are 1

DIRECTIONS: Subtract by using the shortcut. Reduce your answers to lowest terms (1 point for each correct answer).

46. $\frac{1}{2} - \frac{1}{5} =$ _____

54. $\frac{1}{20} - \frac{1}{30} =$ _____

47. $\frac{1}{2} - \frac{1}{3} =$ _____

55. $\frac{1}{5} - \frac{1}{30} =$ _____

48. $\frac{1}{4} - \frac{1}{10} =$ _____

56. $\frac{1}{9} - \frac{1}{12} =$ _____

49. $\frac{1}{6} - \frac{1}{7} =$ _____

57. $\frac{1}{10} - \frac{1}{30} =$ _____

50. $\frac{1}{10} - \frac{1}{15} =$ _____

58. $\frac{1}{6} - \frac{1}{24} =$ _____

51. $\frac{1}{3} - \frac{1}{9} =$ _____

59. $\frac{1}{8} - \frac{1}{16} =$ _____

52. $\frac{1}{5} - \frac{1}{20} =$ _____

60. $\frac{1}{12} - \frac{1}{20} =$ _____

53. $\frac{1}{10} - \frac{1}{12} =$ _____

STUDENT'S SCORE _____

PART E Multiplying and Dividing by 0.1, 0.01, 0.001, etc.

DIRECTIONS: Multiply or divide by moving the decimal point (1 point for each correct answer).

61. 7.01 × 0.1 = _____

64. 3.7291 × 0.001 = _____

62. 31.21 × 0.1 = _____

65. 3,206 × 0.0001 = _____

63. 277 × 0.01 = _____

66. 5 ÷ 0.1 = _____

67. 300 ÷ 0.1 = _____

68. 50.1 ÷ 0.01 = _____

69. 80.2 ÷ 0.001 = _____

70. 7.3 ÷ 0.0001 = _____

STUDENT'S SCORE _____

PART F Multiplying and Dividing by 50 and 25

DIRECTIONS: Multiply or divide by using the shortcut (1 point for each correct answer).

71. 48 × 50 = _____

72. 0.92 × 50 = _____

73. 0.08 × 50 = _____

74. 7.08 × 25 = _____

75. 532 × 25 = _____

76. 40 ÷ 50 = _____

77. 582 ÷ 50 = _____

78. 32.4 ÷ 50 = _____

79. 76 ÷ 25 = _____

80. 32 ÷ 25 = _____

STUDENT'S SCORE _____

PART G Multiplying by 11

DIRECTIONS: Multiply by using the shortcut (1 point for each correct answer).

81. 25 × 11 = _____

82. 32 × 11 = _____

83. 91 × 11 = _____

84. 66 × 11 = _____

85. 47 × 11 = _____

86. 5.7 × 11 = _____

87. 8.8 × 11 = _____

88. 0.26 × 1.1 = _____

89. 0.083 × 11 = _____

90. 6.4 × 1.1 = _____

STUDENT'S SCORE _____

Chapter 5 Estimations, Graphs, and Shortcuts   225

PART H Some Miscellaneous Shortcuts

DIRECTIONS: Add or subtract by using the shortcut (1 point for each correct answer).

91.   64 + 19 = _____   **93.**   90 − 39 = _____

92.   28 + 69 = _____   **94.**   50 − 19 = _____

DIRECTIONS: Multiply from left to right (1 point for each correct answer).

95.   121 × 4 = _____   **97.**   434 × 2 = _____

96.   243 × 2 = _____   **98.**   421 × 3 = _____

DIRECTIONS: Multiply by using the shortcut (1 point for each correct answer).

99.   24
       × 22
       _____

100.   65
        × 33
        _____

STUDENT'S SCORE _____

UNIT 2
APPLICATIONS

6   Keeping a Checking Account

7   Calculating Interest

8   Calculating Time-Payment Plans and Short-Term Loans

9   Purchase Orders and Invoices, Cash Discounts, and Trade Discounts

10 Selling Goods

11 Calculating Gross Pay for Payrolls

12 Inventory Valuation, Cost of Goods Sold, and Depreciation

CHAPTER 6

Keeping a Checking Account

Most businesses and individuals use **checking accounts**. Checking accounts are bank accounts that represent money deposited, and checks may be drawn against these accounts. Making payments with checks is often more convenient and generally safer than paying with cash. As a rule, only enough money to meet expected current expenses is kept in a checking account.

Large companies use computers, accounting machines, or check-writing machines to produce both checks and necessary records in a single operation. Most small businesses, however, use typed or handwritten checks and keep their records on the check stubs. The information on the check stubs is transferred to the business owner's books of account.

Section 6.1 of this chapter applies only to checks and deposits by small businesses or individuals who do not use computers. To keep up with technological developments, however, a brief description of electronic banking is given at the end of this section.

Section 6.2 deals with monthly bank statements and reconciliation statements. These statements have to be prepared regularly to check the accuracy of both the bank's and the depositor's records of a checking account.

SECTION 6.1 Checks and Deposits

A **check** is a written order to a bank to pay a stated amount of money to the individual or business named on the check. Books of blank checks and deposit slips are usually available from the bank where an individual or business has a checking account. Usually, business checkbooks have three or four checks per page. Each check is separated from the others and from its stub by perforations.

Because banks use computers, a magnetized number identifying the bank and the depositor's account appears on each check and deposit ticket. Today, many checks also have the check number preprinted in the upper right corner. With prenumbered checks, there will usually be a corresponding magnetized check number preceding the bank's identification number at the bottom of the check.

PART A Keeping Records of a Checking Account

Suppose you went to a bank to open a checking account, either for a business or for your own personal use. The bank would give you a choice of several styles of checks and methods of recording them. Both businesses and individuals can record checks using check stubs and/or check registers. Examples of both are shown in Figure 6-1.

Today, check stubs are probably used more by businesses than by individuals. When a business checkbook has three or four checks per page, all the stubs are at the left. Usually, they are three-hole punched, so that they may be kept in a binder.

One advantage of the check register is that the *Balance Forward* portion does not have to be

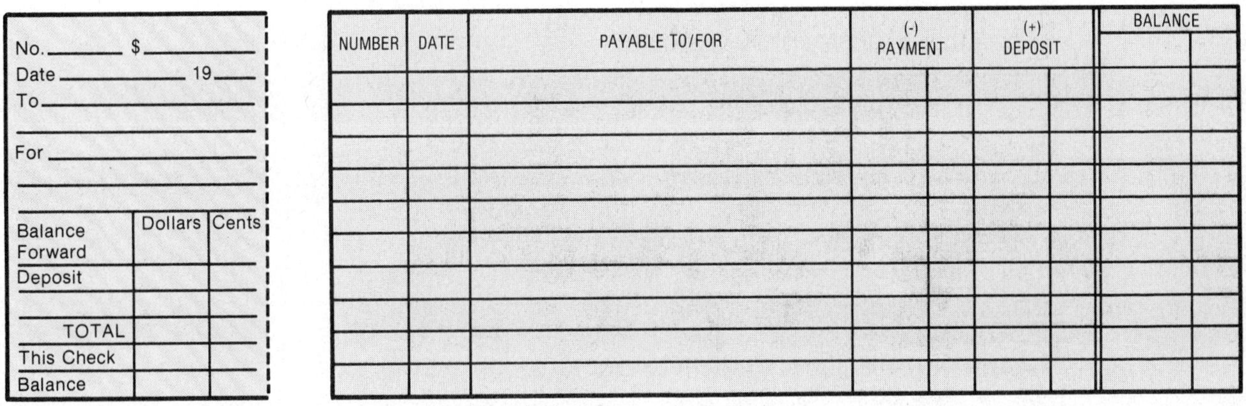

Chapter 6 Keeping a Checking Account

FIGURE 6-1 *A Check Stub and a Check Register*

rewritten for every check. This eliminates a common source of errors—inaccurate copying. In a check register, each deposit has its own line, rather than just a line on a check stub (which is actually the recording of a withdrawal). Although keeping records on check stubs is just as accurate, the check register permits neater recording of deposits and other account adjustments, such as interest, bank service charges, automatic deposits/withdrawals, and correction entries to balance the account.

One style of check register has two lines per transaction; the other style has only one line per transaction. The advantage of having two lines per transaction is that the amount of the transaction can be written directly beneath the *Balance Forward*. This makes the required addition or subtraction easier. The disadvantages of having two lines per check are that it takes twice as much space and that the withdrawal or deposit must be recorded twice, thus allowing another chance for error.

The choice of check stubs or check registers is a matter of preference. The exercises and examples in this book are based primarily on check stubs, because that is what you will most likely find in a business that does not use computer-written checks. Even businesses that use computer-written checks will likely have the ability to write checks manually. For example, suppose someone washes the windows at your business, and you want to pay the window-washer that day with a check. If you have only one computer printer and it is currently printing an invoice, you would not stop the computer and insert blank checks just to print out one check.

To keep accurate records, always complete the check stub whenever you make a deposit and *before* you write a check. When you make a deposit, record the amount and add it to the previous balance, which is usually labeled *Balance Forward* on the check stub. Write the sum in the space labeled *Total*. Before writing a check, be sure to accurately complete all the information on the check stub. After entering the amount of the check, subtract it from the *Total*. The result is the new balance, which you should record and bring forward to the next stub.

Figure 6-2 shows a completed check and its stub. The stub shows that on September 28, Bovio Supply Company had a balance of $2,396.89 (see the *Balance Forward* line). On the same day, John Bovio deposited $1,400 into the account. The new balance was $3,796.89 (see the *Total* line). He also wrote check #1895 for $106.56 to Ladera Plumbing for miscellaneous supplies. After $106.56 was subtracted from $3,796.89, the new balance in the checking account was $3,690.33.

PART B Writing Checks

To write a check, always use a pen. Begin by entering the check number, if it is not preprinted on the check. Then, enter the date. For security reasons, each of the following entries on the check should always begin at the extreme left of the space provided (refer to Figure 6-2):

1. The name of the **payee** (the business or individual to whom the check is payable). If the payee's

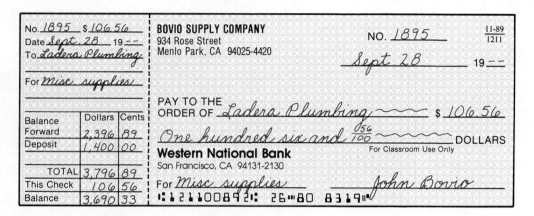

FIGURE 6-2 A Completed Check and Check Stub

name is so short that it does not fill out the *Pay to the order of* line, draw a wavy line to the right of the payee's name. This will help prevent someone from altering the name of the rightful payee.
2. The amount of the check, written in numbers. The first number in the amount should be written as close as possible to the dollar sign.
3. The amount of the check, written in words, on the line labeled *DOLLARS*. The dollar amount is always written in words, but the amount in cents is written as a fraction with a denominator of 100. To prevent confusion, the word *and* should be used to separate the dollars from the cents. It is important to begin at the left and draw the wavy line to the right of the amount to help prevent any alterations.

Some checks have a blank line where you may write the purpose of the check. This is optional, but it makes keeping records easier and more thorough. Finally, be sure to sign the check on the line provided at the bottom right. Remember, *never use a pencil to write any part of a check.*

PART C Endorsing Checks

Before a check can be cashed or sent to the bank for deposit, the payee must **endorse** the check, or sign it on the reverse side. There may be a place reserved on the back of the check for the endorsement. If not, sign across the short side of the check at the end which is against the words *Pay to the order of* on the front. The entire endorsement should fit within the top $1\frac{1}{2}$ inches. Although some banks may request somewhat different limits, the placement of the endorsement in a particular position on the back of the check is the result of a Federal Reserve System regulation included in the Federal Banking Act of 1987.

There are several ways of endorsing a check. In the **blank endorsement** shown in Figure 6-3, only the name of the payee is written. If a check with a blank endorsement is lost or stolen, anybody who has the check may cash it.

Thus, a better way to endorse a check is with a **restrictive endorsement,** which indicates a specific purpose or new payee for the check. The second block in Figure 6-3 shows a restrictive endorsement indicating that this check must be deposited. This endorsement is used when the payee plans to deposit the check or mail it to the bank for deposit. The words *For deposit only* are written above the payee's signature.

In the third block of Figure 6-3 is an even more restrictive endorsement for deposit. Bovio Supply Company wants to ensure that this check is deposited into a specific bank account, possibly because the company has more than one account at this bank.

Most companies, large and small, stamp all incoming checks with a restrictive endorsement as soon as they are received. Thus, if a check is lost or stolen, it cannot be cashed by someone else. An example of such a stamped endorsement is shown in the fourth block of Figure 6-3.

Another type of restrictive endorsement permits the check to be transferred to a different payee, as shown in the fifth block of Figure 6-3. The words *Pay to the order of* are written at the top of the

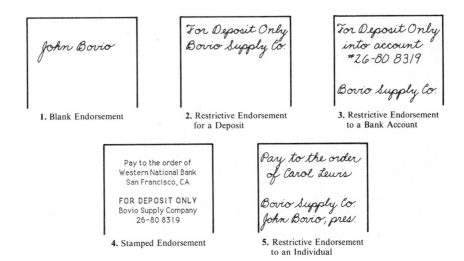

FIGURE 6-3 *Various Forms of Endorsement*

check. Then comes the name of the new payee (an individual or a business). Below that is the signature of the original payee. In this example, Bovio Supply Company is represented by the owner, John Bovio.

Although checks may be transferred in the manner shown in the fifth block of Figure 6-3, it is a poor business practice. Bovio Supply Company will not have a canceled check or a check stub to prove that it ever paid anything to Carol Lewis. A much better practice is to first deposit all checks from customers, and then write new checks to pay the suppliers of goods and services.

PART D Clearing of Checks

After a check has been cashed by the payee, it may take a few days (or longer) for the check to reach a **clearinghouse.** A check is *cleared* when its amount is charged to the bank against which it was drawn. The bank, in turn, subtracts the amount of the check from the depositor's account. After this is done, the check is stamped and becomes a **canceled check.**

The American Bankers Association (ABA), the national organization of banking, has devised a numerical coding system that identifies all banks in the country by location and by official bank number. The **ABA number** for a particular bank is usually printed in the upper right-hand corner of the check. In Figure 6-2, it is 11-89/1211. The magnetized numbers at the bottom are used for routing the checks with computerized equipment and for charging the amount to the depositor's account. The right-hand part of these magnetized numbers is the depositor's checking account number with the bank. In Figure 6-2, it is 26-80 8319.

PART E Making Deposits

When you make a deposit, you complete a deposit ticket detailing the cash and checks you are depositing. Usually, the deposit ticket is made out in duplicate. If you make the deposit in person, a copy of the deposit ticket is handed back to you. If you mail the deposit, the original deposit ticket may be returned to you by the bank, along with your monthly statement and canceled checks.

Figure 6-4 shows a completed deposit ticket. Note that only the totals of currency and coins are listed; the checks are listed separately. Most deposit tickets have room on the reverse side to list additional checks. The total of any checks listed on the back is written on the face of the deposit ticket.

Sometimes, the ABA number of each check deposited is written to the left of the amount. At other times, the person or business that issued the check may be listed.

PART F Transferring Funds Electronically

Although electronic banking eliminates some check writing, it does not eliminate the need to use math

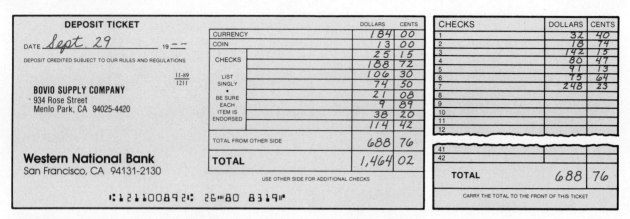

FIGURE 6-4 *A Completed Deposit Ticket (Front and Back)*

to keep accurate account records. You should be familiar with some fundamental aspects of electronic banking.

An **Electronic Funds Transfer (EFT)** system is a new, checkless method of paying for goods or services, making deposits, withdrawing money, etc. It was developed in an effort to reduce the rising costs of processing checks. One of the systems now using EFT is the **Automated Clearinghouse.** Stated simply, information about payments is put on computer storage devices (for example, tapes and disks). The translated data are then run through the bank's computer to complete the desired transfers of funds.

Uses of the Automated Clearinghouse. The federal government, most state governments, and many companies now provide the direct method of paying their employees. Instead of issuing paychecks, a company uses computer media to authorize its bank to transfer money from the company's account to the banks and accounts of its employees.

Another use of the Automated Clearinghouse allows a depositor's bank to pay some of that depositor's regular monthly expenses by electronically transferring funds. Examples of regular monthly expenses are mortgage payments, car installment payments, utility bills, etc. The depositor's monthly bank statement will show all amounts electronically transferred to pay preauthorized expenses.

Automated Teller Machines. One of the first applications of the EFT was the installation of **automated teller machines (ATM)** by banks and savings and loan associations in the early 1970s. The ATM can be used 24 hours a day. It is placed in a lobby, on an outside wall, at a drive-in window, or at other convenient locations, such as a store or a shopping center.

The ATM is a computer terminal that communicates with a bank or savings and loan association. It is activated by the insertion of a plastic card, which has a magnetic strip to identify the individual user. Generally, the user must also enter a secret code number on the keyboard. By doing this, the user can make deposits, withdraw cash, or transfer funds to a different account without using checks or going into a bank. Receipts for the transactions are printed by the ATM. In some cases, the transfer of money can be canceled on three days' notice, which is similar to stopping payment on a check.

A Word of Caution. There are disadvantages to the electronic transfer of money, in spite of strict rules (established by state banking and savings and loan associations) for these new systems. Although EFTs do reduce the total volume of checks, particularly at the consumer level, they do not eliminate the need for checks in the operation of most businesses, nor do they eliminate the need to reconcile accounts.

COMPLETE ASSIGNMENT 21

Name _____

Date _____

ASSIGNMENT 21
SECTION 6.1

Checks and Deposits

	Perfect Score	Student's Score
PROBLEM 1	20	
PROBLEM 2	20	
PROBLEM 3	20	
PROBLEM 4	10	
PROBLEM 5	20	
PROBLEM 6	10	
TOTAL	100	

DIRECTIONS: Using the information given, complete the check stubs and the checks in Problems 1, 2, and 3. Assume that you are John Bovio, and sign his name (10 points for each correctly written check, and 10 points for each correctly written check stub).

1. Check #1896: The date is September 29. The balance brought forward from check #1895 is $3,690.33. The deposit made is $834.47. The amount of this check is $209.50. Make the check out to Ascot and Sons for plastic pipe.

STUDENT'S SCORE _____

2. Check #1905: The date is October 6. The balance brought forward from check #1904 is $2,349.64. The amount of this check is $182.12. Make the check out to Better Roofco for roof repairs.

STUDENT'S SCORE _____

3. Check #1906: The date is October 7. For the balance brought forward from check #1905, see Problem 2. The deposit made is $361.48. The amount of this check is $150. Make the check out to Clean Corp. for janitorial services.

STUDENT'S SCORE _____

4. Bovio Supply Company has received a $48 check from a customer. John Bovio uses this check to pay for $48 worth of office supplies from Office Products, Inc.

   a. In the space provided, show the necessary restrictive endorsement (7 points).

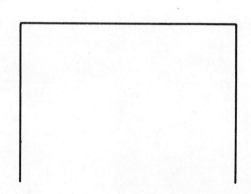

   b. Is this type of endorsement a good business practice (3 points)? _____

STUDENT'S SCORE _____

Chapter 6   Keeping a Checking Account   **235**

5. On October 10, Bovio Supply Company makes a deposit consisting of the following: $127 in currency, $18 in coins, and checks for $97.51, $211.43, $427.50, and $198.25. Complete the deposit ticket (20 points).

DEPOSIT TICKET		DOLLARS	CENTS
	CURRENCY		
DATE _____ 19 ___	COIN		
DEPOSIT CREDITED SUBJECT TO OUR RULES AND REGULATIONS	CHECKS		
11-89 / 1211	LIST SINGLY • BE SURE EACH ITEM IS ENDORSED		
BOVIO SUPPLY COMPANY 934 Rose Street Menlo Park, CA 94025-4420			
	TOTAL FROM OTHER SIDE		
Western National Bank San Francisco, CA 94131-2130	TOTAL		
	USE OTHER SIDE FOR ADDITIONAL CHECKS		

⑆121100892⑆ 26⬝80 8319⑈

STUDENT'S SCORE _____

6. In the spaces provided, show two restrictive endorsements, either of which should be used on all the checks deposited by Bovio Supply Company in Problem 5 (10 points).

STUDENT'S SCORE _____

SECTION 6.2 Bank Statements and Reconciliation Statements

A **monthly bank statement** is sent to each depositor who has a checking account. This monthly statement indicates the depositor's account balance as of a given date. As soon as possible after receiving the bank statement, the depositor should prepare a **reconciliation statement.** This is a statement in which the depositor's current checkbook balance and the balance on the bank statement are compared and adjusted to find the true cash balance in the account as of the current date.

Banks normally return all canceled checks with the monthly bank statement. Some banks also return the original copies of deposit tickets for deposits that were made by mail. Other banks return copies of deposit slips immediately upon receiving the deposits.

PART A Understanding the Monthly Bank Statement

Every bank has its own format for the monthly statement, but the information given on most forms is the same. Shown at either the top or the bottom of the statement is the following *summary information:*

1. The starting balance (or *Balance—Last Statement*).
2. The total deductions (or *Total Amount—Debits*).
3. The total deposits (or *Total Amount—Credits*).
4. The ending balance (or *Balance—This Statement*).

The main body of the statement shows the following items:

1. Checks that have been returned to the bank and subtracted from the depositor's account.
2. Bank charges and other miscellaneous fees that are subtracted from the account.
3. Deposits received by the bank and added to the account.
4. The account balance at the end of each day in which there was some activity.

In the following discussion of each of these items, reference will be made to the monthly statement shown in Figure 6-5.

Listing Order of Checks. The order in which the checks are listed on the monthly statement varies among banks. Some banks list checks in the order in which they are received. However, for the depositor's convenience in reconciling the bank statement, many banks now list the checks in numerical order and show the check number, the date the check was received by the bank, and the amount of the check. With this latter method, missing checks (either destroyed by the depositor or not yet received by the bank) are indicated with a series of asterisks or dashes. Regardless of the method of listing, the ending balance does not include the missing checks.

Bank Charges and Miscellaneous Fees. Banks usually assess a monthly service charge and/or a fee for each check written. In place of these charges, some banks require the daily balance, or perhaps the average daily balance, to exceed some minimum requirement. In Figure 6-5, there is no service charge or check-writing fee because of the large daily balance in the account during the entire month.

Banks usually make additional charges when a bad check is written. Banks make these extra charges because they must do extra work. The writer of a bad check is charged by the bank for **overdrawing** the account. The payee is also charged for depositing a check that comes back marked **NSF (not sufficient funds).** Both the writer of the check and the payee are notified immediately. In Figure 6-5, Bovio Supply Company has deposited two bad checks (for $75 and $89.50) and is charged $10.00 for each check. When notified, John Bovio should immediately reduce his checkbook balance by $85 ($75 + $10) and then by $99.50 ($89.50 + $10).

Deposits Added to the Account. In Figure 6-5, there are five deposits, totaling $2,752.20. A 5 is recorded in the space labeled *Number of Credits,* and $2,752.20 is recorded in the space labeled *Total Amount—Credits.*

Also at the top of the statement, in the space labeled *Number of Debits,* is the number 29—the total of the 25 listed checks, the 2 bad checks returned, and the 2 fees charged (marked X). The total of these debits is $2,418.49, which is recorded in the space labeled *Total Amount—Debits.*

Western National Bank
San Francisco, CA 94131-2130

Checking Account Statement

Date-Last Statement	Number of Debits	Number of Credits		Date-This Statement
8/13/--	29	5		9/13/--

2,435.16	−	2,418.49	+	2,752.20	−	0.00	=	2,768.87
Balance-Last Statement		Total Amount-Debits		Total Amount-Credits		Service Charge		Balance-This Statement

Bovio Supply Company
934 Rose Street
Menlo Park, CA 94025-4420

Account Number
26-80 8319
S-Service charge for previous period unless otherwise noted
X-Charge for overdraft or returned check

DEPOSITS AND OTHER CREDITS		CHECKS AND OTHER DEBITS						DAILY BALANCE	
DATE	AMOUNT	NO.	DATE	AMOUNT	NO.	DATE	AMOUNT	DATE	AMOUNT
8/16	327.44	1852	8/23	132.01	1870	8/27	130.65	8/16	2,689.00
8/26	485.01	1853	8/16	22.76	***			8/17	2,637.57
9/02	659.33	1854	8/17	13.56	1872	9/02	100.00	8/18	2,590.49
9/06	1,080.42	1855	8/18	28.55	1873	8/25	229.78	8/23	2,434.99
9/12	200.00	1856	8/17	37.87	1874	8/27	86.75	8/24	2,398.69
		1857	8/16	50.84	1875	9/02	117.44	8/25	2,083.91
		***			***			8/26	2,422.24
		1859	8/26	47.91	1877	9/13	6.50	8/27	2,033.30
		1860	8/23	8.00	***			8/30	1,763.50
		1861	8/24	23.42	1879	9/06	39.89	9/02	2,205.39
		1862	8/18	18.53	1880	9/13	499.85	9/06	3,174.72
		1863	8/26	98.77	1881	9/06	71.20	9/12	3,275.22
		1864	8/24	12.88	OTHER DEDUCTIONS DEPOSITED CHECKS RETURNED			9/13	2,768.87
		1865	8/23	15.49		8/25	75.00		
		***					10.00X		
		1867	8/30	78.80		9/12	89.50		
		1868	8/27	171.54			10.00X		
		1869	8/30	191.00					

Please examine at once and report any differences to our Auditing Department.

FIGURE 6-5 A Monthly Bank Statement

The amount of $2,768.87 in the space labeled *Balance—This Statement* is calculated by adding the total amount of credits to the balance of the previous statement and then subtracting the total amount of debits from that sum. (For the interest of accounting students, the debits and credits are from the bank's standpoint. The depositor's account is similar to an account payable, as far as the bank is concerned.)

Daily Balance. The daily balance for any date on which some transaction was made is found by adding any deposit made to the previous balance and deducting the total of all checks received by the bank and any bank charges from the previous balance. The last amount in the *Daily Balance* column is the same amount as *Balance—This Statement*.

Note in Figure 6-5 that the notice for the second bad check returned would have probably been sent out one day prior to the statement date. This means that Bovio Supply Company may not yet have reduced the checkbook balance by the total amount of $99.50.

PART B Preparing the Reconciliation Statement

Most banks print a blank reconciliation form on the back of the monthly statement. Generally, however, the depositor first prepares a rough reconciliation statement.

Because banks use computers, they rarely make mistakes on their statements. However, the final balance shown on the monthly statement does not usually agree with the depositor's checkbook balance. This may be due to several reasons, such as: **outstanding checks** (checks that have not yet cleared); deposits added in the checkbook but not yet recorded by the bank; bank service charges or other fees, or even bank interest paid, shown on the monthly statement but not yet recorded by the depositor; checks written by the depositor but not recorded in the checkbook or check stubs; an incorrect deduction by the bank for a check not written by the depositor; or errors in computing the checkbook balance or the bank balance.

Preparing a Rough Reconciliation. There are five items to compare and/or check in the rough reconciliation:

1. Amounts on canceled checks should be compared with amounts listed on the monthly statement.
2. Amounts on canceled checks should be compared with amounts recorded on the check stubs.
3. Amounts of any outstanding checks should be checked.
4. Amounts on returned deposit tickets or the monthly statement should be compared with amounts recorded on the check stubs.
5. Any additions or deductions listed on the monthly statement for which notice has not been received, or which have not yet been entered in the checkbook, should be recorded.

The detailed procedure for checking each of the above items is best illustrated in an example. Assume that you are John Bovio of Bovio Supply Company, whose monthly statement is shown in Figure 6-5 on page 238. Also assume that your checkbook balance on September 13 is $1,905.33. *Typical errors will be introduced intentionally to show how you can correct them.* Before doing any preliminary work, take a separate sheet and label it CORRECTIONS. Under this label, put three subheadings: *Checkbook Balance, Bank Balance,* and *Outstanding Checks.* Then follow these steps:

1. If the amount on each canceled check matches each corresponding amount on the monthly statement, put a small check mark next to the amount on the statement. If these amounts do not match, circle any errors on the statement, and note the date, check number, and amount. In this example, assume that there are no such errors.
2. If the amount on each canceled check matches each corresponding amount on the check stubs, place a check mark after the amount on the stub. If these amounts do not match, circle any errors on the stubs and note the corrections to be made, either additions to or deductions from the checkbook balance. If a canceled check has not been recorded on the check stub, note the check number and the amount, which must be deducted from the checkbook balance.

   In this example, you find two errors. Check stub #1863 shows a deduction of $65.00, but check #1863 in the monthly statement is for $98.77. Furthermore, the check #1863 returned with the statement is not even your check. Thus, your check #1863 is outstanding, and you have been incorrectly charged for $98.77. On the *CORRECTIONS* sheet under *Bank Balance,* write "Incorrect charge: + $98.77" and include $65.00 as an outstanding check. On check stub #1875, you find that you wrote $171.44 instead of $117.44 (the correct amount on the check). This is a common transposition error. On the *CORRECTIONS* sheet under *Checkbook Balance,* write " + $54.00" (the difference between the two numbers).

   If the amount subtracted on the stub is larger than the amount on the check, the difference is an addition under *Checkbook Balance.* If the amount subtracted on the stub is smaller than the amount on the check, the difference is a subtraction under *Checkbook Balance.*
3. On the *CORRECTIONS* sheet under *Outstanding Checks,* list all the check numbers and amounts which have not been checked off on the stubs. Their total will be subtracted from the bank balance on the reconciliation statement.

Outstanding Checks		Reconciliation Statement
No.	Amount	Bovio Supply Company
1858	$286.60	September 13, 19--
1863	65.00	Checkbook balance $1,905.33
1866	514.00	Add:
1871	123.00	Error on #1875 ....+ 54.00
1876	32.12	$1,959.33
1878	37.65	Subtract:
1882	238.39	Returned deposit check and fee...$99.50
1883	52.07	Error on #1880.................. 10.00
1884	43.98	Total ....- 109.50
		Adjusted checkbook balance $1,849.83
		Bank balance $2,768.87
		Add:
		Unrecorded deposit, Sept. 13....$375.00
		Incorrect charge, Aug. 26........ 98.77
		Total ....+ 473.77
		$3,242.64
		Subtract:
		Outstanding checks ....-1,392.81
Total	$ 1,392.81	Adjusted bank balance $1,849.83

FIGURE 6-6 *A Reconciliation Statement*

In this example, you have nine outstanding checks, whose amounts total $1,392.81, as shown in Figure 6-6. If you use an adding machine or printing calculator to calculate any of these totals, save the tape and attach it to the reconciliation statement.

4. Compare each of the five deposits listed on the monthly statement with each of the first five deposits recorded on the check stubs since the previous statement. If they match, check each deposit off on the monthly statement. On the CORRECTIONS sheet under *Bank Balance,* list all deposits which the bank has not recorded. Their total will be added to the bank balance on the reconciliation statement.

In this example, a $375 deposit made on Sept. 13 is not listed on the monthly statement. On the CORRECTIONS sheet under *Bank Balance,* write "Unrecorded deposit: + $375."

5. Note on the CORRECTIONS sheet any additions or deductions listed on the monthly statement of which you have not yet been notified or which have not yet been entered in the checkbook.

In this example, you have just received the notice about the bad check returned and its fee on Sept. 12. On the CORRECTIONS sheet under *Checkbook Balance,* write "Returned check and fee: −$99.50."

From this data, you can now make the following rough reconciliation:

Checkbook balance on 9/13.............	$ 1,905.33
Error on check #1875	+ 54.00
Returned check and fee.....................	− 99.50
Adjusted checkbook balance...............	$ 1,859.83
Bank balance on 9/13	$ 2,768.87
Unrecorded deposit of 9/13	+ 375.00
Incorrect charge	+ 98.77
Outstanding checks	−1,392.81
Adjusted bank balance	$ 1,849.83

Chapter 6 Keeping a Checking Account

The two adjusted balances should be identical, but your rough reconciliation shows that the checkbook balance is $10 more than the bank balance. First, check that all figures were copied correctly. Then, check the arithmetic on the rough reconciliation. You find no errors. Finally, double-check the addition and subtraction on each check stub.

In this example, assume that you find the error on check stub #1880. When you subtracted the check for $499.85, your new balance was $10 more than it should have been. The $10 becomes a subtraction under *Checkbook Balance*. You made a very common error. You borrowed a one in the tens column but did not complete the subtraction properly. If there had been no check stub error, you would check the monthly statement. Notify the bank immediately if you find any errors on the statement.

Form of the Reconciliation Statement. When the two adjusted balances are the same, you can prepare the reconciliation statement. There are various forms, but the one shown in Figure 6-6 is widely used. When there are two or more items to be added or subtracted, they are listed separately (except for outstanding checks). Their total, however, is written in the column on the far right, not directly underneath the items. If there is only one item, the amount is placed directly in the column on the far right.

Correcting the Checkbook. The checkbook balance on September 13 is $1,905.33. After the reconciliation, you know that the correct balance is actually $1,849.83. Thus, you must subtract the difference, $55.50, on the next check stub. The check stub may have an extra blank line which may be used to enter this type of adjustment, as shown in Figure 6-7.

Balance Forward	Dollars	Cents
Balance Forward	1,905	33
Deposit		
Sept. Adj.	-55	50
TOTAL	1,849	83
This Check		
Balance		

No. 1885

FIGURE 6-7 Adjustment Recorded on a Check Stub

You should also keep the reconciliation statement, so you can easily find a complete record of the adjustment. The number on the check stub is 1885, because the last check written was #1884. There is no date entered on the check stub. The date that will be entered is the date that check #1885 is written.

COMPLETE ASSIGNMENT 22

Name _____

Date _____

ASSIGNMENT 22
SECTION 6.2

Bank Statements and Reconciliation Statements

	Perfect Score	Student's Score
PROBLEM 1	25	
PROBLEM 2	25	
PROBLEM 3	50	
TOTAL	100	

1. Using the following information, prepare a bank reconciliation statement for Bovio Supply Company (25 points).

   Checkbook balance, January 14 ....$2,928.31
   Bank balance on statement, January 14 .... 3,881.37
   Deposit, January 14 .... 238.00
   Outstanding checks: #1524 .... 88.25
                        #1541 .... 147.90
                        #1556 .... 72.54
                        #1557 .... 387.25
                        #1558 .... 401.72

The bank statement reported a deposit of $364.51 on January 6, and the same amount was reported on the returned deposit ticket. However, on the check stub, this deposit was recorded as $346.51. Another deposit of $75.40 on December 17 was never entered on any check stub.

STUDENT'S SCORE _____

2. Using the following information, prepare a bank reconciliation statement for Bovio Supply Company (25 points).

Checkbook balance, May 15	$3,107.55
Bank balance on statement, May 15	3,015.28
Deposit, May 15	684.50
Outstanding checks: #1751	472.14
#1770	49.09
#1771	234.41
#1774	122.75
Deposited check returned	195.00
Bank fee for the returned check	10.00

Check #1775 was written for $21.12 but was recorded on the check stub as $21.21. A check (#945) for $81.25 was returned with the statement, but it was not written on the Bovio Supply Company account. It had been signed by John Bovio, but it was actually a check from his personal checking account. This was a bank error.

STUDENT'S SCORE _____

3. The November 14 bank statement received by Bovio Supply Company shows a beginning balance of $3,114.62 and an ending balance of $3,434.85. The deposits, checks, service charge, and ending daily balances from the bank statement are shown on the next page. The company checkbook also shows a starting balance of $3,114.62, but it has an ending balance of $3,252.93. The math from the check stubs is also shown on the next page.

The amounts written on each check and deposit ticket agree with the bank statement. However, two check stubs show incorrect amounts, four checks are still outstanding, and one deposit recorded in the checkbook is not on the bank statement. A rough reconciliation shows that the bank statement and the checkbook do not agree. Further examination of the check stubs reveals three errors in addition and subtraction.

Chapter 6 Keeping a Checking Account **245**

Using the information below, locate all missing checks and deposits, all check stub errors, and the service charge. Adjust the November 14 checkbook balance to reflect the total of all errors and the service charge. Then, prepare the reconciliation statement as of November 14. Use the blank form provided on the next page. The check dates on the bank statement are the dates the checks were received by the bank; the dates on the check stubs are the dates the checks were written (50 points).

DEPOSITS AND OTHER CREDITS		CHECKS AND OTHER DEBITS						DAILY BALANCE	
DATE	AMOUNT	NO.	DATE	AMOUNT	NO.	DATE	AMOUNT	DATE	AMOUNT
10/17	251.34	1913	10/17	412.35	1925	11/10	244.54	10/15	3,114.62
10/20	1,218.74	1914	10/20	125.85	***			10/17	2,953.61
10/29	1,490.00	1915	10/29	465.20	1927	11/13	81.21	10/20	4,046.50
11/05	834.57	1916	10/26	71.00	1928	11/10	197.46	10/24	3,951.69
11/10	676.85	1917	10/24	94.81	1929	11/14	8.27	10/26	3,258.97
		1918	11/04	157.76	1930	11/14	204.40	10/29	3,476.01
		1919	10/29	299.40	***			11/04	3,318.25
		1920	10/29	508.36	***			11/05	4,028.19
		1921	10/26	621.72	1933	11/13	142.89	11/10	4,263.04
		1922	11/05	15.38	1934	11/14	387.67	11/13	4,038.94
		***			OTHER DEDUCTIONS			11/14	3,434.85
		1924	11/05	109.25		11/14	3.75S		

Checkbook Stubs

```
Bal 10/14   3,114|62     1918 10/20  -157|76     1924 10/29  -109|25     1930 11/8   -204|40
1913 10/15  -412|35                   3,266|73                 3,180|45                 3,078|29
            2,702|27     Dep 10/29   1,490|00    1925 11/4    -244|45   Dep 11/10      676|85
Dep 10/17     251|34                  4,756|73                 2,936|00                3,655|14
            2,953|61     1919 10/22   -299|40    1926 11/4    -201|94   1931 11/10    -184|39
1914 10/17   -125|85                  4,457|33                 2,735|06                3,470|75
            2,827|76     1920 10/23   -508|36    1927 11/4     -81|21   1932 11/10     -71|51
1915 10/18   -456|20                  3,958|97                 2,653|85                3,399|24
            2,371|56     1921 10/23   -621|72    Dep 11/5       834|57  1933 11/11    -142|89
1916 10/19    -71|00                  3,337|25                 3,488|42                3,256|35
            2,300|56     1922 10/26    -15|38    1928 11/5    -197|46   1934 11/11    -387|67
Dep 10/20  1,218|74                   3,321|87                 3,290|96                2,868|68
            3,519|30     1923 10/27    -32|17    1929 11/6     -8|27    Dep 11/14      384|25
1917 10/20    -94|81                  3,289|70                 3,282|69                3,252|93
            3,424|49
```

Reconciliation Statement

Bovio Supply Company
November 14, 19--

Checkbook balance..........................$ _____
   Add:
      Errors on
         # _____ $ _____
         # _____   _____
         # _____   _____
         # _____   _____
                   Total................._____
                               $ _____

   Subtract:
      Service charge $ _____
      Errors on
         # _____ $ _____
         # _____   _____
         # _____   _____
         # _____   _____
                   Total................._____

Adjusted checkbook balance................$ _____

Bank balance..............................$ _____
   Add:
      Unrecorded deposit, 11/14............ _____
                                        $ _____

   Subtract:
      Outstanding checks................... _____

Adjusted bank balance.....................$ _____

Outstanding Checks

No.	Amount
___	_____
___	_____
___	_____
___	_____
___	_____
___	_____
___	_____
___	_____
___	_____

Total $ _____

STUDENT'S SCORE _____

CHAPTER 7

Calculating Interest

Interest is a charge for the use of money, just as rent is a charge for the use of property. Banks, savings and loan associations, finance companies, and credit unions are in business primarily to lend money. Many department stores have revolving charge accounts, where interest is charged on balances not paid within 30 days. Most business firms also charge interest on overdue accounts.

In recent years, the cost of borrowing money for individuals and businesses has changed dramatically. Whether you are borrowing or lending money, you should know how interest charges are calculated.

You must consider three elements when calculating interest. These elements are the **principal** (the amount of money borrowed), the **rate** (the percent charged for the use of the principal for one year), and the **time** (the period of days, months, years, or combinations of these for which interest must be paid). The time is also called the **interest period**. The interest rate is almost always given as an annual (yearly) rate, although some businesses also quote monthly rates for some charge accounts.

Section 7.1 explains how to find the due date and the exact time for which interest must be paid. Section 7.2 deals with calculating **simple interest,** which is interest charged on only the original principal. Simple interest may be charged on **short-term loans** (loans made for less than one year). Section 7.3 deals with calculating **compound interest,** which is often charged on **long-term loans** (loans made for one or more years) and may also be charged on short-term loans. When compound interest is charged, the interest for one period is added to the principal for that period, and that total becomes the basis for computing interest for the next period.

SECTION 7.1 The Interest Period

Money may be lent or borrowed for any amount of time that is agreeable to both the borrower and the lender. Whether the time is in years, months, or days, the method of calculating simple interest is the same. *Multiply the principal by the interest rate and then by the time of the interest period.* However, the *interest rate* and the *interest period* must be in agreement. If the interest calculation is based on years, then the interest rate must be an annual rate and the interest period must be expressed in years or fractional parts of years. You need to understand the relationship between these two elements and to know how to determine the length of the interest period.

A borrower obtaining a loan will probably sign a **promissory note.** This is a written promise to repay the amount borrowed, plus interest, within a specified time (the interest period, or loan period). If the interest is calculated on a purchase over time (as may be done with a credit card), the interest rate will almost always be a monthly rate.

The date when the borrower signs the promissory note and receives the money is called the **date of origin.** The date on which the loan must be repaid is called the **due date.** The **loan period** may be defined as the time between these two dates, or it may be defined as a specific amount of time (in days, months, or years) after the date of origin.

Usually, interest rates are stated as annual rates. If the interest period is stated in years, no adjustment is required. If the interest period is stated in months or days, however, it must be converted to an equivalent period in years. To do these conversions, you need to know the number of months and days in a year.

Interest calculations are based on one of two different assumptions about the number of days in one year. For some loans, a year is assumed to have only 360 days. The interest on such loans is called **ordinary simple interest.** For other loans, the year has its actual 365 days (366 in leap years), and the interest on these loans is called **exact interest.**

To calculate exact interest for short-term loans, you must usually express the interest period in the actual number of days, or the **exact time,** of the loan. There are two situations involving ordinary simple interest where you also need to know the exact time of the loan. These are when the loan period is defined with the date of origin and the due date, and when the loan period is stated in months, but the due date falls on a nonbusiness day (Saturday, Sunday, or a holiday).

To count the actual number of days in the interest period, you have to know the number of days in each month and take leap years into consideration. With the exception of February, each month has the same number of days every year. February has 28 days in non-leap years and 29 days in leap years. April, June, September, and November each have 30 days. The other seven months each have 31 days.

Every fourth year is a leap year, with one exception that occurs only once every 100 years. To determine a leap year:

1. Look at the number formed by the *last two digits* of the year. If this number is a multiple of 4, then that year is a leap year. Thus, the years 1768, 1876, 1988, 1992, and 1996 are leap years, because the numbers 68, 76, 88, 92, and 96 are all multiples of 4. Likewise, 1767, 1878, and 1990 are not leap years, because the numbers 67, 78, and 90 are not multiples of 4.

2. When the last two digits of the year are 00, look at the number formed by the *first two digits* of the year. If this number is a multiple of 4, then that year is a leap year. Thus, 1600 and 2000 are leap years, because 16 and 20 are multiples of 4. Likewise, 1700, 1800, and 1900 are not leap years, because 17, 18, and 19 are not multiples of 4.

PART A Finding the Exact Time Between Two Dates

When the loan period for a short-term loan is expressed as the time between two dates, you must count the exact number of days between the two dates to calculate the interest. The starting date is the date of origin, and the ending date is the due date. If both the date of origin and the due date are in the same month, the number of days in the loan period is merely the difference between the two dates.

EXAMPLE:

Find the number of days between April 12 and April 27.

$27 - 12 = 15$ days

If the loan period begins in one month and ends in another month, follow these steps to find the exact time of the loan period:

1. *Beginning month:* Subtract the date of origin from the number of days in the beginning month. This difference is the number of loan period days in the first month.
2. *Intermediate months:* Find the number of days in each of the months between the beginning and ending months of the loan period.
3. *Ending month:* Find the number of loan period days in the ending month. This is the number of days indicated by the due date of the loan.
4. *Total number of days:* Add up the numbers calculated in the three preceding steps.

EXAMPLES:

a. Find the exact time between January 15 and July 21 in a leap year.

Step	Month	Days
1	Jan. (31 − 15)	16
2	Feb.	29
	March	31
	April	30
	May	31
	June	30
3	July	21
4	*Total,* or *Exact Time*	188

Chapter 7 Calculating Interest

b. Find the exact time between June 30 and October 14.

Step	Month	Days
1	June (30 − 30)	0
2	July	31
	Aug.	31
	Sept.	30
3	Oct.	14
4	*Total*, or *Exact Time*	106

PART B Finding the Due Date and Exact Time When the Interest Period is Stated in Days

When the loan period is expressed as a certain number of days from the date of origin, you must find the due date.

If the number of days in the loan period is less than the number of days remaining in the first month of the loan period, the due date can be found in one step. *Add the number of days in the loan period to the date of origin. This sum is the due date.*

EXAMPLE:

Find the due date of a 15-day note, dated November 5.

30 − 5 = 25 days remaining in Nov.

15 (days in the loan period)
 5 (date of origin)
—
20 (due date in Nov.)

If the number of days in the loan period is greater than the number of days remaining in the first month of the loan period, follow these steps to find the due date:

1. *Beginning month:* Subtract the date of origin from the number of days in the beginning month. This difference is the number of loan period days in the first month. Subtract this difference from the number of days in the loan period to get the number of loan period days still remaining.
2. *Ending month:* Compare the number of loan period days remaining (Step 1) with the number of days in the next month.
   a. If the number of loan period days remaining is *less* than the number of days in the next month, then that month is the final, or ending, month of the loan, and it contains the due date.
   b. If the number of loan period days remaining is *greater* than the number of days in the next month, subtract the number of days in that month from the loan period days remaining. This difference is the new number of loan period days remaining. Compare this new number with the number of days in the next month.
   c. Continue this process until the number of days remaining in the loan period is *less than the number of days in the next month*. That next month is the ending month of the loan, and it contains the due date. (*Note:* If the number of days remaining in the loan period after subtraction is 0, then the due date is the final day of the last month subtracted.)
3. *Due date:* The final number of loan period days remaining (the last calculation in *Step 2*) is the actual date in the ending month that the loan is due.

EXAMPLES:

a. Find the due date of a 30-day note dated July 9.

Step	Month	Days	Loan Days Remaining
1	July (31 − 9)	22	30 − 22 = 8
2	Aug.	31	

Since 8 is less than 31 (days in August), the ending month is August.

| 3 | There are 8 loan period days remaining after July, so the due date is August 8. |

b. Find the due date of a 120-day note dated July 20.

Step	Month	Days	Loan Days Remaining
1	July (31 − 20)	11	120 − 11 = 109
2	Aug.	31	109 − 31 = 78
	Sept.	30	78 − 30 = 48
	Oct.	31	48 − 31 = 17
	Nov.	30	

Since 17 is less than 30 (days in November), the ending month is November.

| 3 | There are 17 loan period days remaining after October, so the due date is November 17. |

c. Find the due date of a 90-day note dated June 2.

Step	Month	Days	Loan Days Remaining
1	June (30 − 2)	28	90 − 28 = 62
2	July	31	62 − 31 = 31
	Aug.	31	31 − 31 = 0
3	Since there are no loan period days remaining after August, the ending month is August, and the due date is August 31.		

Due Dates and Nonbusiness Days. Nonbusiness days are Saturdays, Sundays, and legal holidays. Sometimes, a loan period expressed as a certain number of days after the date of origin will lead to a proposed due date that is a nonbusiness day.

EXAMPLE:

Find the due date of a 30-day note dated November 25.

30 − 25 = 5 loan period days in Nov.

Thus, there are 25 loan period days in December. The proposed due date is December 25, which is Christmas Day, a legal holiday.

In actual business practice, loans and notes do not come due on nonbusiness days, because the borrower cannot repay the money on such days. Therefore, the next business day becomes the due date. Since the borrower has the use of the principal for one or more additional days, interest will be charged for those days. Thus, the actual interest period will *not* be exactly 30 days, 60 days, 90 days, etc.

How Legal Holidays are Counted. When a legal holiday falls during the loan period, it is counted as any other day. When a legal holiday falls on a Sunday, the next day becomes the legal holiday.

EXAMPLES:

a. Find the due date and the exact time of the interest period for a 90-day note dated October 3.

Step	Month	Days	Loan Days Remaining
1	Oct. (31 − 3)	28	90 − 28 = 62
2	Nov.	30	62 − 30 = 32
	Dec.	31	32 − 31 = 1
	Jan.	31	
3	The ending month is January. The proposed due date is January 1, a legal holiday. The actual due date is January 2, and the exact time is 91 days.		

b. Find the due date and the exact time of the interest period for a 60-day note dated May 5.

Step	Month	Days	Loan Days Remaining
1	May (31 − 5)	26	60 − 26 = 34
2	June	30	34 − 30 = 4
	July	31	
3	The ending month is July. The proposed due date is July 4, a legal holiday. The actual due date is July 5, and the exact time is 61 days.		

When a legal holiday falls on a Friday, three days must be added to the exact number of days in the interest period, because Saturday and Sunday are also nonbusiness days. Thus, in *Example a,* suppose that January 1 falls on a Friday. The due date would be January 4, and the exact time would be 93 days. In *Example b,* the due date would be July 7, and the exact time would be 63 days.

PART C Finding the Due Date and Exact Time When the Interest Period is Stated in Months

The interest period of a loan may be expressed as a certain number of months from the date of origin. With some exceptions (which will be noted later), you can usually find the due date by following a general rule. *From the beginning month, count forward the given number of months to find the ending month; the due date will generally be the same date in the ending month as the date of origin in the beginning month.*

EXAMPLE:

Find the due date of a four-month note dated March 15.

Count forward four months from March. The due date is July 15.

This general rule does not apply when the proposed due date does not exist in the ending month or when the proposed due date falls on a nonbusiness day. The following examples illustrate these exceptions.

Due Date in February. If the date of origin is the 29th, 30th, or 31st of the beginning month and the ending month is February of a non-leap year, the due date is February 28. In a leap year, the due date is February 29.

EXAMPLE:

Find the due date of a two-month note dated December 31, 1997.

The ending month is February, 1998. Since 1998 is not a leap year, the due date is February 28, 1998.

Due Date in April, June, September, or November. If the date of origin is the 31st of the beginning month and the ending month has only 30 days, the due date is the 30th of the ending month.

EXAMPLE:

Find the due date of a three-month note dated August 31.

Count forward three months from August. The proposed due date is November 31. Since November has only 30 days, the due date is November 30.

Proposed Due Date on a Nonbusiness Day. If the proposed due date is a nonbusiness day, the next business day becomes the due date, as described earlier.

EXAMPLES:

a. Find the due date of a two-month note dated May 4.

The proposed due date is July 4, a legal holiday. Therefore, the actual due date of the note is July 5.

b. Find the due date of a six-month note dated June 25, 1998.

The proposed due date is December 25, 1998, which is both a legal holiday and a Friday. Therefore, the actual due date of the note is Monday, December 28, 1998.

Although the interest period of a short-term promissory note may be expressed in months, the exact time is used to calculate the interest.

EXAMPLES:

a. Find the exact time of a two-month note dated May 4.

Step	Month	Days
1	May (31 − 4)	27
2	June	30
3	July	5
	Total, or Exact Time	62

The actual due date is July 5. The exact time between May 4 and July 5 is 62 days.

b. Find the exact time of a six-month note dated June 25, 1998. (See previous example b.)

Step	Month	Days
1	June (30 − 25)	5
2	July	31
	Aug.	31
	Sept.	30
	Oct.	31
	Nov.	30
3	Dec.	28
	Total, or Exact Time	186

COMPLETE ASSIGNMENT 23

ASSIGNMENT 23
SECTION 7.1

The Interest Period

	Perfect Score	Student's Score
PART A	30	
PART B	40	
PART C	30	
TOTAL	100	

Name _____

Date _____

PART A Finding the Exact Time Between Two Dates

DIRECTIONS: Find the exact time from the first date to the second date. Write your answers on the lines provided (3 points for each correct answer).

	From	To	Days		From	To	Days
1.	July 30	Sept. 30	_____	7.	May 7	July 17	_____
2.	Feb. 16, 1998	Mar. 16, 1998	_____	8.	Jan. 10, 1996	May 10, 1996	_____
3.	Nov. 30	Feb. 28	_____				
				9.	Nov. 15, 1993	Mar. 15, 1994	_____
4.	June 13	Aug. 29	_____				
				10.	May 25	Oct. 25	_____
5.	Jan. 25, 1996	Apr. 10, 1996	_____				
6.	Dec. 9	Dec. 24	_____				

STUDENT'S SCORE _____

PART B Finding the Due Date and Exact Time When the Interest Period is Stated in Days

DIRECTIONS: Find the due date and the exact time in the loan period. Write your answers on the lines provided (3 points for each correct due date, and 2 points for each correct exact time).

	Date of Origin	**Loan Period**	**Due Date**	**Exact Time**
11.	January 8	15 days	_____	_____
12.	February 13, 1996	30 days	_____	_____
13.	June 3	45 days	_____	_____
14.	October 26	60 days	_____	_____
15.	April 5	90 days	_____	_____
16.	January 10, 1994	120 days	_____	_____
17.	August 4	150 days	_____	_____
18.	September 30, 1999	180 days	_____	_____

STUDENT'S SCORE _____

PART C Finding the Due Date and Exact Time When the Interest Period is Stated in Months

DIRECTIONS: Find the due date and the exact time in the loan period. Write your answers on the lines provided (2 points for each correct due date, and 3 points for each correct exact time).

	Date of Origin	**Loan Period**	**Due Date**	**Exact Time**
19.	March 15	1 month	_____	_____
20.	May 4	2 months	_____	_____
21.	January 20, 1994	3 months	_____	_____
22.	August 25	4 months	_____	_____
23.	November 10, 1995	5 months	_____	_____
24.	July 1	6 months	_____	_____

STUDENT'S SCORE _____

Chapter 7   Calculating Interest

SECTION 7.2 Calculating Simple Interest

The two types of interest are *simple interest* and *compound interest*.

Simple interest is important, because it is charged on many loans and is also the foundation for compound interest. As you will see in Section 7.3, compound interest is nothing more than *repeated simple interest*.

The fundamental interest formula is:

$$I = P \times R \times T$$

where I = interest, P = principal, R = rate, and T = time. In other words, *interest equals principal times rate times time,* or simply *I equals PRT.*

As you learned in Section 7.1, when R is an annual interest rate, T must be stated in years or fractional parts of a year. To make a fractional part of a year, divide the number of days in the loan period by the number of days in one year. Recall that the divisor will be 360 days if you are calculating *ordinary interest* and 365 days (366 in a leap year) if you are calculating *exact interest*.

In this text, you are expected to find ordinary interest, unless exact interest is specified. For short-term loans, it will often be necessary to determine the *exact time* of the loan, whether you are calculating ordinary or exact interest.

PART A   Ordinary Interest

In all interest calculations, interest rates expressed as percents must be changed into their decimal equivalents. Then, multiply the principal by the rate and the time (in years). You may use up to six decimal places in the calculations. Round off only the final result.

EXAMPLE:

Find the interest on $900 at 10% for $2\frac{1}{2}$ years.

$I = \$900 \times 0.10 \times 2.5$
$ = (\$900 \times 0.10) \times 2.5$
$ = \$90 \times 2.5$
$ = \$225$

If the time is not already expressed in years, you must change it. If the time is in months, change it to years by dividing the number of months by 12.

EXAMPLE:

Find the interest on $750 at 8% for 21 months.

$I = \$750 \times 0.08 \times \frac{21}{12}$
$ = (\$750 \times 0.08) \times 1.75$
$ = \$60 \times 1.75$
$ = \$105$

If the time is expressed in days, change it to years by dividing the number of days by 360.

EXAMPLE:

Find the interest on $1,200 at 12% for 90 days.

$I = \$1,200 \times 0.12 \times \frac{90}{360}$
$ = (\$1,200 \times 0.12) \times 0.25$
$ = \$144 \times 0.25$
$ = \$36$

Order of Performing the Arithmetic.   Because all the arithmetic operations in a simple interest calculation are either multiplication or division, you may work them in whatever order is most convenient. The preceding examples are repeated in the following examples, but the rate and time are multiplied together first, instead of the principal and rate.

EXAMPLES:

a. Find the interest on $900 at 10% for $2\frac{1}{2}$ years.

$I = \$900 \times 0.10 \times 2.5$
$ = \$900 \times (0.10 \times 2.5)$
$ = \$900 \times 0.25$
$ = \$225$

b. Find the interest on $750 at 8% for 21 months.

$I = \$750 \times 0.08 \times \frac{21}{12}$
$ = \$750 \times (0.08 \times 1.75)$
$ = \$750 \times 0.14$
$ = \$105$

c. Find the interest on $1,200 at 12% for 90 days.

$I = \$1,200 \times 0.12 \times \frac{90}{360}$
$ = \$1,200 \times (0.12 \times 0.25)$
$ = \$1,200 \times 0.03$
$ = \$36$

In *Examples b* and *c*, the fractions $\frac{21}{12}$ and $\frac{90}{360}$ were both changed to decimals to avoid dividing by 12 and 360. However, this may not be important in every problem, especially if you are using a calculator. Simply do all of the multiplication first, and then divide by 12 or 360. In the same examples, the work would be done like this:

b. $I = \$750 \times 0.08 \times \frac{21}{12}$
$= [(\$750 \times 0.08) \times 21] \div 12$
$= (\$60 \times 21) \div 12$
$= \$1,260 \div 12$
$= \$105$

c. $I = \$1,200 \times 0.12 \times \frac{90}{360}$
$= [(\$1,200 \times 0.12) \times 90] \div 360$
$= (\$144 \times 90) \div 360$
$= \$12,960 \div 360$
$= \$36$

Reducing Fractions and Cancellation. When you work with fractions (as in the preceding examples), it is almost always easier to reduce the fractions to lowest terms and to perform all possible cancellations before multiplying. For example, $\frac{21}{12}$ reduces to $\frac{7}{4}$, and $\frac{90}{360}$ reduces to $\frac{1}{4}$. The denominator (4) in the time cancels into either the rate (as in *Example b*) or the principal (as in *Example c*).

b. $I = \$750 \times 0.08 \times \frac{21}{12}$
$= \$750 \times 0.08 \times \frac{7}{4}$
$= \$750 \times \cancel{0.08}^{0.02} \times \frac{7}{\cancel{4}_{1}}$
$= (\$750 \times 0.02) \times 7$
$= \$15 \times 7$
$= \$105$

c. $I = \$1,200 \times 0.12 \times \frac{90}{360}$
$= \$1,200 \times 0.12 \times \frac{1}{4}$
$= \cancel{\$1,200}^{300} \times 0.12 \times \frac{1}{\cancel{4}_{1}}$
$= \$300 \times 0.12$
$= \$36$

Nonbusiness Due Date. If the interest to be calculated is on a short-term promissory note, check the exact time to make sure that the due date does not fall on a nonbusiness day. As mentioned in Section 7.1, the exact time of a note is the number of days from the date of origin to the due date.

EXAMPLE:

Find the interest on a $600, 9%, 60-day note dated October 26 and paid on the due date.

60 days after October 26 is December 25, a legal holiday. December 26 becomes the due date, and the exact time for which interest is charged is 61 days. Thus:

$I = \$600 \times 0.09 \times \frac{61}{360}$

$= \cancel{\$600}^{5} \times \cancel{0.09}^{0.03} \times \frac{61}{\cancel{360}_{\substack{3 \\ 1}}}$

$= \$9.15$

Finding the Maturity Value of a Note. The **maturity value** of a note is the amount to be paid to the lender on the due date (or maturity date). When interest is charged, the maturity value is the sum of the principal and the interest. In the case of non-interest-bearing notes, the maturity value is the same as the principal. These notes are discussed more fully in Chapter 8.

EXAMPLE:

Find the maturity value of a $1,500, 12%, 90-day note dated March 25.

The due date is June 23, which is not a legal holiday. The exact time is 90 days. Thus:

$I = \$1,500 \times 0.12 \times \frac{90}{360}$

$= \$1,500 \times 0.12 \times \frac{1}{4}$

$= \$1,500 \times \cancel{0.12}^{0.03} \times \frac{1}{\cancel{4}_{1}}$

$= \$45$

Maturity value $= \$1,500 + \$45 = \$1,545$

Chapter 7 Calculating Interest

More Cancellation Shortcuts. Perhaps the major reason that 360 days is used (instead of 365 or 366 days) is that it makes several shortcuts possible through cancellation. Most of these shortcuts eliminate the need for a calculator.

The basis for many of these shortcuts is illustrated by the calculations of *6% for 60 days*. 6% is expressed as 0.06, and 60 days is expressed as $\frac{60}{360}$, or $\frac{1}{6}$, year. 0.06 multiplied by $\frac{1}{6}$ equals 0.01. Thus, in any simple ordinary interest problem at 6% for 60 days, simply multiply the principal by 0.01 to find the interest.

EXAMPLE:

Find the ordinary interest on $575 at 6% for 60 days.

$$I = \$575 \times 0.06 \times \frac{60}{360}$$

$$= \$575 \times 0.06 \times \frac{1}{6}$$

$$= \$575 \times 0.06 \times \frac{1}{6}$$

(with 0.06 cancelled to 0.01 and 6 cancelled to 1)

$$= \$575 \times 0.01$$

$$= \$5.75$$

EXAMPLE:

Find the ordinary interest on $955 at 6% for 60 days.

$$I = \$955 \times 0.06 \times \frac{60}{360}$$

$$= \$955 \times 0.01$$

$$= \$9.55$$

There are many other combinations of rate and time that offer useful shortcuts. Three of these shortcuts are illustrated in the following examples:

12% for 30 days:

$$0.12 \times \frac{30}{360} = 0.12 \times \frac{1}{12}$$

$$= 0.12 \times \frac{1}{12}$$

(with 0.12 cancelled to 0.01 and 12 cancelled to 1)

$$= 0.01$$

12% for 60 days:

$$0.12 \times \frac{60}{360} = 0.12 \times \frac{1}{6}$$

$$= 0.12 \times \frac{1}{6}$$

(with 0.12 cancelled to 0.02 and 6 cancelled to 1)

$$= 0.02$$

6% for 120 days:

$$0.06 \times \frac{120}{360} = 0.06 \times \frac{1}{3}$$

$$= 0.06 \times \frac{1}{3}$$

(with 0.06 cancelled to 0.02 and 3 cancelled to 1)

$$= 0.02$$

EXAMPLES:

a. Find the interest on $840 at 12% for 30 days.

$$I = \$840 \times 0.12 \times \frac{30}{360}$$

$$= \$840 \times 0.01$$

$$= \$8.40$$

b. Find the interest on $725 at 12% for 60 days.

$$I = \$725 \times 0.12 \times \frac{60}{360}$$

$$= \$725 \times 0.02$$

$$= \$14.50$$

c. Find the interest on $650 at 6% for 120 days.

$$I = \$650 \times 0.06 \times \frac{120}{360}$$

$$= \$650 \times 0.02$$

$$= \$13.00$$

Because these shortcuts only apply to certain rates and times, they can only be used to find actual final answers to certain problems. Their greatest value is in approximating answers. For example, suppose that you want to calculate the interest on $523 at 6.25% from January 1 to March 1. The exact time is 59 days. Because the rate and time are close to 6% and 60 days, your final answer should be close to $5.23 (or $523 × 0.01). The actual interest is $5.36.

PART B Exact Interest

The formula for simple interest is used to find both ordinary and exact interest:

$$I = P \times R \times T$$

When the time (in years) is a whole number, ordinary interest and exact interest are the same. Exact interest differs from ordinary interest when the time is in days, months, or a fractional part of a year. When this happens, change T to a fraction with a denominator of 365 (366 in a leap year) to find the exact interest.

As with ordinary interest, you may perform the operations in the most convenient order. Also, you can reduce fractions and use cancellation whenever possible.

EXAMPLES:

a. Find the exact interest on $900 at 11% for 60 days.

$I = \$900 \times 0.11 \times \dfrac{60}{365}$
$ = [(\$900 \times 0.11) \times 60] \div 365$
$ = (\$99 \times 60) \div 365$
$ = \$5,940 \div 365$
$ = \$16.273972$ *or* $\$16.27$

b. Find the exact interest on $730 at 16% for 45 days.

$I = \$730 \times 0.16 \times \dfrac{45}{365}$
$ = \$730 \times 0.16 \times \dfrac{9}{73}$
$ = \$\cancel{730}^{10} \times 0.16 \times \dfrac{9}{\cancel{73}_{1}}$
$ = (\$10 \times 0.16) \times 9$
$ = \$1.60 \times 9$
$ = \$14.40$

c. Find the exact interest on $1,000 at 9% from February 15, 1996 to March 15, 1996.

1996 is a leap year, so the exact time is 29 days.

$I = \$1,000 \times 0.09 \times \dfrac{29}{366}$

$ = \$\cancel{1,000}^{500} \times \cancel{0.09}^{0.03} \times \dfrac{29}{\cancel{366}_{\cancel{122}_{61}}}$

$ = (\$500 \times 0.03) \times \dfrac{29}{61}$
$ = (\$15 \times 29) \div 61$
$ = \$435 \div 61$
$ = \$7.131148$ *or* $\$7.13$

When a calculator is not available, calculating exact interest requires more time and effort than ordinary interest, because neither 365 nor 366 reduces nicely.

$$365 = 5 \times 73$$
$$366 = 2 \times 3 \times 61$$

However, 360 does reduce nicely.

$$360 = 2 \times 2 \times 2 \times 3 \times 3 \times 5$$

Therefore, 360 has many more divisors than 365 or 366 and provides more opportunities for cancellation. Of course, the calculations for exact and ordinary interest are equally easy with a calculator.

If you had calculated the ordinary interest for the preceding examples, you would have used a denominator of 360. The results would have been as follows:

a. $I = \$900 \times 0.11 \times \dfrac{60}{360}$
$ = \$16.50$ (ordinary interest)

b. $I = \$730 \times 0.16 \times \dfrac{45}{360}$
$ = \$14.60$ (ordinary interest)

c. $I = \$1,000 \times 0.09 \times \dfrac{29}{360}$
$ = \$7.25$ (ordinary interest)

In each example, the exact interest is slightly smaller than the corresponding ordinary interest.

Chapter 7 Calculating Interest

For the same principal, rate, and time, the exact interest will always be slightly smaller than the ordinary interest. Based upon this, a lender should prefer to use ordinary interest, and a borrower should prefer to use exact interest.

PART C Monthly Interest Rates

If both borrower and lender agree, the simple interest calculation can be done in terms of months. The same formula ($I = P \times R \times T$) is used, but both R and T are stated in months. R is a monthly rate, and T is some number of months.

EXAMPLE:

Find the simple interest on $200 at 1.5% per month for 3 months.

$I = \$200 \times 0.015 \times 3$
$= \$9$

To change a monthly rate into an annual rate, multiply the monthly rate by 12. In this example, the annual rate is 18% ($1.5\% \times 12$). To change 3 months into years, divide 3 by 12: $\frac{3}{12} = \frac{1}{4} = 0.25$ years. In terms of years, this example is calculated as follows:

$I = \$200 \times 0.18 \times \frac{3}{12}$
$= \$9$

Notice that the answers are identical, which will always be the case with simple interest. If the rate is monthly and the time is in years, the interest can be calculated in terms of either years or months. To change years into months, *multiply* the number of years by 12. The following example is done both in years and in months.

EXAMPLE:

Find the simple interest on $500 at 1.25% per month for 2 years.

Years: 1.25% (monthly) $\times$ 12 = 15% (annually)

$I = \$500 \times 0.15 \times 2$
$= \$150$

Months: 2 (years) $\times$ 12 = 24 (months)

$I = \$500 \times 0.0125 \times 24$
$= \$150$

If the rate is an annual rate and the time is in months, the interest can be calculated in terms of either years or months. To change an annual rate into a monthly rate, *divide* the annual rate by 12. To change months into years, divide the number of months by 12. The following example is done both in years and in months.

EXAMPLE:

Find the simple interest on $250 at 12% per year for 9 months.

Years: 9 (months) $\div$ 12 = 0.75 (year)

$I = \$250 \times 0.12 \times 0.75$
$= \$22.50$

Months: 12% (annually) $\div$ 12 = 1% (monthly)

$I = \$250 \times 0.01 \times 9$
$= \$22.50$

There are two things you should know about this material. First, when the interest rate is not specified as annual, monthly, etc., *assume that it is annual*. Second, this section describes how to calculate *simple interest* over a time period measured in months. This is *not* the same as calculating interest on an unpaid balance every month, as is done with a credit card or a charge account. That topic will be discussed in Section 8.1 of the next chapter.

COMPLETE ASSIGNMENT 24

Name _____

Date _____

ASSIGNMENT 24
SECTION 7.2

Calculating Simple Interest

	Perfect Score	Student's Score
PART A	44	
PART B	40	
PART C	16	
TOTAL	100	

PART A Ordinary Interest

DIRECTIONS: *Solve the following problems. Show your work, and write your answers on the lines provided (3 points for each correct answer).*

1. Find the ordinary interest on $800 at 7% for 2 years.

   _____

2. Find the ordinary interest on $700 at 8% for 4 years.

   _____

3. Find the ordinary interest on $1,200 at 10% for 18 months.

   _____

4. Find the ordinary interest on $1,500 at 12% for 15 months.

   _____

5. Find the ordinary interest on $750 at 9% for 60 days.

   _____

6. Find the ordinary interest on $2,500 at 12% for 75 days.

   _____

7. Find the maturity value of a $3,000, 13%, 120-day note dated May 10 and paid on the due date.

8. Find the total payment made on a note for $2,000 at 12%, dated July 10 and paid on September 20.

DIRECTIONS: Multiply the interest rate (expressed as a decimal) by the time (expressed in years, with a denominator of 360 days). Then, multiply that product by the principal to find the ordinary interest (1 point for each correct answer).

	Principal	Rate	Time	Rate × Time	Interest
9.	$1,750	12%	30 days		
10.	$1,200	12%	60 days		
11.	$2,000	12%	90 days		
12.	$750	6%	60 days		
13.	$900	6%	120 days		
14.	$1,100	18%	60 days		
15.	$1,250	8%	45 days		
16.	$800	8%	90 days		
17.	$2,275	10%	36 days		
18.	$1,000	15%	120 days		

STUDENT'S SCORE _____

PART B Exact Interest

DIRECTIONS: Find the interest, as specified. Show your work, and write your answers on the lines provided (5 points for each correct answer).

19. Find the exact interest on $1,460 at 10% for 120 days.

_____

20. Find the exact interest on $3,660 at 9% for 60 days in a leap year.

_____

21. Find the exact interest on a $730, 11%, 4-month note, dated May 10.

_____

22. Find the exact interest of a $1,000, 12%, 2-month note, dated May 4.

_____

23. Find the exact interest on $1,500 at 10% from August 19 to October 31.

_____

24. Find the exact interest on $2,500 at 12% from June 21 to November 14.

_____

25. By how much is the ordinary interest greater than the exact interest on $2,044 at 12% for 60 days?

_____

26. By how much is the ordinary interest greater than the exact interest on $3,000 at 15% for 73 days?

_____

STUDENT'S SCORE _____

Chapter 7   Calculating Interest

PART C   Monthly Interest Rates

DIRECTIONS: Use either monthly or annual computations to find the simple interest. Show your work, and write your answers on the lines provided (2 points for each correct answer).

Principal	Rate	Time	Interest
27. $1,000	1.2% per month	6 months	_____
28. $750	12% per year	1.5 years	_____
29. $600	1.3% per month	$\frac{1}{3}$ year	_____
30. $800	18% per year	8 months	_____
31. $1,200	1% per month	21 months	_____
32. $1,600	1.25% per month	$\frac{1}{2}$ year	_____
33. $2,000	9% per year	15 months	_____
34. $800	1.5% per month	$\frac{1}{4}$ year	_____

STUDENT'S SCORE _____

SECTION 7.3 Calculating Compound Interest

With simple interest (whether time is measured in years, months, or days), only one calculation is necessary: I = P × R × T. The simple interest on $400 at 12% for 3 years is $144 ($400 × 0.12 × 3). *The simple interest for the three-year loan is $144*.

When interest is compounded annually, the interest is calculated *every year*. The annual interest is added to the principal for subsequent calculations. To find interest compounded annually on a three-year loan, you must perform three separate interest calculations. The principal changes each year. When the interest for one year is calculated, it is added to the principal for that year. That sum, called the **compound amount** (usually labeled "A"), becomes the principal for the next year. The *total compound interest* for the three-year loan is the difference between the final compound amount and the original principal.

PART A Compounding Interest at Different Periods

Interest may be compounded at any convenient interval, such as annually, semiannually, quarterly, monthly, or daily.

Interest Compounded Annually. To compound interest annually, find the ordinary interest for the first year. Add this interest to the original principal to obtain the compound amount at the end of the first year. This amount becomes the new principal, on which interest for the second year is based. Repeat this procedure once for each year in the loan period. Do not round off until you have the final answer.

EXAMPLE:

Find the final compound amount (A) and the compound interest at the end of three years on $400 invested at 12%, compounded annually.

Year 1: I = $400 × 0.12 × 1
= $48

A = P + I
= $400 + $48
= $448

Year 2: I = $448 × 0.12 × 1
= $53.76

A = P + I
= $448.00 + $53.76
= $501.76

Year 3: I = $501.76 × 0.12 × 1
= $60.2112

A = P + I
= $501.76 + $60.2112
= $561.9712

Since the third year is the final year, the final compound amount (A) is rounded off to $561.97. The total compound interest is the difference between the final compound amount and the original principal:

$561.97 − $400.00 = $161.97

In each of these calculations, the time was equal to 1. It is not necessary to show the multiplication by 1, because it does not affect the product. *Any number multiplied by 1 is equal to itself.*

As you can see from the preceding example, the number of calculations increases greatly when the number of compounding periods is large. Today, almost everyone uses special tables and/or a calculator to find compound interest. Using these tables is the subject of Part B of this section. The use of calculators to find compound interest is described in Part C.

Interest Compounded Semiannually. To compound interest *semiannually* (twice a year), follow these steps:

1. Multiply the given number of years by 2. This product is the number of compounding periods.
2. Divide the annual interest rate by 2; this quotient is the periodic (semiannual) interest rate.
3. Follow the procedure for compounding interest annually.

EXAMPLE:

Find the final compound amount (A) at the end of two years on $400 invested at 12%, compounded semiannually.

Compounding periods = 2 (years) × 2
= 4 (half-years)

Periodic rate = 12% (annually) ÷ 2
= 6% (per half-year)

Period 1 (first half-year):

A = $400 + ($400 × 0.06)
 = $400 + $24
 = $424

Period 2 (second half-year):

A = $424 + ($424 × 0.06)
 = $424.00 + $25.44
 = $449.44

Period 3 (third half-year):

A = $449.44 + ($449.44 × 0.06)
 = $449.44 + $26.9664
 = $476.4064

Period 4 (fourth half-year):

A = $476.4064 + ($476.4064 × 0.06)
 = $476.4064 + $28.584384
 = $504.990784 *or* $504.99

The final compound amount at the end of two years on $400 invested at 12%, compounded semiannually, is $504.99. Compare this amount to the compound amount ($501.76) at the end of the second year in the preceding example, where interest was compounded annually. In both examples, the original principal was $400, and the annual interest rate was 12%. The only difference was in the frequency of the compounding. If everything else is the same, the final compound amount (and, therefore, the final compound interest) will be greater when interest is compounded more often.

Interest Compounded Quarterly. To compound interest *quarterly,* follow these steps:

1. To find the number of *quarterly* compounding periods, multiply the given number of years by 4.
2. To find the *quarterly* interest rate, divide the given annual rate by 4.
3. Follow the procedure for compounding interest annually.

Interest Compounded Monthly. To compound interest *monthly,* follow these steps:

1. To find the number of *monthly* compounding periods, multiply the given number of years by 12.
2. To find the *monthly* interest rate, divide the given annual rate by 12.
3. Follow the procedure for compounding interest annually.

Interest Compounded Daily. To compound interest *daily,* follow these steps:

1. To find the number of *daily* compounding periods, multiply the given number of years by 365 (or 366).
2. To find the *daily* interest rate, divide the given annual rate by 365 (or 366).
3. Follow the procedure for compounding interest annually.

Banks that compound interest daily can easily find the daily compound amounts with computers.

PART B Using Compound-Interest Tables

Compound-interest tables eliminate all of the individual steps performed in calculating the final compound amount. In the first example (on page 267) in Part A, it took six calculations to get the final compound amount of $561.97. It would have taken 20 steps if the time had been 10 years.

Finding compound interest is easier with a calculator, but you still have to perform each step separately. In actual practice, most people use either comprehensive compound-interest tables or financial calculators. Here, we will discuss the use of interest tables. Each concept presented here can also be applied with a financial calculator.

Table 7-1 on pages 269 and 270 shows the compound amounts for a principal of $1.00 at various periodic interest rates and various numbers of compounding periods. The column labeled *Period* shows the numbers of compounding periods. The row of interest rates across the top shows the periodic rates. Within the table, each number is in a row and column and represents the compound amount for a principal of $1.00 at that rate for that number of periods.

Using Tables When Interest is Compounded Annually. To find the compound amount when interest is compounded annually, follow these steps:

1. Find the number of years in the *Period* column, and go across that row to the column for the given rate. Where the *Period* row and the Rate column meet is a number called the **compound amount factor.**

Chapter 7 Calculating Interest

Period	1.00	1.50	2.00	3.00	4.00	5.00	6.00	Period
1	1.010 000	1.015 000	1.020 000	1.030 000	1.040 000	1.050 000	1.060 000	1
2	1.020 100	1.030 225	1.040 400	1.060 900	1.081 600	1.102 500	1.123 600	2
3	1.030 301	1.045 678	1.061 208	1.092 727	1.124 864	1.157 625	1.191 016	3
4	1.040 604	1.061 364	1.082 432	1.125 509	1.169 859	1.215 506	1.262 477	4
5	1.051 010	1.077 284	1.104 081	1.159 274	1.216 653	1.276 282	1.338 226	5
6	1.061 520	1.093 443	1.126 162	1.194 052	1.265 319	1.340 096	1.418 519	6
7	1.072 135	1.109 845	1.148 686	1.229 874	1.315 932	1.407 100	1.503 630	7
8	1.082 857	1.126 493	1.171 659	1.266 770	1.368 569	1.477 455	1.593 848	8
9	1.093 685	1.143 390	1.195 093	1.304 773	1.423 312	1.551 328	1.689 479	9
10	1.104 622	1.160 541	1.218 994	1.343 916	1.480 244	1.628 895	1.790 848	10
11	1.115 668	1.177 949	1.243 374	1.384 234	1.539 454	1.710 339	1.898 299	11
12	1.126 825	1.195 618	1.268 242	1.425 761	1.601 032	1.795 856	2.012 196	12
13	1.138 093	1.213 552	1.293 607	1.468 534	1.665 074	1.885 649	2.132 928	13
14	1.149 474	1.231 756	1.319 479	1.512 590	1.731 676	1.979 932	2.260 904	14
15	1.160 969	1.250 232	1.345 868	1.557 967	1.800 944	2.078 928	2.396 558	15
16	1.172 579	1.268 986	1.372 786	1.604 706	1.872 981	2.182 875	2.540 352	16
17	1.184 304	1.288 020	1.400 241	1.652 848	1.947 900	2.292 018	2.692 773	17
18	1.196 147	1.307 341	1.428 246	1.702 433	2.025 817	2.406 619	2.854 339	18
19	1.208 109	1.326 951	1.456 811	1.753 506	2.106 849	2.526 950	3.025 599	19
20	1.220 190	1.346 855	1.485 947	1.806 111	2.191 123	2.653 298	3.207 135	20
21	1.232 392	1.367 058	1.515 666	1.860 295	2.278 768	2.785 963	3.399 564	21
22	1.244 716	1.387 564	1.545 980	1.916 103	2.369 919	2.925 261	3.603 537	22
23	1.257 163	1.408 377	1.576 899	1.973 586	2.464 716	3.071 524	3.819 750	23
24	1.269 735	1.429 503	1.608 437	2.032 794	2.563 304	3.225 100	4.048 935	24
25	1.282 432	1.450 945	1.640 606	2.093 778	2.665 836	3.386 355	4.291 871	25
26	1.295 256	1.472 710	1.673 418	2.156 591	2.772 470	3.555 673	4.549 383	26
27	1.308 209	1.494 800	1.706 886	2.221 289	2.883 369	3.733 456	4.822 346	27
28	1.321 291	1.517 222	1.741 024	2.287 928	2.998 703	3.920 129	5.111 687	28
29	1.334 504	1.539 980	1.775 845	2.356 565	3.118 651	4.116 136	5.418 388	29
30	1.347 849	1.563 080	1.811 362	2.427 262	3.243 397	4.321 942	5.743 491	30

TABLE 7-1 Compound Amounts When the Principal is $1

Period	8.00	9.00	10.00	12.00	14.00	16.00	18.00	Period
1	1.080 000	1.090 000	1.100 000	1.120 000	1.140 000	1.160 000	1.180 000	1
2	1.166 400	1.188 100	1.210 000	1.254 400	1.299 600	1.345 600	1.392 400	2
3	1.259 712	1.295 029	1.331 000	1.404 928	1.481 544	1.560 896	1.643 032	3
4	1.360 489	1.411 582	1.464 100	1.573 519	1.688 960	1.810 639	1.938 778	4
5	1.469 328	1.538 624	1.610 510	1.762 342	1.925 415	2.100 342	2.287 758	5
6	1.586 874	1.677 100	1.771 561	1.973 823	2.194 973	2.436 396	2.699 554	6
7	1.713 824	1.828 039	1.948 717	2.210 681	2.502 269	2.826 220	3.185 474	7
8	1.850 930	1.992 563	2.143 589	2.475 963	2.852 586	3.278 415	3.758 859	8
9	1.999 005	2.171 893	2.357 948	2.773 079	3.251 949	3.802 961	4.435 454	9
10	2.158 925	2.367 364	2.593 742	3.105 848	3.707 221	4.411 435	5.233 836	10
11	2.331 639	2.580 426	2.853 117	3.478 550	4.226 232	5.117 265	6.175 926	11
12	2.518 170	2.812 665	3.138 428	3.895 976	4.817 905	5.936 027	7.287 593	12
13	2.719 624	3.065 805	3.452 271	4.363 493	5.492 411	6.885 791	8.599 359	13
14	2.937 194	3.341 727	3.797 498	4.887 112	6.261 349	7.987 518	10.147 244	14
15	3.172 169	3.642 482	4.177 248	5.473 566	7.137 938	9.265 521	11.973 748	15
16	3.425 943	3.970 306	4.594 973	6.130 394	8.137 249	10.748 004	14.129 022	16
17	3.700 018	4.327 633	5.054 470	6.866 041	9.276 464	12.467 685	16.672 247	17
18	3.996 019	4.717 120	5.559 917	7.689 966	10.575 169	14.462 514	19.673 251	18
19	4.315 701	5.141 661	6.115 909	8.612 762	12.055 693	16.776 517	23.214 436	19
20	4.660 957	5.604 411	6.727 500	9.646 293	13.743 490	19.460 759	27.393 035	20
21	5.033 834	6.108 808	7.400 250	10.803 848	15.667 578	22.574 481	32.323 781	21
22	5.436 540	6.658 600	8.140 275	12.100 310	17.861 039	26.186 398	38.142 061	22
23	5.871 464	7.257 874	8.954 302	13.552 347	20.361 585	30.376 222	45.007 632	23
24	6.341 181	7.911 083	9.849 733	15.178 629	23.212 207	35.236 417	53.109 006	24
25	6.848 475	8.623 081	10.834 706	17.000 064	26.461 916	40.874 244	62.668 627	25
26	7.396 353	9.399 158	11.918 177	19.040 072	30.166 584	47.414 123	73.948 980	26
27	7.988 061	10.245 082	13.109 994	21.324 881	34.389 906	55.000 382	87.259 797	27
28	8.627 106	11.167 139	14.420 994	23.883 866	39.204 493	63.800 444	102.966 560	28
29	9.317 275	12.172 182	15.863 093	26.749 930	44.693 121	74.008 514	121.500 541	29
30	10.062 657	13.267 678	17.449 402	29.959 922	50.950 158	85.849 877	143.370 638	30

TABLE 7-1 Compound Amounts When the Principal is $1 (continued)

Chapter 7 Calculating Interest

2. Multiply the compound amount factor by the principal to get the final compound amount.

In the first example (on page 267) in Part A, $400 was invested for 3 years at 12%, compounded annually. Looking at Period 3 in the 12% column, you find the compound amount factor 1.404928. Multiply $400 by 1.404928, and you get $561.9712. This answer is exactly the same as the answer obtained on page 267. Here, it took only one step after you found the compound amount factor in the table.

EXAMPLES:

a. Find the final compound amount and the compound interest on $1,500 at 9%, compounded annually for 12 years.

Find Period 12 in the 9% column. The compound amount factor is 2.812665.

A = $1,500 × 2.812665
  = $4,218.9975 *or* $4,219.00

Compound interest = $4,219.00 − $1,500
  = $2,719

b. Find the final compound amount and the compound interest on $3,000 at 14%, compounded annually for 21 years.

Find Period 21 in the 14% column. The compound amount factor is 15.667578.

A = $3,000 × 15.667578
  = $47,002.734 *or* $47,002.73

Compound interest = $47,002.73 − $3,000.00
  = $44,002.73

Using Tables When Interest is Compounded Semiannually. To find the compound amount when interest is compounded semiannually, follow these steps:

1. Find the number of compounding periods by multiplying the number of years by 2.
2. Find the periodic interest rate by dividing the annual rate by 2.
3. Find the compound amount factor in the table, and multiply it by the original principal.

In the example on page 267 in Part A, $400 was invested for 2 years at 12%, compounded semiannually. There were 4 compounding periods (half-years), and the periodic interest rate was 6% (12% ÷ 2).

Looking at Period 4 in the 6% column, you find the compound amount factor 1.262477. Multiply $400 by 1.262477, and you get $504.9908. This is very close to the answer obtained in the example, $504.990784. The reason for the small discrepancy is that the values in Table 7-1 are rounded off to six decimal places. The exact compound amount factor is 1.26247696. When this exact factor is multiplied by $400, the result is exactly $504.990784.

EXAMPLE:

Find the compound amount and the compound interest on $2,000 at 10%, compounded semiannually for 9 years.

Compounding periods = 9 (years) × 2
  = 18 (half-years)

Periodic rate = 10% (annually) ÷ 2
  = 5% (per half-year)

Find Period 18 in the 5% column. The compound amount factor is 2.406619.

A = $2,000 × 2.406619
  = $4,813.238 *or* $4,813.24

Compound interest = $4,813.24 − $2,000.00
  = $2,813.24

Using Tables When Interest is Compounded Quarterly. To find the compound amount when interest is compounded quarterly, follow these steps:

1. Find the number of compounding periods by multiplying the number of years by 4.
2. Find the periodic interest rate by dividing the annual rate by 4.
3. Find the compound amount factor in the table, and multiply it by the original principal.

EXAMPLE:

Find the compound amount and the compound interest on $2,500 at 8%, compounded quarterly for 6 years.

Compounding periods = 6 (years) × 4
= 24 (quarters)

Periodic rate = 8% (annually) ÷ 4
= 2% (per quarter)

Find Period 24 in the 2% column. The compound amount factor is 1.608437.

A = $2,500 × 1.608437
= $4,021.0925 *or* $4,021.09

Compound interest = $4,021.09 − $2,500.00
= $1,521.09

Using Tables When Interest is Compounded Monthly. To find the compound amount when interest is compounded monthly, follow these steps:

1. Find the number of compounding periods by multiplying the number of years by 12.
2. Find the periodic interest rate by dividing the annual rate by 12.
3. Find the compound amount factor in the table, and multiply it by the original principal.

EXAMPLE:

Find the compound amount and the compound interest on $4,200 at 18%, compounded monthly for $1\frac{1}{2}$ years.

Compounding periods = $1\frac{1}{2}$ (years) × 12
= 18 (months)

Periodic rate = 18% (annually) ÷ 12
= 1.5% (per month)

Find Period 18 in the 1.5% column. The compound amount factor is 1.307341.

A = $4,200 × 1.307341
= $5,490.8322 *or* $5,490.83

Compound interest = $5,490.83 − $4,200.00
= $1,290.83

PART C Calculating Compound Amount Factors

Table 7-1 is limited, because it contains only a few sample periodic interest rates and the number of periods only goes up to 30. However, there is a simple procedure for calculating any compound amount factor.

Look back at the two examples that were solved in Part A. Both of these were also solved in Part B with Table 7-1. Study both solutions for each example. Recall the numbers of periods, the periodic interest rates, and the compound amount factors. Then, study the following examples.

EXAMPLE:

Find the final compound amount at the end of 3 years on $400 invested at 12%, compounded annually.

In Part A, the final compound amount after three calculations with the annual interest rate of 12% was $561.9712.

In Part B, the compound amount factor was 1.404928, because there were 3 periods and a periodic interest rate of 12%. The final compound amount was $561.9712.

The compound amount factor (rounded to six decimal places) can be calculated as follows:

$$1.404928 = 1.12 \times 1.12 \times 1.12$$

EXAMPLE:

Find the final compound amount at the end of 2 years on $400 invested at 12%, compounded semiannually.

In Part A, the final compound amount after four calculations with the semiannual interest rate of 6% was $504.990784.

In Part B, the compound amount factor was 1.262477, because there were 4 periods and a periodic interest rate of 6%. The final compound amount was $504.9908.

The compound amount factor (rounded to six decimal places) can be calculated as follows:

$$1.262477 = 1.06 \times 1.06 \times 1.06 \times 1.06$$

To use Table 7-1, you only need to know the number of periods and the periodic interest rate. If you know both of these, you can always find any compound amount factor by following these steps:

1. Determine the number of periods (n).
2. Express the periodic interest rate as a decimal (r).

Chapter 7 Calculating Interest

3. Add 1 to r to obtain $1 + r$.
4. Multiply $1 + r$ by itself n times. This product is the compound amount factor.

EXAMPLES:

Find the compound amount factors. Check your answers against Table 7-1.

a. For $n = 3$ and $r = 0.05$, the compound amount factor is 1.157625 ($1.05 \times 1.05 \times 1.05$).

b. For $n = 4$ and $r = 0.02$, the compound amount factor is 1.082432 ($1.02 \times 1.02 \times 1.02 \times 1.02$).

c. For $n = 5$ and $r = 0.10$, the compound amount factor is 1.610510 ($1.10 \times 1.10 \times 1.10 \times 1.10 \times 1.10$).

If n is a large number, there will be many multiplications to perform, and it is easy to lose count. For example, if 12% is compounded monthly for only 3 years, 1.01 will be multiplied by itself 36 times (12% ÷ 12 = 1%, and 3 years × 12 = 36 months).

Calculators that quickly find the compound amount factor are available at a modest cost. These calculators are called *scientific, business,* or *financial calculators*. They will likely have a key labeled Y^x or X^y. Using this key, you can find the compound amount factors for the preceding examples as follows:

EXAMPLES:

a. $n = 3$ and $r = 0.05$

Key-enter	Display
1.05	1.05
Y^x	1.05
3	3.
=	1.157625

b. $n = 4$ and $r = 0.02$

Key-enter	Display
1.02	1.02
Y^x	1.02
4	4.
=	1.082432

c. $n = 5$ and $r = 0.10$

Key-enter	Display
1.10	1.10
Y^x	1.10
5	5.
=	1.610510

With this method, the number of periods (n) is called an **exponent**. In *Example a*, the exponent is 3; the problem is read as *1.05 raised to the third power* and is written as 1.05^3. If the key on the calculator is labeled Y^x, the x is the exponent.

COMPLETE ASSIGNMENT 25

ASSIGNMENT 25
SECTION 7.3

Name _____

Date _____

Calculating Compound Interest

	Perfect Score	Student's Score
PART A	24	
PART B	64	
PART C	12	
TOTAL	100	

PART A Compounding Interest at Different Periods

DIRECTIONS: Without using a table, find the final compound amount and the compound interest in the following problems. Show your work, and write your answers on the lines provided (10 points for each correct final compound amount, and 2 points for each correct compound interest).

1. Find the final compound amount and the compound interest on $1,000 at 7%, compounded annually for 3 years.

   Amount _____ Interest _____

2. Find the final compound amount and the compound interest on $800 at 18%, compounded monthly for 3 months.

   Amount _____ Interest _____

STUDENT'S SCORE _____

PART B Using Compound-Interest Tables

DIRECTIONS: Use Table 7-1 on pages 269–270 to find the final compound amount and the compound interest in the following problems. Show your work, and write your answers on the lines provided (6 points for each correct final compound amount, and 2 points for each correct compound interest).

3. Find the final compound amount and the compound interest on $1,250 at 9%, compounded annually for 25 years.

   Amount _____ Interest _____

4. Find the final compound amount and the compound interest on $600 at 10%, compounded semiannually for 11 years.

   Amount _____ Interest _____

5. Find the final compound amount and the compound interest on $1,500 at 12%, compounded quarterly for 5 years.

   Amount _____ Interest _____

6. Find the final compound amount and the compound interest on $2,500 at 18%, compounded monthly for 2 years.

   Amount _____ Interest _____

7. Find the final compound amount and the compound interest on $2,000 at 10%, compounded annually for 15 years.

   Amount _____ Interest _____

8. Find the final compound amount and the compound interest on $900 at 16%, compounded semiannually for $11\frac{1}{2}$ years.

   Amount _____ Interest _____

9. Find the final compound amount and the compound interest on $1,600 at 8%, compounded quarterly for $7\frac{1}{4}$ years.

   Amount _____ Interest _____

10. Find the final compound amount and the compound interest on $3,000 at 12%, compounded monthly for $2\frac{1}{2}$ years.

    Amount _____ Interest _____

STUDENT'S SCORE _____

Chapter 7 Calculating Interest

PART C Calculating Compound Amount Factors

DIRECTIONS: Calculate the compound amount factors for the given rates and time periods. Show your work. Round off your answers to six decimal places, and compare your answers with the values given in Table 7-1 on pages 269–270 (2 points for each correct answer).

11. Find the compound amount factor for 5%, compounded annually for 5 years.

14. Find the compound amount factor for 10%, compounded quarterly for 1 year.

12. Find the compound amount factor for 7.5%, compounded annually for 3 years.

15. Find the compound amount factor for 18%, compounded monthly for 4 months.

13. Find the compound amount factor for 12%, compounded quarterly for 1 year.

16. Find the compound amount factor for 15%, compounded monthly for 3 months.

STUDENT'S SCORE _____

CHAPTER 8

Calculating Time-Payment Plans and Short-Term Loans

Many retailers offer *credit plans* to "buy now and pay later." These plans allow a buyer to immediately receive goods or services and pay for them over an extended period. Usually, the deferred payment price consists of the cash price plus a **finance charge**. The finance charge covers such items as interest on the loan of money to the customer in the form of goods or services, bookkeeping expenses for recording payments, losses (if payments are not made), and insurance on the goods. Three kinds of credit plans for deferred payments are:

1. A *time-payment plan*, in which the buyer makes equal monthly or weekly payments, called **installments**, for a specified number of months or weeks. A down payment is often required.
2. A *revolving charge account*, in which the buyer has two options: paying the full amount due within 30 days without a finance charge, or making minimum monthly payments with a finance charge based on the average daily balance. The payments will vary with the balance in the account.
3. A *90-day period for payment* with a **service charge** instead of a finance charge.

Consumer loans (personal, family, or household loans) are often repaid with time-payment plans. The monthly or weekly payments cover the amount of the loan plus a finance charge. Section 8.1 deals with the calculation of finance charges, amounts of the installments, and annual percentage rates in time-payment plans.

As opposed to a consumer loan, a short-term **commercial loan** is made to a business. Usually, the entire amount of a commercial loan plus interest is repaid at the end of the loan period; in some cases, however, it can be repaid in installments, just like consumer loans. Section 8.2 discusses short-term commercial loans (both interest-bearing and non-interest-bearing) and what it means to discount them.

SECTION 8.1 Time-Payment Plans

The Consumer Credit Protection Act, better known as the Truth in Lending Act, went into effect on July 1, 1969. Businesses extending credit on an installment basis must give the consumer a **disclosure statement** before the sale or loan is completed. The disclosure statement contains the following information in writing:

1. The cash price
2. The down payment required
3. The total amount of the unpaid balance (which is the cash price minus the down payment)
4. Any additional charges to be financed
5. The total amount to be financed
6. The actual amount of the finance charge
7. The amount of each installment payment
8. The charges for late payments
9. The true annual percentage rate (APR) based on the reduced balances

Violators of the Truth in Lending Act are subject to penalties.

The term *finance charge* is used, rather than *interest charge*, because time-payment plans often contain charges other than an interest rate that might be quoted. For example, a fee for bookkeeping may be included. This is a cost of financing that the buyer is required to pay.

The **annual percentage rate (APR)** is essentially an annual interest rate based on the total

279

finance charge, even though the interest amounts may be calculated monthly. The purpose of the APR is to help consumers compare different financing options that may be unclear or confusing to them.

First of all, *1% per month* may simply sound better to a consumer than *12% per year*. As discussed in Chapter 7, these two statements are identical if you are calculating *simple interest*. However, most time-payment plans calculate interest based on the unpaid balance each month, and this actually makes the interest *compound* instead of *simple*. The following two examples, using 1.5% per month (or 18% per year), illustrate the difference.

EXAMPLE:

Find the simple interest on $800 at a monthly rate of 1.5% for a period of 2 months.

Months: I = $800 × 0.015 × 2
             = $24

Years: 1.5% (monthly) × 12 = 18% (annually)
         2 (months) ÷ 12 = $\frac{1}{6}$ (year)
         I = $800 × 0.18 × $\frac{1}{6}$
           = $24

Computed in either months or years, the amount of *simple interest* is $24. Compare that result with monthly interest based on the unpaid balance each month, as shown in the following example.

EXAMPLE:

A store offers charge accounts to its good customers. The cost to the customer is 1.5% per month on the unpaid balance. Suppose someone has an unpaid balance of $800 and makes neither any more purchases nor any payments at all for two months. What will be the new unpaid balance after the two months? What is the total interest charge for the two months?

First month: I = $800 × 0.015 × 1
                 = $12

Unpaid balance = $800 + $12
               = $812

When the next monthly statement arrives, the unpaid balance has increased to $812. The customer makes no payment. The amount of $812 is used to calculate the interest in the second month.

Second month: I = $812 × 0.015 × 1
                 = $12.18

Unpaid balance = $812.00 + $12.18
               = $824.18

Total interest = $824.18 − $800.00
               = $24.18

The interest for the two months is $24.18 ($12 for the first month, and $12.18 for the second month). Notice that this is larger than the simple interest calculated in the preceding example. When nothing was paid in the first month, the unpaid balance increased from $800 to $812. This caused the interest in the second month to be greater than in the first month ($12.18 instead of $12).

The customer had to pay *interest on the interest* in the second month. This is the concept of compound interest. The APR includes the effect of compound interest and thus allows consumers to compare interest rates.

Usury is an interest charge that is above the legal limit. All states have laws (called **usury laws**) which establish the maximum annual interest rate that can be charged to a consumer. Because the usury laws are stated in terms of annual rates, an annual percentage rate is needed to determine if the interest charges are actually within the legal limits.

PART A Finding the Amount of the Finance Charge

To find the total price of goods bought on a time-payment plan, multiply the number of payments by the amount of the required installment and then add the down payment (if there is one) to that product. The finance charge (FC) is the difference between the total time-payment price and the cash price.

EXAMPLE:

Valley Discount Center advertises a video camera for $985 cash or $85 down and $85 a month for 12 months. What is the finance charge?

Chapter 8 Time-Payment Plans and Short-Term Loans

Total installments: 12 × $85 = $1,020
Down payment: +   85
Total time-payment price: $1,105
Cash price: –   985
Finance charge: $ 120

With a consumer loan, the total amount that the borrower repays is the product of the number of installment payments and the amount of each payment. The finance charge is the difference between the total amount repaid and the amount of the loan.

EXAMPLE:

A $900 loan is repaid in 9 monthly installments of $109. What is the finance charge?

Total payments: 9 × $109 = $981
Amount of loan: – 900
Finance charge: $ 81

Most time-payment plans are based on monthly payments. Therefore, the examples and problems in this section involve only monthly payments. If the payments were weekly, bimonthly, etc., the same procedure could be used to find the total finance charge.

PART B Finding the Amount of the Monthly Installment

Each monthly installment consists of two parts: the finance charge for the month, and the payment on the unpaid balance. The amount of the monthly installment is most easily and accurately calculated with special tables, a financial calculator, or a computer. Here, we will illustrate a single method that gives a reasonable approximation. The two separate steps in this method are:

1. Find the total amount of the finance charge.
2. Find the monthly installment.

Computing the Total Finance Charge. Since the finance charge for each month is based on the unpaid balance for that month, the finance charge is different each month. You may compute the exact total finance charge by finding the monthly finance charge for each month and adding them together. However, this requires many separate calculations and is time-consuming. An approximation of the total finance charge can be calculated with the following formula:

$$FC = \frac{\text{1st month's charge} \times (\text{no. of payments} + 1)}{2}$$

EXAMPLE:

East Bay Appliance sells a giant screen television for $1,800 cash or a 5% down payment, with the balance to be paid in 18 equal monthly installments. The finance charge is 1.5% per month on the reduced balances. Find the total finance charge.

Step 1. Find the unpaid cash balance.
Cash price: $1,800
Down payment (5% ×
  $1,800):   –   90
Unpaid balance: $1,710

Step 2. Find the first month's finance charge by using the formula
I = P × R × T. Refer to Chapter 7 to review monthly interest.
I = $1,710 × 0.015 × 1
  = $25.65

Step 3. Find the total finance charge.
$$FC = \frac{\$25.65 \times (18 + 1)}{2}$$
  = $243.675 or $243.68

Computing the Monthly Installment Amount. The total amount to be repaid is equal to the sum of the unpaid cash balance and the total finance charge. Divide this sum by the number of monthly payments to find the monthly installment amount.

In the East Bay Appliance example, the unpaid cash balance after the down payment is $1,710. The total finance charge is $243.68. The total amount to be repaid is $1,953.68 ($1,710.00 + $243.68). Therefore, the amount of each monthly installment is $108.54 ($1,953.68 ÷ 18 = $108.5378).

Two items should be mentioned. First, 18 payments of $108.54 equal $1,953.72. This is 4¢ more than the total amount to be repaid ($1,953.68). Therefore, the final payment will be decreased by 4¢. There will be 17 payments of $108.54 and a final payment of $108.50.

Second, remember that the total finance charge of $243.68 is only an *approximation*. Therefore, the monthly payment of $108.54 is also only an approximation. The actual installment amount can be determined, but it requires a complete set of financial tables, a financial calculator, or a computer.

EXAMPLES:

a. A freezer is advertised for $450 cash or 10% down, with the balance to be paid in 6 monthly installments. The finance charge is 1.2% per month on the reduced balances. Find the monthly payment.

Step 1. Find the unpaid cash balance.

Cash price:	$450
Down: (10% × $450)	− 45
Unpaid balance:	$405

Step 2. Find the first month's charge.
$405 × 0.012 × 1 = $4.86

Step 3. Find the total FC.

$$FC = \frac{\$4.86 \times (6 + 1)}{2}$$
$$= \$17.01$$

Step 4. Find the sum of the unpaid balance and the finance charge.

Unpaid cash balance:	$405.00
Finance charge:	+ 17.01
Total time-payment price:	$422.01

Step 5. Find the amount of the monthly payment.
Monthly payment = $422.01 ÷ 6
= $70.335 *or*
$70.34

Since 6 payments of $70.34 equal $422.04, the last payment will be reduced by 3¢, to $70.31.

b. A loan of $1,800 is to be repaid in 24 equal monthly installments. The finance charge is 1.25% per month on the reduced cash balances. Find the monthly installment.

Step 1. The unpaid cash balance is $1,800.
Step 2. Find the first month's charge.

$1,800 × 0.0125 × 1 = $22.50

Step 3. Find the total FC.

$$FC = \frac{\$22.50 \times (24 + 1)}{2}$$
$$= \$281.25$$

Step 4. Find the sum of the unpaid cash balance and the finance charge.

Unpaid cash balance:	$1,800.00
Finance charge:	+ 281.25
Total of payments:	$2,081.25

Step 5. Find the amount of the monthly payment.
Monthly payment = $2,081.25 ÷ 24
= $86.71875 *or*
$86.72

Since 24 payments of $86.72 equal $2,081.28, the last payment will be reduced by 3¢, to $86.69.

PART C Finding the Annual Percentage Rate (APR)

The formula for finding simple interest rates, $R = I \div (P \times T)$, cannot be used to find the true annual percentage rate, because the unpaid principal is reduced by each payment. The APR can be accurately determined with tables, a financial calculator, or a computer. There are also methods of estimating the APR. The following formula is easy to use with monthly installment payment plans and gives a reasonably accurate approximation.

$$APR = \frac{24 \times \text{total finance charge}}{\text{unpaid cash bal.} \times (\text{no. of payments} + 1)}$$

The number 24 in this formula represents *2 times the number of months in one year:* $24 = 2 \times 12$. If the installments are weekly, use 104 instead of 24, because 104 represents *2 times the number of weeks in one year:* $104 = 2 \times 52$. In this section, however, only monthly payment plans will be considered.

Before the formula can be applied, the unpaid cash balance (the cash price minus the down payment, if there is one) and the total finance charge must be found for goods bought on an installment basis. In the case of consumer loans, the unpaid cash balance is the amount of the loan.

To find the annual percentage rate, follow these five steps:

Chapter 8   Time-Payment Plans and Short-Term Loans

1. Find the total dollar amount of the payments (the monthly installment amount times the number of payments).
2. Find the total time-payment price (the total amount of installments plus the down payment). For consumer loans, there is no down payment, so this will be the amount found in Step 1.
3. Find the total finance charge (the total time-payment price minus the cash price, or the total time-payment price minus the amount of the loan).
4. Find the unpaid cash balance (the cash price minus the down payment, or the amount of the loan). Note that the unpaid cash balance does *not* include the finance charge.
5. Find the APR by substituting the correct figures in the formula.

$$APR = \frac{24 \times \text{total finance charge}}{\text{unpaid cash bal.} \times (\text{no. of payments} + 1)}$$

EXAMPLES:

a. Freeway Cycle Shop advertises a motorcycle for $2,799 cash or $99 down and $99 a month for 36 months. Find the annual percentage rate.

   Step 1. Find the total amount of the payments.

Monthly payment:	$99
Number of payments:	× 36
Total of payments:	$3,564

   Step 2. Find the total time-payment price.

Total of payments:	$3,564
Down payment:	+ 99
Total time-payment price:	$3,663

   Step 3. Find the total finance charge.

Total time-payment price:	$3,663
Cash price:	−2,799
Finance charge:	$864

   Step 4. Find the unpaid cash balance.

Cash price:	$2,799
Down payment:	− 99
Unpaid cash balance:	$2,700

   Step 5. Find the annual percentage rate.

   $$APR = \frac{24 \times \$864}{\$2,700 \times (36 + 1)}$$
   $$= 0.2076 \text{ or } 20.8\%$$

b. A loan of $1,800 is to be repaid in 24 payments: 23 equal installments of $86.72 each, and a final payment of $86.69. Find (approximate) the annual percentage rate.

   Step 1. Find the total amount of the payments.

Monthly payment:	$ 86.72
Number of payments:	× 23
Total of 23 payments:	$1,994.56
24th payment:	+ 86.69
Total of payments:	$2,081.25

   Step 2. Find the total time-payment price.
   $2,081.25 (It is a loan.)

   Step 3. Find the total finance charge.

Total time-payment price:	$2,081.25
Cash price:	−1,800.00
Finance charge:	$ 281.25

   Step 4. Find the unpaid cash balance.
   $1,800 (The amount of the loan)

   Step 5. Find the annual percentage rate.

   $$APR = \frac{24 \times \$281.25}{\$1,800 \times (24 + 1)}$$
   $$= 0.15 \text{ or } 15\%$$

COMPLETE ASSIGNMENT 26

Name _____

Date _____

ASSIGNMENT 26
SECTION 8.1

Time-Payment Plans

	Perfect Score	Student's Score
PART A	22	
PART B	42	
PART C	36	
TOTAL	100	

PART A Finding the Amount of the Finance Charge

DIRECTIONS: The first group of the following problems is a set of time-payment purchases, and the second group is a set of short-term loans. Find the missing items in each group and write the answers on the lines provided (1 point for each correct answer).

	Cash Price	Down Payment	Monthly Payment	Number of Payments	Total Time-Payment Price	Finance Charge
1.	$1,000	$150	$ 85.75	12	_____	_____
2.	$850	$ 50	$101.50	9	_____	_____
3.	$1,500	$225	$ 84.50	18	_____	_____
4.	$2,200	20%	$ 87.25	24	_____	_____
5.	$3,650	10%	$250.00	15	_____	_____

	Amount of Loan	Monthly Payment	Number of Payments	Total Amount Repaid	Finance Charge
6.	$3,000	$375.00	9	_____	_____
7.	$2,500	$187.50	15	_____	_____
8.	$1,200	$225.60	6	_____	_____
9.	$800	$ 39.25	24	_____	_____
10.	$4,200	$275.00	18	_____	_____
11.	$1,650	$160.00	12	_____	_____

STUDENT'S SCORE _____

PART B Finding the Amount of the Monthly Installment

DIRECTIONS: Find the first month's interest, the total finance charge (approximation), and the monthly payment. Write your answers on the lines provided. Put an asterisk (*) after any monthly payment for which the last payment must be adjusted (2 points for each correct answer).

	Cash Price/ Loan Amount	Down Payment	No. of Months	Monthly Rate	1st Month's Interest	Finance Charge	Monthly Payment
12.	$1,250	$250	12	1%			
13.	$1,900	$300	6	1.5%			
14.	$3,500	20%	21	2%			
15.	$2,000	10%	15	1.75%			
16.	$6,000	—	24	1.5%			
17.	$2,200	—	9	1%			
18.	$3,200	—	36	1.25%			

STUDENT'S SCORE _____

PART C Finding the Annual Percentage Rate (APR)

DIRECTIONS: Find the total time-payment price, the total finance charge, and the APR (approximation) to the nearest 0.1%. Write your answers on the lines provided (2 points for each correct answer).

	Cash Price/ Loan Amount	Down Payment	Monthly Payment	No. of Months	Total Time-Payment Price	Finance Charge	APR
19.	$2,400	$600	$110	18			
20.	$2,900	$500	$115	24			
21.	$3,500	20%	$213	15			
22.	$800	10%	$127	6			
23.	$1,800	—	$170	12			
24.	$3,500	—	$135	30			

STUDENT'S SCORE _____

Chapter 8 Time-Payment Plans and Short-Term Loans

SECTION 8.2 Short-Term Commercial Loans

A commercial loan is made for business use, just as a consumer loan is made for personal use. Commercial loans can be *short-term* (for less than one year) or *long-term* (for one or more years). Long-term loans (bonds, debentures, mortgages, etc.) vary greatly and often have complicated terms of repayment. Therefore, they are not discussed in this section.

Short-term commercial loans often require a note that is **negotiable,** or transferable to another person, to confirm the debt. The promissory note (defined in Chapter 7) is an example of a negotiable instrument. A simple promissory note is illustrated in Figure 8-1. Most lending agencies, however, use a much longer form with detailed legal specifications. Notes may also be required from purchasers who are late in settling their accounts. The bank to which payment is to be made is often specified on these notes. Other negotiable instruments used in connection with short-term commercial loans are *trade acceptances, bills of exchange,* and *drafts.*

When a commercial loan is made, the borrower or signer of the note is called the **maker;** the lender is called the **payee.** The amount of the loan is the **face value,** or **face,** of the note. Sometimes, the maker is required to furnish **collateral,** which is something of value (such as stocks, equipment, accounts receivable, real estate, etc.). The collateral is evidence of the maker's ability to pay, and the lender has the right to sell the collateral if the loan is not repaid.

A note may be **non-interest-bearing** or **interest-bearing.** As mentioned in Chapter 7, the maturity value of a non-interest-bearing note is equal to the face value of that note. When a note is interest-bearing, the interest rate must be shown on the note. The maturity value of an interest-bearing note is equal to the face value plus the interest.

PART A Finding the Interest Charge, Due Date, and Maturity Value

Interest rates on short-term commercial loans vary considerably and often depend on the circumstances or uses for each loan. The **prime rate** is the rate charged by leading banks to their best, low-risk customers. As a rule of thumb, the greater the risk is, the higher the interest rate will be. To compute the interest on either a 360- or 365-day basis, the exact number of days of the loan is used. Even if the time of the loan is stated in weeks or months, the exact number of days is still used. For ease of computation, **banker's interest** (based on 360 days) is used in all examples and problems in this section.

In consumer loans repaid in installments, the finance charge (which includes the interest charge) is part of each payment. In contrast, the interest on short-term commercial loans is usually payable on the due date of the note. For this reason, the interest on short-term commercial loans is often called **interest due at maturity, interest after date,** or **interest to follow.** The maturity value of the note is paid on the due date.

$2,000.00   San Francisco, CA March 3, 19--

For value received *sixty days* after date *I* promise to pay to the order of *Western National Bank*

Two thousand and no/100 —————————— DOLLARS

with annual interest at the rate of *15* percent.

No. *374*   Due *May 2, 19--*   *John Bovio*

FIGURE 8-1 *A Promissory Note*

Finding the due date, the interest charge, and the maturity value of the notes in the following examples should serve as a review of what you learned about calculating interest in Chapter 7.

EXAMPLES

a. Find the due date, the interest charge, and the maturity value of a 120-day note, dated June 25, for $1,500, with interest at 12% annually.

Month	Days	Loan Days Remaining
June	30 − 25 = 5	120 − 5 = 115
July	31	115 − 31 = 84
Aug.	31	84 − 31 = 53
Sept.	30	53 − 30 = 23
Oct.	31	

Since the final number of loan days remaining (23) is less than the number of days in October (31), the due date is October 23.

Due date: October 23

$$\text{Interest} = \$1{,}500 \times 0.12 \times \frac{120}{360}$$
$$= \$60$$

Maturity value = $1,500 + $60
$\phantom{\text{Maturity value }}= \$1{,}560$

b. Find the due date, the interest charge, and the maturity value of a three-month note, dated April 17, for $1,850, with interest at 15% annually.

Due date: July 17
Number of days:
April	30 − 17 =	13
May		31
June		30
July		17
Total		91 days

$$\text{Interest} = \$1{,}850 \times 0.15 \times \frac{91}{360}$$
$$= \$70.1458 \text{ or } \$70.15$$

Maturity value = $1,850.00 + $70.15
$\phantom{\text{Maturity value }}= \$1{,}920.15$

Remember that the exact number of days is used to calculate interest on these loans. In *Example b*, the exact time of the 3-month loan is 91 days. Also, recall that we are using a 360-day year for all problems in this section.

PART B Discounting Non-Interest-Bearing Notes

When a person borrows money from another for a short period of time, the interest is normally paid, along with the principal, on the due date of the loan. When a person borrows from a bank, however, the bank may collect the interest at the time the loan is made. The bank merely deducts the amount of interest from the face value of the note, and the borrower receives the difference. This kind of interest, paid in advance, is called **bank discount,** and the note is said to be **discounted.** The amount of money actually received by the borrower is called the **proceeds** of the note. Because the interest has already been paid, the note can be considered non-interest-bearing.

There are two kinds of short-term, non-interest-bearing notes which can be discounted: a note payable signed by the maker, with the bank as the payee (as previously described), and a note receivable which a company has received from a customer in settlement of an account. From the accounting point of view, a note is called a **note payable** by the maker of the note and a **note receivable** by the lender.

Finding the Proceeds of Discounted Notes Payable. Calculating simple bank discount is done by using a concept similar to the one used in calculating simple interest:

Discount = Face value × Rate × Time
Proceeds = Face value − Discount

The percentage rate charged by the bank is called the **discount rate** and the time is called the **discount period.** The maturity value is the face value of the note. It is very important to know that the discount rate is *not* the interest rate, and that the face value is *not* the principal.

Chapter 8 Time-Payment Plans and Short-Term Loans

EXAMPLES:

a. On March 10, Daryl Watson borrowed money from his bank. He signed a 90-day note for $800, which the bank discounted at 11%. Find the proceeds of the note.

Bank discount = $800 \times 0.11 \times \dfrac{90}{360}$
= $22

Proceeds = $800 − $22
= $778

b. On October 19, Beverly Wong borrowed money from her bank. She signed a $1,200, 60-day note, which the bank discounted at 14%. Find the proceeds.

Bank discount = $1,200 \times 0.14 \times \dfrac{60}{360}$
= $28

Proceeds = $1,200 − $28 = $1,172
= $1,172

Discount Rates vs. Interest Rates (Optional). In the first preceding example, Daryl Watson signed a 90-day note for $800, which the bank discounted at a rate of 11%. Watson received $778 (the proceeds) and had to repay $800 (the face value) to the bank at the end of 90 days. The bank earned a profit of $22 (the discount amount). The discount was calculated as follows:

Discount = Face value × Discount rate × Time
= $800 \times 0.11 \times \dfrac{90}{360}$
= $22

Another way to visualize this example is to view the $778 that Watson received from the bank as the principal of a loan. The $22 is interest. After 90 days, Watson repays $800 ($778 principal plus $22 interest). Recall from Chapter 7 that the fundamental formula for simple interest is:

$$I = P \times R \times T$$

or

Interest = Principal × Rate × Time

If you know the interest amount, the principal, and the time, you can calculate the interest rate by using the following formula:

$$R = \dfrac{I}{P \times T}$$

Substituting $778, $22, and $\dfrac{90}{360}$ year (90 days) into this formula, you get:

$$R = \dfrac{\$22}{\$778 \times \dfrac{90}{360}}$$

= 0.1131 or 11.31%

Note that the interest rate is 11.31% and the discount rate is only 11%. The 11.31% may sound less attractive to some borrowers. However, the actual dollar cost of borrowing in either case is the same—$22 to borrow $778 for 90 days.

By basing the calculation of interest on the face value ($800) instead of the actual amount borrowed ($778), the lender can use a lower percentage rate—11% instead of 11.31%. This always occurs with discounting, and it can be confusing for a borrower if it is not explained clearly.

Finding the Proceeds of Discounted Notes Receivable. Conceivably, the payee of a note may need cash before the due date of the note that he or she possesses. The payee can sell the note to a third party, which is often a bank. To make a profit on the transaction, the bank does not pay the entire maturity value of the note. The bank discounts the note at some percentage rate. The amount that the bank pays for the note is the proceeds.

If the note is discounted on the same day that it was written, the discount period is the same as the time of the note. More often, however, the discount date is some time *after* the date of origin. In this case, the discount period runs from the date of discount to the due date of the note. There is no charge for the days preceding the date of discount.

To find the proceeds of a non-interest-bearing note receivable which has been discounted, follow these four steps:

1. Find the due date if it does not appear on the note.

2. Find the number of days in the discount period. Remember that the discount period starts on the date that the note was discounted. In most cases, this is not the same as the date of origin.
3. Find the bank discount by multiplying the face value by the discount rate by the discount period.
4. Find the proceeds by subtracting the bank discount from the face value.

EXAMPLES:

a. Martha Stewart accepted a 90-day, non-interest-bearing note for $400 on August 9 in settlement of a customer's account. She had it discounted at her bank on September 23 at 15%. Find the proceeds.

Step 1. Find the due date.

Month	Days	Loan Days Remaining
Aug.	31 − 9 = 22	90 − 22 = 68
Sept.	30	68 − 30 = 38
Oct.	31	38 − 31 = 7
Nov.	30	

Due date: November 7

Step 2. Find the discount period.

Sept.	30 − 23 =	7
Oct.		31
Nov.		7
Discount period:		45 days

Step 3. Find the bank discount.

$$\$400 \times 0.15 \times \frac{45}{360} = \$7.50$$

Step 4. Find the proceeds.
$400.00 − $7.50 = $392.50

b. Wallace Giles accepted a 60-day, non-interest-bearing note for $750 on May 17 in settlement of a customer's account. He had it discounted at his bank on June 27 at 13%. Find the proceeds.

Step 1. Find the due date.

Month	Days	Loan Days Remaining
May	31 − 17 = 14	60 − 17 = 43
June	30	43 − 30 = 13
July	31	

Due date: July 13

Step 2. Find the discount period.

June	30 − 27 =	3
July		13
Discount period:		20 days

Step 3. Find the bank discount.

$$\$750 \times 0.13 \times \frac{20}{360} = \$5.4167 \text{ or } \$5.42$$

Step 4. Find the proceeds.
$750.00 − $5.42 = $744.58

PART C   Discounting Interest-Bearing Notes

In addition to non-interest-bearing notes, banks also buy interest-bearing notes. Suppose that one company, or one person, holds a note from a second party. The second party promises to pay the face value of the note, plus interest, on the due date. The total amount paid on the due date is the maturity value of the note.

The party holding the note may be able to sell the note to a bank sometime between the date of origin and the due date. The bank will have to evaluate the risk involved, because the person who actually wrote the note may not be a customer of that bank. If the bank buys the note, the bank will receive the entire maturity value on the due date. To make a profit, the bank will pay less than the maturity value for the note; in other words, the bank *discounts* the note. The amount that the bank pays for the note is the proceeds; the bank's profit is the bank discount.

The bank discount is based on the maturity value of the note, because this is the amount that the bank will collect on the due date. To find the amount of interest, use the face value of the note. To find the proceeds, subtract the bank discount from the maturity value, *not* from the face value.

To find the proceeds of an interest-bearing note receivable, follow these five steps:

1. Find the due date if it does not appear on the note.
2. Find the maturity value of the note. If the time of the note is expressed in months, the exact number of days between the date of origin and the due date *must* be used in calculating the interest. For example, a two-month note which is dated June 21 is due on August 21; the exact time of interest is 61 days.
3. Find the number of days in the discount period. This is the exact time from the date of discount to the due date. Remember that there is no charge for the time before the date of discount.
4. Find the bank discount. Multiply the maturity value by the rate of discount by the discount period.
5. Find the proceeds by subtracting the bank discount from the maturity value.

EXAMPLES:

a. Ed North accepted a 90-day, 12% interest-bearing note for $900 on November 15. His bank discounted the note on December 29 at 15%. Find the proceeds.

Step 1. Find the due date.

Month	Days	Loan Days Remaining
Nov.	30 − 15 = 15	90 − 15 = 75
Dec.	31	75 − 31 = 44
Jan.	31	44 − 31 = 13
Feb.	28	

Due date: February 13

Step 2. Find the maturity value.

$$\text{Interest} = \$900 \times 0.12 \times \frac{90}{360}$$
$$= \$27$$
$$\text{Maturity value} = \$900 + \$27 = \$927$$

Step 3. Find the discount period.

Dec.	31 − 29 =	2
Jan.		31
Feb.		13
Discount period:		46 days

Step 4. Find the bank discount.

$$\$927 \times 0.15 \times \frac{46}{360} = \$17.7675 \text{ or } \$17.77$$

Step 5. Find the proceeds.

$$\$927.00 - \$17.77 = \$909.23$$

b. Sandra Klein accepted a four-month, 10% interest-bearing note for $650 on April 25. She took it to her bank on June 10 for discounting at 12%. Find the proceeds.

Step 1. Find the due date.

Due date: Aug. 25

Step 2. Find the maturity value. First, calculate the interest period.

April	30 − 25 =	5
May		31
June		30
July		31
Aug.		25
Interest period:		122 days

$$\text{Interest} = \$650 \times 0.10 \times \frac{122}{360}$$
$$= \$22.0278 \text{ or } \$22.03$$
$$\text{Maturity value} = \$650.00 + \$22.03 = \$672.03$$

Step 3. Find the discount period.

June	30 − 10 =	20
July		31
Aug.		25
Discount period:		76 days

Step 4. Find the bank discount.

$672.03 × 0.12 × $\frac{76}{360}$ = $17.0248 *or* $17.02

Step 5. Find the proceeds.

$672.03 − $17.02 = $655.01

COMPLETE ASSIGNMENT 27

ASSIGNMENT 27
SECTION 8.2

Name

Date

Short-Term Commercial Loans

	Perfect Score	Student's Score
PART A	30	
PART B	46	
PART C	24	
TOTAL	100	

PART A Finding the Interest Charge, Due Date, and Maturity Value

DIRECTIONS: The interest is due at maturity on the following notes. Find the due date, the amount of interest, and the maturity value for each. Write your answers on the lines provided (1 point for each correct answer).

	Face Value	Date of Note	Interest Period	Interest Rate	Due Date	Amount of Interest	Maturity Value
1.	$1,000	June 17	60 days	12%			
2.	$800	Dec. 16	45 days	13%			
3.	$1,250	Jan. 15	30 days	8%			
4.	$2,000	Oct. 27	120 days	11%			
5.	$1,600	Aug. 17	90 days	12%			
6.	$750	Apr. 23	150 days	10%			
7.	$1,850	Dec. 12	75 days	15%			
8.	$1,400	Sept. 1	60 days	14%			
9.	$2,500	July 8	3 months	9%			
10.	$500	May 24	4 months	11%			

STUDENT'S SCORE _____

PART B Discounting Non-Interest-Bearing Notes

DIRECTIONS: The following non-interest-bearing notes payable are signed by the maker with a bank as the payee and are discounted on the dates of origin at the given rates. Find the bank discount and the proceeds for each. Write your answers on the lines provided (1 point for each correct answer).

	Face Value	Date of Note	Discount Period	Discount Rate	Bank Discount	Proceeds
11.	$2,200	Nov. 18	75 days	14%		
12.	$800	June 7	150 days	12%		
13.	$1,400	Mar. 10	90 days	9%		
14.	$1,625	Jan. 30	30 days	15%		
15.	$2,500	Dec. 15	120 days	8%		
16.	$1,000	Apr. 10	60 days	10%		
17.	$1,800	May 25	45 days	13%		
18.	$3,600	Feb. 28	120 days	16%		
19.	$1,500	Oct. 12	60 days	14%		
20.	$900	Sept. 9	2 months	10%		
21.	$2,000	July 16	3 months	11%		

DIRECTIONS: The following non-interest-bearing notes receivable are discounted on the given dates at the given rates. Find the due date, the discount period, the bank discount, and the proceeds for each (1 point for each correct answer).

22. 90-day note, dated March 22, for $1,200; discounted on May 21 at 11%

   Due date: _____
   Discount period: _____
   Bank discount: _____
   Proceeds: _____

23. 120-day note, dated May 25, for $2,600; discounted on July 1 at 8%

   Due date: _____
   Discount period: _____
   Bank discount: _____
   Proceeds: _____

Chapter 8 Time-Payment Plans and Short-Term Loans

24. 75-day note, dated August 9, for $2,000; discounted on Sept. 12 at 12%

   Due date: _____
   Discount period: _____
   Bank discount: _____
   Proceeds: _____

26. 30-day note, dated January 28, for $800; discounted on Feb. 12 at 9%

   Due date: _____
   Discount period: _____
   Bank discount: _____
   Proceeds: _____

25. 60-day note, dated June 5, for $1,800; discounted on June 15 at 10%

   Due date: _____
   Discount period: _____
   Bank discount: _____
   Proceeds: _____

27. 3-month note, dated April 30, for $1,400; discounted on June 15 at 13%.

   Due date: _____
   Discount period: _____
   Bank discount: _____
   Proceeds: _____

STUDENT'S SCORE _____

PART C Discounting Interest-Bearing Notes

DIRECTIONS: The following interest-bearing notes receivable are discounted on the given dates at the given rates. Find the due date, the interest amount, the maturity value, the discount period, the bank discount, and the proceeds for each. Write your answers on the lines provided (1 point for each correct answer).

28. 90-day, 10% interest-bearing note, dated March 15, for $1,500; discounted on May 29 at 12%

   Due date: _____
   Interest amount: _____
   Maturity value: _____
   Discount period: _____
   Bank discount: _____
   Proceeds: _____

29. 60-day, 12% interest-bearing note, dated July 1, for $1,900; discounted on July 31 at 18%

   Due date: _____
   Interest amount: _____
   Maturity value: _____
   Discount period: _____
   Bank discount: _____
   Proceeds: _____

30. 120-day, 8% interest-bearing note, dated August 31, for $2,250; discounted on November 1 at 10%

Due date: _____
Interest amount: _____
Maturity value: _____
Discount period: _____
Bank discount: _____
Proceeds: _____

31. 3-month, 9% interest-bearing note, dated June 25, for $3,200; discounted on August 25 at 13%

Due date: _____
Interest amount: _____
Maturity value: _____
Discount period: _____
Bank discount: _____
Proceeds: _____

STUDENT'S SCORE _____

CHAPTER 9

Purchase Orders and Invoices, Cash Discounts, and Trade Discounts

Chapter 9 introduces some fundamental concepts about an important feature of our economy—the sales transaction. Over the centuries, somewhat uniform procedures and documents have developed for sales transactions. In a properly recorded transaction, both the buyer and the seller prepare separate documents. Section 9.1 discusses the *purchase order* (the buyer's document) and the *invoice* (the seller's document). An understanding of these documents is increasingly important because of the increasing use of computers to prepare them.

In a competitive economy, there will always be more than one seller, or **vendor**, for the same products. To try to increase sales and profits, vendors may offer several types of discounts to buyers. Section 9.2 is about **cash discounts**, which sellers use to encourage buyers to pay promptly. Section 9.3 is about **trade discounts**, which enable a seller to give different prices to different types of customers without printing several different catalogs.

SECTION 9.1 Purchase Orders and Invoices

Businesses increase their profits by increasing revenues and decreasing costs. Those businesses that purchase many items can better control costs by having one department handle all purchasing. In a small company, purchasing might be handled by only one person. This person would attempt to verify that all purchases are legitimate, of the proper quality, and obtained at the best price.

In most large companies, a person who needs materials to be purchased completes a form called a **purchase requisition**. Additional signatures may be required to verify the need for these materials. The requisition then goes to the purchasing department, which prepares a *purchase order* and sends it to a vendor.

PART A Preparing and Checking Purchase Orders

An order for goods may be given directly to a supplier's representative, placed by telephone or telegram, or sent by mail. Many companies use their own printed forms for ordering goods.

A **purchase order** shows the particular goods wanted and specifies quantities and unit prices. It also includes details on how and when the order is to be shipped, as well as the terms and conditions to which the vendor must agree before accepting the order. The purchase order must then be signed by the purchasing agent or another responsible person. Several copies of it are prepared and kept for future reference.

A purchase order is illustrated in Figure 9-1. Note that the purchaser, the Coastal Garden Stores, uses its own printed form. After Coastal Garden's purchasing department received the approved Requisition No. 87921, it prepared Purchase Order No. 58830 on April 14.

Purchase Order No. 58830 shows that the order is to be shipped via United Parcel Service (UPS) by April 28. Coastal Garden's purchasing agent is Jan Mason. The order is to be shipped to the company's store in Palo Alto, California. The vendor to whom the purchase order is sent is the Bovio Supply Company.

```
┌─────────────────────────────────────────────────────────────────────────┐
│                          PURCHASE ORDER                                 │
│                     Coastal Garden Stores            NO. 58830          │
│                                                 SHOW PURCHASE ORDER NUMBER │
│                     1814 West Holmby, Suite 200    ON PACKAGES, INVOICES, AND │
│                     Los Angeles, CA 90024-2418        CORRESPONDENCE    │
└─────────────────────────────────────────────────────────────────────────┘
```

REQUISITION NO.	DATE	SHIP VIA	SHIP BY	REFER INQUIRES TO BUYER
87921	4/14/--	UPS	4/28/--	Jan Mason

Bovio Supply Company
934 Rose Street
Menlo Park, CA 94025-4420

SHIP TO:

1729 Anderson Avenue
Palo Alto, CA 94303-3870

PLEASE SUPPLY THE FOLLOWING

QUANTITY	MODEL NO.	DESCRIPTION	PRICE
30	VA-196-B	Sprinkler valve (brass)	$6.85 ea
4 doz	SP-003-B	Impulse sprinkler head (brass)	5.70 ea
500	B-1201-B	Shrub bubbler head (brass)	85.00/C
75 doz	B-419-P	Shrub bubbler head (plastic)	4.60/doz

Per telephone quote 4/10/--

PLEASE ACKNOWLEDGE RECEIPT OF THIS ORDER IMMEDIATELY. MAIL INVOICES IN DUPLICATE. IN ACCEPTING THIS ORDER YOU AGREE TO ALL TERMS AND CONDITIONS STATED ON THE REVERSE SIDE.

FIGURE 9–1 A Purchase Order

The quantities, descriptions, and unit prices of the items are listed in the main part of the purchase order. The printed statement at the bottom of the purchase order states that the vendor must acknowledge receipt of the order and that two copies of the invoice are to be sent. It also calls attention to the terms and conditions (printed on the back) to which the vendor must agree if the order is accepted.

When the order is delivered, a clerk in the receiving department checks the goods against a copy of the purchase order. If the vendor has included a **bill of lading** (or a packing slip, or a shipping ticket), which is a description of the shipment without any prices, that is also checked. The clerk then makes out a **receiving record** of the goods delivered and informs the purchasing and accounting departments if any items are missing or damaged or if the wrong items or quantities have been delivered.

When the invoice for the goods is received, an accounting clerk checks it for any errors reported by the receiving clerk and then checks all extensions and the total. If there are any errors either in the order itself or on the invoice, the vendor is notified immediately, so that the necessary adjustments can be made as soon as possible.

PART B Preparing and Checking Invoices

Many companies use printed invoice forms designed to fit their particular needs. In a small company, invoices may be prepared on a billing machine or a typewriter, or they may be done by hand. In a large firm, they are often produced by a computer.

A distinction is sometimes made between an invoice and a bill. An **invoice** is a statement for goods that have been sold, while a **bill** is a statement for services that have been performed.

An invoice is also different from an order form. An **order form** is the vendor's internal document, just as the purchase requisition is the purchaser's internal document. The invoice results from the order form, just as the purchase order results from the purchase requisition.

The **monthly statement** is the vendor's summarization of the invoices (or bills) and the cash payments that a purchaser has made each month. This section is concerned with purchase orders and invoices only.

Although designs of individual invoices vary considerably, the information that they contain is quite standard. The upper part of the invoice shows the vendor's name and address, the purchaser's name and address, the invoice number, the order number, and the invoice date. In addition, it may show the terms of payment, shipping details, the salesperson's name or number, the purchase order number, and other information that may be necessary.

The lower part of the invoice has separate columns for the quantity, description, unit price, amount, and other information needed about each item. Most manufacturers and distributors identify each product or item with a separate stock number. On some invoices, the stock number may be included in the description column.

Invoices prepared by computers or billing machines may omit the dollar signs on prices and amounts. For the most part, handwritten or typewritten invoices do include dollar signs for the first price and amount in a column and for the total.

Figure 9–2 shows Invoice No. 90027 of Bovio Supply Company, which was sent to the administrative offices of Coastal Garden Stores in Los Angeles, California. This invoice was prepared from an internal order form (Order No. 49613) of Bovio Supply Company. Order No. 49613 was the result of Coastal Garden's Purchase Order No. 58830 and was taken by salesperson LJ on April 17.

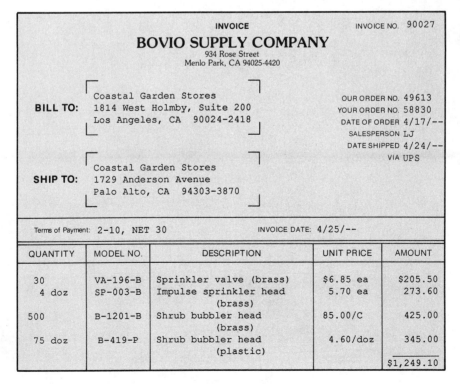

FIGURE 9–2 An Invoice

The invoice was prepared on April 25, and it shows that the shipment was to be sent via UPS, with payment terms of 2-10, NET 30. (Terms of payment will be discussed in Section 9.2.) The merchandise was shipped on April 24 to Coastal Garden Stores in Palo Alto.

Note: For receipt of the invoice, many businesses have an address called the *bill-to-address*; for receipt of goods, they have another address (or several other addresses) called the *ship-to-address*.

Generally, the purchaser will express the quantities in the same units of measure that the vendor used in the price quotation. If the purchaser expresses the quantity in a different unit of measure, the vendor may change it on the order form, on the invoice, or in the extension. If the invoice is prepared by a computer, the quantities will be expressed in a unit of measure that agrees with the one used for the unit price.

The extensions are price calculations for each item ordered. The prices are then listed in the *Amount* column. Before calculating the extensions in Figure 9–2, you may wish to review the discussion about units of measure in Chapter 4. In particular, review such measures as unit, dozen, gross, hundred, and hundredweight.

The extensions for Figure 9–2 are calculated as follows:

1. 30 × $6.85 = $205.50
2. 4 doz = 4 × 12 = 48 units
   48 × $5.70 = $273.60
3. 500 units ÷ 100 = 5C
   5 × $85.00 = $425.00
4. 75 × $4.60 = $345.00

Note that the total is placed under the last extension. On some invoices, there is a separate column for the total.

Note: While some companies package their small items in lots of 10, 25, 50, 100, or more, with the unit price per package, others continue to use unit prices per dozen or per gross.

From the invoice in Figure 9–2, you can see that Bovio Supply Company uses both order forms and invoices. Some companies, however, do not use both forms. They may simply use one copy of the invoice as their order form. In addition, instead of preparing a separate bill of lading, they may use a copy of the invoice with the prices marked out.

COMPLETE ASSIGNMENT 28

_____ **ASSIGNMENT 28**
Name   **SECTION 9.1**

	Perfect Score	Student's Score
PART A	30	
PART B	70	
TOTAL	100	

_____ Purchase Orders and
Date   Invoices

PART A Preparing and Checking Purchase Orders

1. With the information given, complete Purchase Order No. 59747, dated May 23. Use the blank form provided (30 points).

   Purchase Requisition No. 89032 contained the items listed below. The buyer, Jan Mason, verified the prices by telephone on May 20 with salesperson Lars Jensen. The order is to be shipped by June 15 via IME (Inter-Mountain Express) to the store at 1729 Anderson Avenue, Palo Alto, CA, 94303-3870. The vendor is Bovio Supply Company, 934 Rose Street, Menlo Park, CA, 94025-4420.

Quantity	Model No.	Description	Price
300 units	EJ-046-B	Elbow joint 5/8'' (brass)	$76.00/C
20 doz	S-1321-P	Pop-up sprinkler (plastic)	$8.50/doz
40 units	T-512-BD	Water-Miser timer	$12.75/ea
5 doz	N-2004-B	Water nozzle (brass)	$3.17/ea

   PURCHASE ORDER

   Coastal Garden Stores

   1814 West Holmby, Suite 200
   Los Angeles, CA 90024-2418

   NO.
   SHOW PURCHASE ORDER NUMBER ON PACKAGES, INVOICES, AND CORRESPONDENCE

REQUISITION NO.	DATE	SHIP VIA	SHIP BY	REFER INQUIRES TO BUYER

   SHIP TO:

   PLEASE SUPPLY THE FOLLOWING

QUANTITY	MODEL NO.	DESCRIPTION	PRICE

   PLEASE ACKNOWLEDGE RECEIPT OF THIS ORDER IMMEDIATELY. MAIL INVOICES IN DUPLICATE. IN ACCEPTING THIS ORDER YOU AGREE TO ALL TERMS AND CONDITIONS STATED ON THE REVERSE SIDE.

STUDENT'S SCORE _____

PART B Preparing and Checking Invoices

2. The purchase order in Problem 1 was received by Bovio Supply Company on May 27. The vendor created Order Form No. 50284 and shipped the order on June 10. Complete Invoice No. 90659, dated June 11, for this order, with terms of 2-10, NET 30 (30 points).

INVOICE

BOVIO SUPPLY COMPANY
934 Rose Street
Menlo Park, CA 94025-4420

INVOICE NO.

BILL TO:

SHIP TO:

OUR ORDER NO.
YOUR ORDER NO.
DATE OF ORDER
SALESPERSON
DATE SHIPPED
VIA

Terms of Payment: INVOICE DATE:

QUANTITY	MODEL NO.	DESCRIPTION	UNIT PRICE	AMOUNT

DIRECTIONS: *Complete the extensions and find the totals. Write your answers on the lines provided (2 points for each correct answer).*

3.
- 11 units $41.50 ea   _____
- 4 doz   2.50 ea   _____
- 9 doz   0.28 ea   _____
- 60 units   14.20/doz   _____
- 3 gross   0.15 ea   _____
- 180 units   28.50/doz   _____
- 35 units   13.00 ea   _____
- 29 units   3.86 ea   _____
- 96 units   4.25/doz   _____
- TOTAL

4.
- 5,000 units $28.90/M   _____
- 4 gross   3.25/doz   _____
- 60 doz   54.00/gross _____
- 3,500 lb   9.40 cwt   _____
- 720 units   12.50/gross _____
- 315 units   0.12 ea   _____
- 250 lb   8.40/lb   _____
- 30 units   84.90/doz   _____
- 90 ft   1.56/yd   _____
- TOTAL

STUDENT'S SCORE _____

SECTION 9.2 Cash Discounts

The **terms of payment** (or terms) for a business transaction state the time within which the purchaser must pay the total amount of the invoice to avoid a penalty. The terms are normally found in the upper part of an invoice. Often, the purchaser will have 30 days to pay an invoice without a penalty, but the time may vary, depending on company policy. If the terms are **COD (cash on delivery)**, payment must be made upon delivery of the goods. The payment for purchased goods is often called the **remittance**.

If the purchaser does not pay the invoice within the specified period, the vendor will declare the account **past due** and charge the penalty. The penalty is charged because the purchaser is using the vendor's merchandise (and therefore the vendor's money). The penalty is actually an interest charge for the use of the merchandise. Often, the vendor's company policy regarding the penalty is printed at the bottom of the invoice. The following are some typical policies:

1. 6% charge on past due accounts.
2. Interest is charged at the rate of 1% per month on past due accounts.
3. Any invoice not paid by the 25th of the month following the date of the invoice will be subject to a $1\frac{1}{2}\%$ finance charge. This is an annual percentage rate of 18%.

Sometimes, the terms of payment include a provision for a cash discount. The discount is stated as two numbers (for example, 2-10). The first number is a percent (the rate of the discount), and the second number is the number of days within which the purchaser must pay the invoice to receive the discount. The length of the discount period is always less than the length of the payment period. For a typical 30-day payment period, the discount period might be 10 days. The primary purpose of the discount period is to encourage the purchaser to pay the invoice promptly. As a result, the discount reduces the cost of the goods to the purchaser; of course, it also reduces the revenue to the vendor.

PART A The Discount Periods for Cash Discounts

There are several ways to express the discount periods for payment terms that provide cash discounts.

Ordinary Dating. In **ordinary dating**, the rate and number of days in the cash-discount period are followed by the total number of days the purchaser can pay the invoice without penalty. For example, *2-10, n-30* means that 2% of the net amount of the invoice may be deducted if the complete payment is made on or before the 10th day after the invoice date; the net amount (n), or *net*, is due within 30 days after the invoice date. Other ways to write these terms are:

$$2/10, \text{ n}/30$$
$$2/10, \text{ N-30}$$
$$2\%\text{-}10\text{-}30$$
$$2\text{-}10, \text{ NET } 30$$

Some companies offer two or more possible dates for taking advantage of cash discounts, such as *5-10, 2-20, n-60*. These terms mean that 5% of the net may be deducted if the complete payment is made within 10 days, 2% of the net may be deducted if the complete payment is made within 20 days, and the net (total) amount must be paid within 60 days to avoid a penalty.

When the last day of the cash-discount period or the payment date of the net amount falls on a nonbusiness day, the next business day is used. The extra days are not counted. And if a shipment is delayed, the purchaser can use the date of receipt of the goods, rather than the invoice date, as the beginning of the discount period.

EXAMPLES:

Find the time between the invoice date and the date of payment. Indicate the percent of discount given.

a. 2-10, n-30; date of invoice is January 30; date of payment is February 7

The time between dates is 8 days; the cash discount is 2%.

b. $\frac{1}{2}\%$-10-30; date of invoice is June 24; date of payment is July 5

The last day of the discount period is July 4, a legal holiday. The time between dates is 11 days. Since the extra day is allowed, the cash discount is $\frac{1}{2}\%$.

c. 2/10, 1/20, n/30; date of invoice is August 19; date of payment is September 5

The time between dates is 17 days; the cash discount is 1%.

d. 2-10, NET 30; date of invoice is October 7; date of payment is November 4

The time between dates is 28 days; there is no cash discount and no penalty.

e. 3/10, n/30; date of invoice is January 28; date of payment is February 28

The time between dates is 31 days; there is no cash discount. The invoice is past due and subject to penalty.

Extra Dating. When goods are sold before they are in season, the period allowed for the cash discount may be extended. This is called **extra dating**. Thus, *2/10-90 extra* means that a 2% cash discount can be taken if the invoice is paid within 100 days (10 + 90), from the date of the invoice. Instead of *extra*, the symbol *X* is often used; for example, 2/10-90X. If the cash discount is not taken, the net amount is assumed to be due 20 days after the last day of the discount period. Due dates vary among vendors. However, each vendor will make the due date explicitly clear to a purchaser.

EXAMPLE:

The terms on an invoice dated March 15 are 2/10-60X. Find the last day of the discount period and the last day for paying the invoice without penalty.

The discount period is 10 days plus 60 days, or 70 days. The last day of the discount period is May 24, which is 70 days after March 15. The invoice is due on June 13, which is 20 days after May 24.

ROG Dating. When a vendor knows that shipments may be delayed, **receipt-of-goods (ROG) dating** is often used instead of ordinary dating. For example, under the terms 3-10 ROG, the date of receipt of the goods, not the invoice date, is used as the first day of the discount period. Usually, the net period is not stated but is assumed to be 30 days after the receipt of goods.

EXAMPLES:

Find the number of days between the first day of the discount period and the date of payment. Indicate the percent of discount given.

a. 2-10 ROG (received April 24); date of invoice, March 9; date of payment, May 4

The time between April 24 and May 4 is exactly 10 days; the cash discount is 2%.

b. 2/10 ROG (received Dec. 28); date of invoice, November 30; date of payment, January 25

The time between December 28 and January 25 is 28 days; there is no cash discount and no penalty.

EOM Dating. In **end-of-month (EOM) dating**, the discount period starts immediately after the end of the month in which the invoice is dated. The usual discount period is 10 days. For example, on an invoice dated March 2 with terms of 2-10 EOM, the last day for taking advantage of the cash discount is April 10, which is 10 days after March 31. If the cash discount is not taken, the net amount is usually due 30 days after the end of the invoice month. In this example, the due date is April 30, which is 30 days after March 31.

If the invoice date is the 26th or any later date in the month, some companies allow an extra month for the discount period. For example, on an invoice dated March 26 with terms of 2-10 EOM, the discount period ends on May 10. As previously noted, if the discount is not taken, the business usually allows an additional 20 days for payment without penalty. In this case, the due date without penalty is May 30, which is 20 days after May 10. To simplify their terms, some companies that use EOM dating will specify the due date for payment without penalty as the last day of the discount month. Thus, in this example, the due date of the net amount would be May 31.

Instead of *EOM*, the term **prox.** is sometimes used. It stands for the Latin *proximo mense*, which means *in the next month*.

EXAMPLES:

Determine if a cash discount can be taken. Indicate the percent of discount given.

a. 3-10 EOM; date of invoice, May 31; date of payment, June 9

June 9 is 9 days after May 31; the cash discount of 3% can be taken.

Chapter 9   Purchase Orders, Invoices, and Discounts

b. 2-10 EOM; date of invoice, September 27; date of payment, November 10

Because the invoice date is after the 26th, an extra month is allowed for the discount period. The cash discount of 2% can be taken.

c. 1-20 prox.; date of invoice, May 5; date of payment, July 5

June 20 is the last day for taking advantage of the cash discount, and June 30 is the due date of the net amount. Therefore, a penalty may be charged.

Other Ways of Expressing Terms of Payment. In the last few years, there has been a trend to eliminate the formal terms of payment on the top part of the invoice. Instead, the cash discount and the last day to take it are included in a printed statement at the bottom of the invoice.

EXAMPLES:

a. IF PAID BY THIS DATE   10/10/--,
   TAKE THIS DISCOUNT $4.25

b. Cash discount as marked, if paid by the 10th of the month.

   TERMS: Net if paid by the last date of the month.

   Add 1% per month service charge if paid thereafter.

   IF INVOICE IS PAID BEFORE 10th OF NEXT MONTH, YOU SAVE $3.18.

Note that the terms in *Examples a* and *b* are similar to EOM, since the 10th of the month following the invoice month is used as the last day of the discount period.

PART B   Calculating Cash Discounts

When the amount due on an invoice is paid within the cash-discount period, the purchaser deducts the discount amount when making the remittance. To find the amount of the cash discount, multiply the cash discount percent by the net amount of the invoice.

EXAMPLES:

Find the remittance for each of the following invoices.

a. Invoice for $750; dated March 26; terms of 2/10, NET 30; paid on April 5

Cash discount = $750 × 2%
             = $15.00

Remittance = $750.00 − $15.00
           = $735.00

b. Invoice for $832.50; dated April 27; terms of 1/10-90X; paid on August 4

Cash discount = $832.50 × 1%
             = $8.325

Remittance = $832.50 − $8.325
           = $824.175 *or* $824.18

c. Invoice for $1,325.71; dated October 6; terms of 2/10 ROG; paid on January 3; goods received on December 26

Cash discount = $1,325.71 × 2%
             = $26.5142

Remittance = $1,325.71 − $26.5142
           = $1,299.1958 *or* $1,299.20

d. Invoice for $479.80; dated January 15; terms of 3/10 EOM; paid on February 10

Cash discount = $479.80 × 3%
             = $14.394

Remittance = $479.80 − $14.394
           = $465.406 *or* $465.41

Note that in each of the four preceding examples, no numbers were rounded off until the final answer. This is a good practice. In some problems, a mistake in rounding off an intermediate answer becomes even larger as you continue solving the problem. Likewise, the following calculator procedures do not include any intermediate rounding off, either.

On calculators with a percent key %, the net payment can usually be calculated in a single sequence of entries, without any intermediate steps or clearing. Unfortunately, this key does not work the same way on every calculator. The particular sequence of keystrokes will depend on the brand and model of the calculator.

Three common sequences are shown, where AMT is the total net amount of the invoice and RATE is the discount rate. If, on your calculator, one of these sequences results in the net payment, then the other two will probably not result in the net payment. Some sequences may also display the amount of the cash discount after some step, but this also varies among brands and models of calculators.

1. [AMT] [−] [RATE] [%] [=]

After the [%] key is pressed, the amount displayed is the amount of the cash discount. After the [=] key is pressed, the amount displayed is the net payment.

2. [AMT] [−] [RATE] [%]

After the [%] key is pressed, the amount displayed is the net payment.

3. [AMT] [×] [RATE] [%] [−]

After the [%] key is pressed, the amount displayed is the amount of the cash discount. After the [−] key is pressed, the amount displayed is the net payment.

It is important that you practice this procedure on your own calculator and that you read the instruction manual that comes with it. When you have determined the appropriate sequence of keystrokes for your calculator, go back and rework *Examples a, b, c,* and *d* with the calculator.

When a Portion of the Goods Is Returned. A cash discount is only allowed on goods actually purchased. If some goods included in the invoice total will be returned, subtract them from the invoice total before calculating the cash discount.

EXAMPLE:

An invoice for $564.28, with terms of 2-10, NET 30, includes returned goods worth $87.58. Find the remittance if the invoice is paid within the discount period.

Actual purchases = $564.28 − $87.58
 = $476.70

Cash discount = $476.70 × 2%
 = $9.534

Remittance = $476.70 − $9.534
 = $467.166 *or* $467.17

When Prepaid Freight Charges Are Paid by the Purchaser. A vendor sometimes includes prepaid freight charges in the invoice total. Remember that a cash discount is offered by the vendor, *not* by the freight company. This discount applies only to the merchandise, *not* to the freight charges. The freight charges are on the invoice because the vendor has already paid them, and the purchaser now owes the vendor that entire amount.

Deduct any prepaid freight charges from the invoice total before calculating the cash discount. Then, calculate the cash discount, based on the remainder of the invoice amount, and subtract it from that remainder. Finally, add back the freight charge to get the amount of the remittance.

EXAMPLE:

An invoice for $965.83, with terms of 1-10 EOM, includes a freight charge of $71.33. Find the remittance if the invoice is paid within the discount period.

Value of goods = $965.83 − $71.33
 = $894.50

Cash discount = $894.50 × 1%
 = $8.945

Remittance = $894.50 − $8.945 + $71.33
 = $956.885 *or* $956.89

When Only a Portion of the Invoice Is Paid Within the Discount Period. Sometimes, a purchaser does not have sufficient funds available during the discount period to pay the entire invoice but is able to pay part of it. The cash discount applies only to that amount paid within the discount period. The purchaser must decide both how much to leave as a remaining balance and how much of an actual remittance to mail to the vendor. Compare the following examples:

Chapter 9   Purchase Orders, Invoices, and Discounts

EXAMPLES:

a. An invoice for $1,800 has terms of 2-10, NET 30. The purchaser wants to pay enough during the discount period to have a remaining balance of $1,000. Find the cash discount and the remittance.

The purchaser wants to receive a total credit of $800 ($1,800 − $1,000) on the account. The discount will be 2% of the $800.

Cash discount = $800 × 2%
             = $16.00

Remittance = $800.00 − $16.00
           = $784.00

Since 2% of the $800 is the amount of the cash discount, the remittance is 98% of the $800. *The amount of the remittance is 98% of the amount credited ($800).*

Remittance = $800 × 98%
           = $784.00

b. An invoice for $1,800 has terms of 2-10, NET 30. The purchaser wants to mail the vendor a check for $800 during the discount period. Find the unpaid balance of the invoice and the amount of the cash discount.

The purchaser wants to make an actual remittance of $800. *The amount of the remittance ($800) is 98% of the amount credited.* Stated like the problems in Chapter 3, *$800 is 98% of what amount?* This problem is solved by dividing $800 by 98%.

Remittance = $800

Amount credited = $800 ÷ 98%
               = $816.3265 *or* $816.33

Unpaid balance = $1,800.00 − $816.33
              = $983.67

Cash discount = $816.33 − $800.00
             = $16.33

COMPLETE ASSIGNMENT 29

ASSIGNMENT 29
SECTION 9.2

Cash Discounts

	Perfect Score	Student's Score
PART A	40	
PART B	60	
TOTAL	100	

Name _____

Date _____

PART A The Discount Periods for Cash Discounts

DIRECTIONS: Given the terms, the date of invoice, and the date of remittance in the first three columns, place a check mark on the lines provided to show if a cash discount was allowed, if the net amount was due, or if the net amount was past due (2 points for each correct answer).

	Terms	Date of Invoice	Date of Remittance	Cash Discount	Net Due	Past Due
1.	2/10, NET 30	Mar. 30	Apr. 9			
2.	3/10, 1/20, n/60	Sept. 21	Oct. 11			
3.	2%-10-30	Oct. 17	Nov. 17			
4.	2-10, 1-30, n-60	Dec. 14	Feb. 13			
5.	2-10, 60 extra	Mar. 15	June 10			
6.	3/10, 90X	Apr. 1	July 14			
7.	2/10 ROG (received Nov. 18)	Sept. 15	Dec. 18			
8.	3-10 ROG (received July 11)	June 14	July 20			
9.	1/20 ROG (received July 27)	May 7	Aug. 13			
10.	2/10 EOM	Jan. 18	Feb. 10			
11.	4-10 EOM	Apr. 15	May 12			
12.	1/20 EOM	Aug. 12	Sept. 19			
13.	1/20, n/60	July 10	Sept. 8			
14.	1/10, 2/20, n/30	Mar. 25	Apr. 14			
15.	3/20, 90X	Jan. 10	Apr. 30			
16.	1/10 prox.	Oct. 26	Nov. 9			
17.	3/15, 1/30, n/60	July 25	Aug. 10			
18.	3/10 prox.	June 19	July 19			
19.	2-10, 1-20, n-30	Aug. 25	Sept. 25			
20.	5/10, 3/20, n/30	May 13	June 3			

STUDENT'S SCORE _____

PART B Calculating Cash Discounts

DIRECTIONS: Assume that payment on the invoice is to be made on the last day of the discount period. Calculate the answers for the blanks below. The column labeled Amount Credited means the sum of the remittance and the cash discount (2 points for each correct answer).

Terms	Invoice Total	Amount Credited	Date of Invoice	Date of Remittance	Cash Discount	Amount of Remittance
21. 1/10, NET 30	$1,250.00	$1,250.00	Aug. 28			
22. 2%-20-60	$875.77 (includes $48.32 for freight)	$ 875.77	Oct. 23			
23. 3/15, n/30	$654.73 (includes return of $184.40)	Total due	Dec. 24			
24. 2-10, 1-20, n-30	$2,425.00	$1,500.00	Jan. 22			
25. 1-10 ROG	$743.90	Total due	July 12 (received Aug. 27)			
26. 2/10 ROG	$828.15		May 8 (received June 25)			$ 400.00
27. ½-20 ROG	$2,218.00 (includes $124.56 for freight)	$2,218.00	Aug. 18 (received Nov. 21)			
28. 2-10 EOM	$952.75	$ 450.00	May 19			
29. 1-20 EOM	$1,765.86		Feb. 26			$1,000.00
30. 3/15 prox.	$3,256.48	$2,500.00	Oct. 13			

STUDENT'S SCORE _____

Chapter 9 Purchase Orders, Invoices, and Discounts

SECTION 9.3 Trade Discounts

Retailers quote the prices of their merchandise in dollars and cents. Many, if not most, manufacturers and distributors do likewise. Some of the latter group, however, quote their prices in terms of a suggested retail price *less* one or more *trade discounts*. The result is a *wholesale price*.

A distributor may offer more than one trade discount because he or she might have several different wholesale prices for the same item, depending on the customer.

Trade discounts are not the same as cash discounts. Trade discounts are given to a customer before the invoice is even written. The customer qualifies for the trade discount by being in a specific business or by selling specific quantities of products at the retail level. As you learned earlier, cash discounts are reductions that are subtracted from the invoice. A customer qualifies for the cash discount by making an early payment of the invoice.

Manufacturers and distributors usually send retailers catalogs containing pictures, descriptions, and stock or model numbers of their products. The **suggested retail price**, also called the **catalog** or **list price**, may also be shown in a catalog. In addition, manufacturers and distributors send retailers separate sheets which show the trade discount rates offered on the products. For example, a refrigerator may be listed at $498 in the catalog, with a trade discount of 40% on the discount sheet. The **net price** to the retailer can be found by deducting the amount of the trade discount from the list price: $498.00 − $199.20 = $298.80.

Using the terms *net cost* or *net price* may depend on whether you are the buyer or the seller. If you are the buyer, you would be more likely to think of the price as *your cost*. However, the price is certainly not the *cost* to the seller. In this section, we will use the term *net price* from the seller's viewpoint.

There are advantages to quoting unit prices in catalogs and mailing out separate discount sheets. For example, retailers can find their net cost by applying the trade discounts. For items which they do not carry, retailers can show the catalog to their customers without disclosing the net cost of these items.

Prices that retailers must pay fluctuate throughout the year. If the actual prices to the retailer were printed in a catalog, then a new catalog would be necessary whenever the prices changed. However, the catalog can remain the same until the actual list of merchandise changed if separate sheets for discount rates, and even for list prices themselves, are printed. By doing this, distributors save on the cost of printing catalogs and can make changes more quickly. Also, separate discount sheets can be prepared periodically for only those items that are out-of-date or unpopular.

To increase the price to retailers, the distributor will decrease the trade discount rate. To decrease the price to retailers, the distributor will increase the trade discount rate.

PART A Finding the Net Price with a Single Trade Discount

There are two methods of finding the net price when a single trade discount is offered.

Method 1. The steps for the first method are as follows:

1. Multiply the list price by the trade discount rate to find the amount of the trade discount.
2. Subtract the amount of the trade discount from the list price to find the net price.

EXAMPLES:

Find the net price.

a. A television set is listed at $359.50, less a trade discount of 30%.

   Trade discount = $359.50 × 30%
                  = $107.85

   Net price = $359.50 − $107.85
             = $251.65

b. A power saw is listed at $189.90, less a trade discount of 40%.

   Trade discount = $189.90 × 40%
                  = $75.96

   Net price = $189.90 − $75.96
             = $113.94

On calculators with a percent key %, the net price can usually be calculated in a single sequence of entries, without any intermediate steps or clearing. As indicated in Section 9.2, the particular sequence of steps will depend on the brand and model of the calculator.

Three common sequences are shown, where LP is the list price, or the total net amount, of the invoice and RATE is the discount rate. If one of these sequences results in the net price, then the other two will probably not result in the net price. Some sequences may also display the amount of the trade discount after some step, but this also varies among brands and models of calculators.

1. [LP] [−] [RATE] [%] [=]

After the [%] key is pressed, the amount displayed is the amount of the trade discount After the [=] key is pressed, the amount displayed is the net price.

2. [LP] [−] [RATE] [%]

After the [%] key is pressed, the amount displayed is the net price.

3. [LP] [×] [RATE] [%] [−]

After the [%] key is pressed, the amount displayed is the amount of the trade discount. After the [−] key is pressed, the amount displayed is the net price.

Method 2. The steps for the second method of finding the net price when there is a single trade discount are as follows:

1. Subtract the trade discount rate from 100%. The difference is called the **complement** of the discount rate.
2. Multiply the complement of the discount rate by the list price.

Use this method to find the net price in the two preceding examples.

EXAMPLES:

Find the net price.

a. A television set is listed at $359.50, less a trade discount of 30%.

$$\text{Complement of } 30\% = 100\% - 30\%$$
$$= 70\%$$

$$\text{Net price} = \$359.50 \times 70\%$$
$$= \$251.65$$

b. A power saw is listed at $189.90, less a trade discount of 40%.

$$\text{Complement of } 40\% = 100\% - 40\%$$
$$= 60\%$$

$$\text{Net price} = \$189.90 \times 60\%$$
$$= \$113.94$$

When a group of items has the same trade discount rate, the rate is shown next to the last item in the group on the discount sheet. The net price is found by multiplying the complement of the rate by the total of the list prices.

The invoice in Table 9-1 is designed specifically to show list prices, list totals, trade discount rates,

Quantity	Description	List	Total	Discount	Net
4	Grass trimmer #2450	$55.95 ea	$223.80		
10	Hedge clipper #6160	29.50 ea	295.00	40%	
			518.80		$311.28
720	Plastic gloves #G-25	2.50/doz	150.00	25%	112.50
					$423.78

TABLE 9-1 Sample Invoice with Trade Discounts

Chapter 9 Purchase Orders, Invoices, and Discounts

and the net price of each group of items. The first trade discount of 40% is taken on the total of the first two items, and their total net price is placed in the *Net* column.

PART B Finding the Net Price with a Chain of Discounts

Trade discounts are sometimes offered as a series of discounts, called **chain discounts**. They may be written in several ways:

$$40\%, 20\%, 10\%$$
$$40\%\text{-}20\%\text{-}10\%$$
$$40/20/10$$

Only the first format is used in Section 9.3.

Chain discounts provide a means of changing a net price to retailers. Chain discounts are also useful when the discounts allowed depend on the size of the order. For example, 40% may be given on orders under $100; 40%, 20% may be given for orders from $100 to $500; and 40%, 20%, 10% may be given for orders more than $500.

Like finding the net price when a single discount is offered, there are two methods of finding the net price when chain discounts are offered.

Method 1. The first method consists of the following steps:

1. Multiply the list price by the first discount rate and subtract the product from the list price to get the *first net price*.
2. Multiply the first net price by the second discount rate and subtract the product from the first net price. Repeat this procedure until you have deducted all the discounts.
3. To find the total amount of the trade discount, subtract the final net price from the list price.

Do not round off until you obtain the final answer. Since the order of multiplication does not matter, you may take the discounts in any convenient order, not necessarily as listed. The same net price will result.

EXAMPLES:

Find the net price and the total trade discount.

a. A gas-powered weed cutter is listed at $79.95, less discounts of 40% and 20%.

1st discount = $79.95 × 40%
 = $31.98

1st net price = $79.95 − $31.98
 = $47.97

2nd discount = $47.97 × 20%
 = $9.594

Final net price = $47.97 − $9.594
 = $38.376 *or* $38.38

Total discount = $79.95 − $38.38
 = $41.57

b. A power lawn mower is listed at $142.50, less discounts of 30%, 20%, and 10%.

1st discount = $142.50 × 30%
 = $42.75

1st net price = $142.50 − $42.75
 = $99.75

2nd discount = $99.75 × 20%
 = $19.95

2nd net price = $99.75 − $19.95
 = $79.80

3rd discount = $79.80 × 10%
 = $7.98

Final net price = $79.80 − $7.98
 = $71.82

Total discount = $142.50 − $71.82
 = $70.68

On a calculator with a percent key %, extend the procedure for a single trade discount (see page 311) to the remaining discounts.

Method 2. The second method of finding the net price when chain discounts are offered consists of the following steps:

1. Find the complement of each discount rate.
2. Multiply the list price by each of the complements in succession. The final product is the final net price.

The two preceding examples are used here so that you can compare the results of the two methods.

EXAMPLES:

Find the net price.

a. A gas-powered weed cutter is listed at $79.95, less discounts of 40% and 20%.

Complement of 40% = 100% − 40%
= 60%
Complement of 20% = 100% − 20%
= 80%

1st net price = $79.95 × 60%
= $47.97
Final net price = $47.97 × 80%
= $38.376 *or* $38.38

b. A power lawn mower is listed at $142.50, less discounts of 30%, 20%, and 10%.

Complement of 30% = 100% − 30%
= 70%
Complement of 20% = 100% − 20%
= 80%
Complement of 10% = 100% − 10%
= 90%

1st net price = $142.50 × 70%
= $99.75
2nd net price = $99.75 × 80%
= $79.80
Final net price = $79.80 × 90%
= $71.82

PART C Using Net Price Factors for Chain Discounts

When the same chain discounts occur repeatedly, you can save time by using the **net price factor** for the series. The net price factor is the product of the complements of the discount rates, expressed as a decimal. Actually, the net price factor is equal to $1.00 less the series of discounts. The net price factor is also called the *net cost factor*.

EXAMPLE:

Show that $1.00 less 30% and 20% is the same as the product of the complements of 30% and 20%.

Complement of 30% = 100% − 30%
= 70%
Complement of 20% = 100% − 20%
= 80%
Net price factor = 70% × 80%
= 56% *or* 0.56

$1.00 × 30% = $0.30
$1.00 − $0.30 = $0.70
$0.70 × 20% = $0.14
$0.70 − $0.14 = $0.56

To find the net price factor for a series of trade discounts, first find the complement of each rate. Then, find the product of the complements and change that product to its decimal form.

Using the Net Price Factor When One Item Is Listed with Chain Discounts. To find the net price of an item listed with chain discounts, multiply the list price by the net price factor for the series.

EXAMPLE:

Find the net price of a gas-powered weed cutter listed at $79.95, less discounts of 40% and 20%, by using the net price factor.

Complement of 40% = 100% − 40%
= 60%
Complement of 20% = 100% − 20%
= 80%

Net price factor = 60% × 80%
= 48% *or* 0.48

Net price = $79.95 × 0.48
= $38.376 *or* $38.38

Note that this result is identical to the answers in *Example a* of Part A and Part B in this section.

Using the Net Price Factor When Several Items Have the Same Series of Discounts. When a number of items have the same series of discounts, multiply the total list price of the items by the net price factor for the series. On an invoice, the chain discounts appear under the last item in the group. The net price of the group is put in the *Net* column.

Chapter 9 Purchase Orders, Invoices, and Discounts

EXAMPLE:

Find the net price on the following invoice by using the net price factor for the series.

Qty.	Description	Unit Price	Total	Net
25	Gate valve	$6.78	$169.50	
15	Globe valve	7.39	110.85	
20	Ball valve	7.89	157.80	
10	Sink faucet	5.75	57.50	
			$495.65	
Less 40%, 20%, 10%				$214.12

Net price factor = $0.6 \times 0.8 \times 0.9$
                 = 0.432

Net price = 495.65×0.432
          = $214.1208 or $214.12

Finding the Equivalent Single Discount of a Series.

Occasionally, a vendor may want to know the single discount which is equivalent to a series. To find the **equivalent single discount** of a series, follow these steps:

1. Find the net price factor for the series and change it to a percent.
2. Subtract that percent from 100%.

Thus, the equivalent single discount of the 40%, 20%, 10% chain used in the preceding example is calculated as follows:

Net price factor = $0.6 \times 0.8 \times 0.9$
                 = 0.432 *or* 43.2%

Equivalent single discount = 100% − 43.2%
                           = 56.8%

Using a Net Price Factor Table. Prior to the widespread use of calculators, net price factor (NPF) tables were often used to compute the net price of an item on which chain discounts were offered. Table 9-2 shows a partial net price factor table. These tables are still used, because some companies find it easier to train their employees to use the table than to train them to find the complements and multiply them together.

In a net price factor table, a single discount heads each column, and various combinations of discounts are listed in the column at the extreme left. To find the NPF for a series, first locate any one of the series discount rates in the top row. Then, in the column at the extreme left, find the combination that contains the remaining discount rates. Go across that row to the column under the first discount rate you located. The number at the intersection of that row and column is the NPF. Remember that the order of the discounts is unimportant, since multiplication can occur in any order.

EXAMPLES:

a. Find the net price factor for 25%, 10%, 5%.

   Go down the 25% column and across the 10–5 row. The NPF is 0.64125.

b. Find the net price of an air compressor listed at $375, less 20%, 10%, 10%.

   NPF = 0.648 (from Table 9–2)

   Net price = 375×0.648
             = $243

Rate %	5%	10%	15%	20%	25%
5	0.9025	0.855	0.8075	0.76	0.7125
5–5	0.85738	0.81225	0.76713	0.722	0.67688
10		0.81	0.765	0.72	0.675
10–5		0.7695	0.72675	0.684	0.64125
10–10		0.729	0.6885	0.648	0.6075

TABLE 9-2 *Partial Net Price Factor Table for $1.00*

COMPLETE ASSIGNMENT 30

Name _____

Date _____

ASSIGNMENT 30
SECTION 9.3

Trade Discounts

	Perfect Score	Student's Score
PART A	30	
PART B	30	
PART C	40	
TOTAL	100	

PART A Finding the Net Price with a Single Trade Discount

DIRECTIONS: Find the net price by using the method indicated. Show your calculations, and write your answers on the lines provided (2 points for each correct answer).

Multiply by the rate:

	List Price	Trade Discount	Amount of Discount	Net Price
1.	$475.00	20%	_____	_____
2.	$250.60	25%	_____	_____
3.	$ 84.75	40%	_____	_____
4.	$842.80	35%	_____	_____
5.	$168.00	45%	_____	_____

Multiply by the complement of the rate:

	List Price	Trade Discount	Net Price
6.	$421.40	35%	_____
7.	$ 84.75	40%	_____
8.	$900.00	20%	_____
9.	$750.00	45%	_____
10.	$125.30	25%	_____

STUDENT'S SCORE _____

PART B Finding the Net Price with a Chain of Discounts

DIRECTIONS: Find the net price by using the method indicated. Show your calculations, and write your answers on the lines provided (2 points for each correct answer).

Multiply by the discounts:

	List Price	Chain Discounts	Amount of Trade Discount	Net Price
11.	$425.00	30%, 20%	_____	_____
12.	$74.50	25%, 10%	_____	_____
13.	$138.80	30%, 20%, 10%	_____	_____
14.	$48.20	30%, 15%, 5%	_____	_____
15.	$240.00	40%, 20%, 10%, 5%	_____	_____

Multiply by the complements of the discounts:

	List Price	Chain Discounts	Net Price
16.	$49.00	25%, 10%	_____
17.	$112.50	30%, 20%	_____
18.	$196.40	30%, 15%, 5%	_____
19.	$66.40	30%, 20%, 10%	_____
20.	$500.00	40%, 20%, 10%, 5%	_____

STUDENT'S SCORE _____

Chapter 9 Purchase Orders, Invoices, and Discounts

PART C Using Net Price Factors for Chain Discounts

DIRECTIONS: Find the net price factors and the net prices. Use Table 9-2 where possible. Show your calculations, and write your answers on the lines provided (1 point for each correct answer).

	List Price	Chain Discounts	Net Price Factor	Net Price
21.	$150.00	20%, 10%	_____	_____
22.	$75.50	25%, 20%	_____	_____
23.	$110.00	10%, 10%	_____	_____
24.	$280.00	10%, 5%	_____	_____
25.	$24.60	15%, 5%	_____	_____
26.	$200.00	50%, 25%, 10%	_____	_____
27.	$500.00	40%, 20%, 10%	_____	_____
28.	$1,200.00	20%, 10%, 5%	_____	_____
29.	$640.00	40%, 30%, 20%	_____	_____
30.	$480.00	25%, 20%, 10%	_____	_____

Unit 2   Applications

DIRECTIONS: Extend each item, take the trade discounts indicated, round to the nearest cent, and find the net invoice due. Write your answers on the lines provided (1 point for each correct answer).

31.

Quantity	Description	List Price	Amount	Net
5 doz	PVC cap 1/2"	$ 0.89	_____	
3 doz	PVC plug 1/2"	0.69	_____	
10 doz	PVC threaded coupling 1/2"	0.45	_____	
	(Subtotal)		_____	
	Less 25%, 10%, 10%..			_____
12	PVC Pipe cutter	12.95	_____	
8	Pipe wrench 10"	26.75	_____	
	(Subtotal)		_____	
	Less 30%, 10%			_____
	Net Invoice Due			_____

32.

Quantity	Description	List Price	Amount	Net
8	M-81D metal snips 12"	$ 18.50	_____	
4	C-15P tubing cutter	13.75	_____	
2	BC-8K bolt cutter 24"	95.00	_____	
	(Subtotal)		_____	
	Less 20%, 10%, 5%			_____
2	S-34T belt sander	138.85	_____	
6	B-374R elec drill 3/8"	127.50	_____	
	(Subtotal)		_____	
	Less 40%, 20%..			_____
	Net Invoice Due..			_____

STUDENT'S SCORE _____

CHAPTER 10

Selling Goods

Manufacturers, wholesale distributors, and retailers are in business to sell their products at a profit. Therefore, their total sales revenue must be high enough to cover three elements: the cost of the goods sold, the expenses related to selling them, and the desired net profit.

Markup is the difference between the cost of the goods and their selling price. Markup is also called **markon, margin on sales**, or **gross profit**. A markup may be expressed as a dollar amount. It may also be written as a percent (called the *markup rate*) of either the cost or the selling price of the merchandise.

Section 10.1 deals with the relationships among selling price (SP), cost (C), markup (M), and markup rate (MR). The relationships among these elements are presented in a basic equation, or formula. This equation can be used whether the markup rate is based on cost or on selling price.

After the cost of the goods is subtracted from sales revenue, the amount remaining is the markup. All of the **operating expenses** (such as salaries, advertising, office supplies, selling expenses, utilities, etc.) are deducted from the markup. The amount remaining after all operating expenses are deducted is the **net profit**. If the total operating expenses exceed gross profit or markup, then the business has a **net loss**.

Section 10.2 deals with the relationships among selling price (SP), cost (C), operating expenses (OE), and net profit (NP) or net loss (NL). These relationships are presented in expanded forms of the basic equation introduced in Section 10.1.

SECTION 10.1 Selling Price, Cost, and Markup Rate

As stated earlier, markup can be expressed as either a dollar amount or a percent, called the markup rate. When markup is expressed as an amount, the basic relationship among markup (M), cost (C), and selling price (SP) is given by the following equation:

$$SP = C + M$$

This equation states, for example, that an item that costs $100 (C) and is marked up by $25 (M) will sell for $125 (SP).

When the markup rate (MR) is used, it is expressed as a percent of a base. The base can be either the selling price or the cost. Many businesses, especially retailers, use selling price as the base for markup. Businesses that might use cost as the base for markup include some wholesale distributors, manufacturers, and drug companies. Retail distributors of large or expensive products, such as furniture, appliances, jewelry, etc., may also use cost as the base for markup.

To find the amount of markup (M) whether the base is cost or selling price, multiply the base (C or SP) by the markup rate (MR). In equation form, these relationships are as follows:

$$M = C \times MR \text{ (based on cost)}$$

or

$$M = SP \times MR \text{ (based on selling price)}$$

The first equation states that if an item costs $100 (C) and is marked up by 25% (MR) of cost, then the amount of the markup (M) is $25.

$$\begin{aligned} M &= C \times MR \\ &= \$100 \times 25\% \\ &= \$100 \times 0.25 \\ &= \$25 \end{aligned}$$

Likewise, the second equation states that if an item sells for $125 (SP) and is marked up by 20% (MR) of selling price, then the amount of the markup (M) is $25.

$$\begin{aligned} M &= SP \times MR \\ &= \$125 \times 20\% \\ &= \$125 \times 0.20 \\ &= \$25 \end{aligned}$$

PART A Cost as the Base for Markup

The equation $SP = C + M$ is a relationship among three terms. If any two of the terms are known, the third term can always be calculated in a minimum of two steps. As you will see, it is sometimes possible to reduce the number of steps by combining two equations.

Finding the Selling Price. When the cost and the markup rate based on cost are known, the selling price can be found in two steps:

1. Find the amount of markup by multiplying the cost by the markup rate.

$$M = C \times MR$$

2. Find the selling price by adding the amount of markup to the cost.

$$SP = C + M$$

EXAMPLES:

Find the selling price if the markup is based on cost.

a. The cost is $50; the markup rate is 30%.

$$\begin{aligned} \text{Step 1.}\ M &= C \times MR \\ &= \$50 \times 30\% \\ &= \$15 \end{aligned}$$

$$\begin{aligned} \text{Step 2.}\ SP &= C + M \\ &= \$50 + \$15 \\ &= \$65 \end{aligned}$$

b. The cost is $75; the markup rate is 40%.

$$\begin{aligned} \text{Step 1.}\ M &= C \times MR \\ &= \$75 \times 40\% \\ &= \$30 \end{aligned}$$

$$\begin{aligned} \text{Step 2.}\ SP &= C + M \\ &= \$75 + \$30 \\ &= \$105 \end{aligned}$$

The arithmetic from Steps 1 and 2 can be combined as follows:

$$\begin{array}{ll} M = C \times MR & SP = C + M \\ \$30 = \$75 \times 40\% \quad \text{and} \quad & \$105 = \$75 + \$30 \end{array}$$

If you substitute $75 \times 40\%$ for the $30 on the right, the equation becomes

$$\$105 = \$75 + (\$75 \times 40\%)$$

In Chapter 3, you learned that $100\% = 1$; therefore, $\$75 = \$75 \times 100\%$. Substituting $75 \times 100\%$ for $75 in the preceding equation, you can write:

$$\$105 = (\$75 \times 100\%) + (\$75 \times 40\%)$$

This equation can be rewritten as:

$$\$105 = \$75 \times (100\% + 40\%)$$

or

$$\$105 = \$75 \times 140\%$$

Using symbols in place of the numbers, you get the following equation:

$$SP = C \times (100\% + MR)$$

This equation is useful when the selling price is the unknown term and you do not need to know the actual dollar amount of markup. Use this equation to find the selling price in *Example a*, where cost was $50 and the markup rate was 30%:

$$\begin{aligned} SP &= C \times (100\% + MR) \\ &= \$50 \times (100\% + 30\%) \\ &= \$50 \times 130\% \\ &= \$50 \times 1.3 \\ &= \$65 \end{aligned}$$

Finding the Markup Rate. Sometimes, a business owner bases the selling price of an item on a competitor's selling price for a similar item. In this situation, the cost and the selling price are both known, and the markup rate must be calculated. This can be done in two steps.

1. Find the amount of markup by subtracting the cost from the selling price.

$$M = SP - C$$

2. Find the markup rate by dividing the amount of markup by the cost and converting that quotient to a percent.

$$MR = \frac{M}{C} \text{ or } MR = M \div C$$

EXAMPLES:

Find the markup rate based on cost.

a. The cost is $12; the selling price is $18.

Step 1. $M = SP - C$
$= \$18 - \$12$
$= \$6$

Step 2. $MR = M \div C$
$= \$6 \div \$12$
$= 0.5 \text{ or } 50\%$

b. The cost is $140; the selling price is $280.

Step 1. $M = SP - C$
$= \$280 - \$140$
$= \$140$

Step 2. $MR = M \div C$
$= \$140 \div \$140$
$= 1.0 \text{ or } 100\%$

Note that when the amount of markup is equal to the cost (as in *Example b*), the markup rate (based on cost) is 100%.

c. The cost is $72; the selling price is $180.

Step 1. $M = SP - C$
$= \$180 - \$72$
$= \$108$

Step 2. $MR = M \div C$
$= \$108 \div \$72$
$= 1.5 \text{ or } 150\%$

Note that when the amount of markup is greater than the cost (as in *Example c*), the markup rate (based on cost) is greater than 100%.

Finding the Cost. Occasionally, it may be necessary or useful to calculate the cost of an item. Suppose that you are in a business with a 25% average markup rate. Assume that you have a competitor who is selling an item for $60. You can estimate your competitor's cost for the item and compare it with your cost for the same item.

Or, suppose that you have lost your record of an article's cost. If you know the selling price and your markup rate, you can work backward and determine the cost. The calculation is based on the equation that was previously developed, $SP = C \times (100\% + MR)$, and it takes two steps:

1. Add 100% and the known markup rate.

$$100\% + MR$$

2. Find the cost by dividing the selling price by the total percent obtained in *Step 1*.

$$C = \frac{SP}{100\% + MR}$$

or

$$C = SP \div (100\% + MR)$$

Note that the equation in *Step 2* is simply a variation of the earlier equation, $SP = C \times (100\% + MR)$.

This type of problem can be checked by using the reverse logic. Multiply the cost by the markup rate to find the amount of markup. Then, add the amount of markup to the cost. The result should be the given selling price.

EXAMPLES:

Find the cost if the markup is based on cost.

a. The selling price is $80; the markup rate is 60%, based on cost.

Step 1. $100\% + MR = 100\% + 60\%$
$= 160\%$

Step 2. $C = SP \div (100\% + MR)$
$= \$80 \div 160\%$
$= \$80 \div 1.6$
$= \$50$

Check: $C \times MR = M$
$\$50 \times 60\% = \$30$
$C + M = SP$
$\$50 + \$30 = \$80 \text{ (given SP)}$

b. The selling price is $100; the markup rate is 150%, based on cost.

Step 1. $100\% + MR = 100\% + 150\%$
$= 250\%$

Step 2. $C = SP \div (100\% + MR)$
$= \$100 \div 250\%$
$= \$100 \div 2.5$
$= \$40$

Check: $C \times MR = M$
$\$40 \times 150\% = \$60$

$C + M = SP$
$\$40 + \$60 = \$100$ (given SP)

PART B Selling Price as the Base for Markup

In Part A, two equations were used. The fundamental relationship was expressed by $SP = C + M$, and the formula for markup based on cost was $M = C \times MR$. In Part B, the fundamental relationship is still the same. However, a formula for markup based on selling price will be used:

$$M = SP \times MR$$

Recall that if any two of the three terms in the equation are known, then the third term can always be calculated in a minimum of two steps.

Finding the Cost. When the selling price and the markup rate based on the selling price are known, the cost can be found in two steps:

1. Find the amount of markup by multiplying the selling price by the markup rate.

$$M = SP \times MR$$

2. Find the cost by subtracting the markup from the selling price.

$$C = SP - M$$

EXAMPLES:

Find the cost if the markup is based on selling price.

a. The selling price is $50; the markup rate is 30%.

Step 1. $M = SP \times MR$
$= \$50 \times 30\%$
$= \$15$

Step 2. $C = SP - M$
$= \$50 - \$15$
$= \$35$

b. The selling price is $75; the markup rate is 40%.

Step 1. $M = SP \times MR$
$= \$75 \times 40\%$
$= \$30$

Step 2. $C = SP - M$
$= \$75 - \$30$
$= \$45$

On a calculator, the same procedure used to find cash or trade discounts and net prices can be used to find the markup and the cost.

As in Part A, the arithmetic from Steps 1 and 2 can be combined as follows:

$M = SP \times MR$ and $C = SP - M$
$\$30 = \$75 \times 40\%$ $\$45 = \$75 - \$30$

If you substitute $75 × 40% for the $30 on the right, the equation becomes:

$$\$45 = \$75 - (\$75 \times 40\%)$$

Because $100\% = 1$, $\$75 = \$75 \times 100\%$. Substituting $75 × 100% for $75 in the preceding equation, you can write:

$$\$45 = (\$75 \times 100\%) - (\$75 \times 40\%)$$

This equation can be rewritten as:

$$\$45 = \$75 \times (100\% - 40\%)$$

or

$$\$45 = \$75 \times 60\%$$

Using symbols in place of the numbers, you get the following equation:

$$C = SP \times (100\% - MR)$$

This equation is useful when the cost is the unknown term and you do not need to know the actual dollar amount of markup. Use this equation to find the cost in *Example a*, where selling price was $50 and the markup rate was 30%.

$C = SP \times (100\% - MR)$
$= \$50 \times (100\% - 30\%)$
$= \$50 \times 70\%$
$= \$35$

Finding the Markup Rate. When the cost and the selling price are known, the markup rate based on the selling price can be found in two steps:

1. Find the amount of markup by subtracting the cost from the selling price.

$$M = SP - C$$

2. Find the markup rate by dividing the amount of markup by the selling price and converting that quotient to a percent.

$$MR = \frac{M}{SP} \quad or \quad MR = M \div SP$$

EXAMPLES:

Find the markup rate based on selling price.

a. The cost is $12; the selling price is $18.

Step 1. M = SP − C
= $18 − $12
= $6

Step 2. MR = M ÷ SP
= $6 ÷ $18
= 0.333 or $33\frac{1}{3}$%

b. The cost is $140; the selling price is $280.

Step 1. M = SP − C
= $280 − $140
= $140

Step 2. MR = M ÷ SP
= $140 ÷ $280
= 0.5 or 50%

c. The cost is $72; the markup is $108.

Step 1. SP = C + M
= $72 + $108
= $180

Step 2. MR = M ÷ SP
= $108 ÷ $180
= 0.6 or 60%

When markup is based on cost, the markup rate can be greater than 100%. However, when markup is based on selling price, the markup rate can *never* be more than 100%. This is because the markup is a part of the selling price (which is always 100% in the basic equation SP = C + M). In other words, when markup is based on selling price, it is impossible to have a selling price smaller than the markup.

Finding the Selling Price. In retailing, the selling price is generally used as the base for markup, based on the theory that there can be no real profit until the goods are sold. In business, **markup tables** or **markup wheels** are used. They show what the selling price should be for a given unit cost and a given markup rate (based on the selling price).

Without markup tables or markup wheels, you can find the selling price by following these two steps:

1. Find the difference between 100% and the markup rate.

$$100\% - MR$$

2. Find the selling price by dividing the cost by the percent obtained in Step 1.

$$SP = \frac{C}{100\% - MR}$$

or

$$SP = C \div (100\% - MR)$$

You do not find the dollar amount of markup by using this equation. It can be found only after the selling price has been determined.

This type of problem is checked by using reverse logic. Multiply the answer (the selling price) by the markup rate based on the selling price. Then, subtract the markup from the selling price. The result should be the given cost.

EXAMPLES:

Find the selling price if the markup is based on selling price.

a. The cost is $80; the markup rate is 60%, based on the selling price.

Step 1. 100% − MR = 100% − 60%
= 40%

Step 2. SP = C ÷ (100% − MR)
= $80 ÷ 40%
= $80 ÷ 0.4
= $200

Check: SP × MR = M
$200 × 60% = $120
SP − M = C
$200 − $120 = $80 (given cost)

b. The cost is $100; the markup rate is 50%, based on the selling price.

Step 1. 100% − MR = 100% − 50%
= 50%

Step 2. SP = C ÷ (100% − MR)
= $100 ÷ 50%
= $100 ÷ 0.5
= $200

Check: SP × MR = M
$200 × 50% = $100
SP − M = C
$200 − $100 = $100 (given cost)

PART C Converting Markup Rates from One Base to Another

In certain situations, a businessperson may want to change the base on which the markup rate is calculated. More likely, he/she will want to know both of the markup rates—the one based on selling price and the one based on cost. Parts A and B explained how to calculate markup rates based on cost and selling price, respectively, when both the selling price and the cost are known. Part C describes how to find one markup rate directly from the other markup rate by using either a table or a formula. Both methods will be demonstrated.

Converting a Markup Rate on Selling Price to a Markup Rate on Cost. The percents in the extreme left column and the top row of Table 10-1 represent markup rates based on selling price. The equivalent markup rates based on cost are represented by the numbers in the other 10 columns.

For example, a 9% markup rate on selling price is represented by the 9% in the top row of percents. The equivalent markup rate on cost is located at the intersection of the 0% row and the 9% column, which is 9.89%. A 15% markup rate on selling price is found by locating 10% in the extreme left column and 5% in the top row. The equivalent markup rate on cost is located at the intersection of the 10% row and the 5% column, which is 17.65%.

	0%	1%	2%	3%	4%	5%	6%	7%	8%	9%
0%	0.00	1.01	2.04	3.09	4.17	5.26	6.38	7.53	8.70	9.89
10%	11.11	12.36	13.64	14.94	16.28	17.65	19.05	20.48	21.95	23.46
20%	25.00	26.58	28.21	29.87	31.58	33.33	35.14	36.99	38.89	40.85
30%	42.86	44.93	47.06	49.25	51.52	53.85	56.25	58.73	61.29	63.93
40%	66.67	69.49	72.41	75.44	78.57	81.82	85.19	88.68	92.31	96.08
50%	100.00	104.08	108.33	112.77	117.39	122.22	127.27	132.56	138.10	143.90
60%	150.00	156.41	163.16	170.27	177.78	185.71	194.12	203.03	212.50	222.58
70%	233.33	244.83	257.14	270.37	284.62	300.00	316.67	334.78	354.55	376.19
80%	400.00	426.32	455.56	488.24	525.00	566.67	614.29	669.23	733.33	809.09
90%	900.00	1011.11	1150.00	1328.57	1566.67	1900.00	2400.00	3233.33	4900.00	9900.00

Note: The percents in the extreme left column and upper row of Table 10-1 (in boldface) represent markup rates based on selling price. The equivalent markup rates based on cost are represented within the table itself.

TABLE 10-1 Markup Rates Based on Selling Price and Equivalent Markup Rates Based on Cost

EXAMPLE:

Use Table 10-1 to convert a 23% markup rate on selling price to the equivalent markup rate on cost. Locate 20% in the extreme left column and 3% in the top row. The equivalent markup rate on cost is located at the intersection of the 20% row and the 3% column, which is 29.87%.

To make the same change *without* using Table 10-1, assume that the original base of the markup rate is $100. Then, proceed as follows.

$$SP = \$100$$
$$M = SP \times MR$$
$$= \$100 \times 23\%$$
$$= \$23$$
$$C = SP - M$$
$$= \$100 - \$23$$
$$= \$77$$

$$MR \text{ on } C = M \div C$$
$$= \$23 \div \$77$$
$$= 0.2987 \text{ or } 29.87\%$$

Converting a Markup Rate on Cost to a Markup Rate on Selling Price. The percents in the extreme left column and the top row of Table 10-2 represent markup rates based on cost. The equivalent markup rates based on selling price are represented by the numbers in the other 10 columns.

For example, a 9% markup rate on cost is represented by the 9% in the top row. The equivalent markup rate on selling price is located at the intersection of the 0% row and the 9% column, which is 8.26%. A 15% markup rate on cost is found by locating 10% in the extreme left column and 5% in the top row. The equivalent markup rate on selling price is located at the intersection of the 10% row and the 5% column, which is 13.04%.

	0%	1%	2%	3%	4%	5%	6%	7%	8%	9%
0%	0.00	0.99	1.96	2.91	3.85	4.76	5.66	6.54	7.41	8.26
10%	9.09	9.91	10.71	11.50	12.28	13.04	13.79	14.53	15.25	15.97
20%	16.67	17.36	18.03	18.70	19.35	20.00	20.63	21.26	21.88	22.48
30%	23.08	23.66	24.24	24.81	25.37	25.93	26.47	27.01	27.54	28.06
40%	28.57	29.08	29.58	30.07	30.56	31.03	31.51	31.97	32.43	32.89
50%	33.33	33.77	34.21	34.64	35.06	35.48	35.90	36.31	36.71	37.11
60%	37.50	37.89	38.27	38.65	39.02	39.39	39.76	40.12	40.48	40.83
70%	41.18	41.52	41.86	42.20	42.53	42.86	43.18	43.50	43.82	44.13
80%	44.44	44.75	45.05	45.36	45.65	45.95	46.24	46.52	46.81	47.09
90%	47.37	47.64	47.92	48.19	48.45	48.72	48.98	49.24	49.49	49.75

Note: The percents in the extreme left column and upper row of Table 10-2 (in boldface) represent markup rates based on cost. The equivalent markup rates based on selling price are represented within the table itself.

TABLE 10-2 *Markup Rates Based on Cost and Equivalent Markup Rates Based on Selling Price*

EXAMPLE:

Use Table 10-2 to convert a 25% markup rate on cost to the equivalent markup rate on selling price.

Locate 20% in the extreme left column and 5% in the top row. The equivalent markup rate on selling price is located at the intersection of the 20% row and the 5% column, which is 20.00%, or simply 20%.

To make the same change *without* using Table 10-2, assume that the original base of the markup rate is $100. Then, proceed as follows.

$$C = \$100$$
$$M = C \times MR$$
$$= \$100 \times 25\%$$
$$= \$25$$
$$SP = C + M$$
$$= \$100 + \$25$$
$$= \$125$$
$$MR \text{ on } SP = M \div SP$$
$$= \$25 \div \$125$$
$$= 0.2 \text{ or } 20\%$$

COMPLETE ASSIGNMENT 31

ASSIGNMENT 31
SECTION 10.1

Name

Date

Selling Price, Cost, and Markup Rate

	Perfect Score	Student's Score
PART A	40	
PART B	40	
PART C	20	
TOTAL	100	

PART A Cost as the Base for Markup

DIRECTIONS: In the following problems, markup is based on cost. Find the items indicated by the column headings. Show your work, and write your answers on the lines provided (1 point for each correct answer).

	Cost	MR	Amount of Markup	Selling Price
1.	$150.00	40%	_____	_____
2.	$80.60	75%	_____	_____
3.	$183.79	100%	_____	_____
4.	$240.00	175%	_____	_____
5.	$151.60	225%	_____	_____

	Cost	Selling Price	Amount of Markup	MR
6.	$64.00	$80.00	_____	_____
7.	$112.95	$180.72	_____	_____
8.	$9.54	$19.08	_____	_____
9.	$88.40	$198.90	_____	_____
10.	$248.50	$869.75	_____	_____

Unit 2   Applications

DIRECTIONS: In the following problems, markup is based on cost. The selling price and the markup rate are given; find the cost. Show your work, and write your answers on the lines provided (2 points for each correct answer).

	Selling Price	MR	Cost
11.	$11.10	25%	_____
12.	$17.32	$33\frac{1}{3}\%$	_____
13.	$82.25	40%	_____
14.	$127.47	50%	_____
15.	$192.50	$66\frac{2}{3}\%$	_____
16.	$429.45	75%	_____
17.	$342.38	100%	_____
18.	$938.75	150%	_____
19.	$426.64	$166\frac{2}{3}\%$	_____
20.	$939.00	200%	_____

STUDENT'S SCORE _____

Chapter 10   Selling Goods

PART B Selling Price as the Base for Markup

DIRECTIONS: In the following problems, markup is based on selling price. Find the items indicated by the column headings. Show your work, and write your answers on the lines provided (1 point for each correct answer).

	Selling Price	MR	Amount of Markup	Cost
21.	$6.52	25%	_____	_____
22.	$27.60	40%	_____	_____
23.	$49.88	50%	_____	_____
24.	$120.00	$66\frac{2}{3}\%$	_____	_____
25.	$298.96	75%	_____	_____

	Cost	Selling Price	Amount of Markup	MR
26.	$23.60	$ 29.50	_____	_____
27.	$22.89	$ 32.70	_____	_____
28.	$61.08	$152.70	_____	_____
29.	$47.57	$237.85	_____	_____
30.	$49.50	$495.00	_____	_____

DIRECTIONS: In the following problems, markup is based on selling price. The cost and the markup rate are given; find the selling price. Show your work, and write your answers on the lines provided (2 points for each correct answer).

	Cost	MR	Selling Price
31.	$5.08	20%	_____
32.	$11.22	25%	_____
33.	$12.25	30%	_____
34.	$18.00	$33\frac{1}{3}$%	_____
35.	$38.85	40%	_____
36.	$41.49	50%	_____
37.	$49.70	60%	_____
38.	$54.20	$66\frac{2}{3}$%	_____
39.	$50.00	75%	_____
40.	$40.00	80%	_____

STUDENT'S SCORE _____

PART C Converting Markup Rates from One Base to Another

DIRECTIONS: For each of the following problems, find the missing markup rate. You may either use Tables 10-1 and 10-2 on pages 326–327 or do the calculation by formula (2 points for each correct answer).

	Markup Rate Based on SP	Markup Rate Based on C		Markup Rate Based on C	Markup Rate Based on SP
41.	25%	_____	46.	25%	_____
42.	40%	_____	47.	40%	_____
43.	50%	_____	48.	50%	_____
44.	$66\frac{2}{3}$%	_____	49.	$66\frac{2}{3}$%	_____
45.	75%	_____	50.	75%	_____

STUDENT'S SCORE _____

Chapter 10 Selling Goods

SECTION 10.2 Profit and Loss

When selling goods, you should know how the net profit (NP), net loss (NL), and the marking *up* or *down* of goods are related. Three elements determine whether there is a net profit or a net loss on an item: the selling price (SP), the cost (C), and that item's share of operating expenses (OE).

To determine whether there is a net profit or a net loss on an item, compare the selling price (SP) with the *sum* of the cost of the merchandise and the operating expenses (C + OE). If the selling price is larger than this sum, there is a net profit equal to the difference between the selling price (SP) and the *sum* of the cost and the operating expenses (C + OE). In equation form, this is:

$$NP = SP - (C + OE)$$

If the selling price is smaller than this sum, there is a net loss equal to the difference between the *sum* of the cost and the operating expenses (C + OE) and the selling price (SP). In equation form, this is:

$$NL = (C + OE) - SP$$

PART A Net Profit

Recall from Section 10.1 that the markup is added to the cost to determine the selling price of goods. This was expressed as the basic equation:

$$SP = C + M$$

When establishing the selling price, a businessperson plans for the markup (M) to cover two things: operating expenses (OE), and the desired net profit (NP). With symbols, this is expressed as follows:

$$M = OE + NP$$

Taken together, these relationships state that the operating expenses (OE) and the net profit (NP) are both added to the cost (C) to determine the selling price. The expanded equation for selling price becomes:

$$SP = C + OE + NP$$

Compare this equation with the equation for net profit that was given previously: NP = SP − (C + OE). These two equations are simply two different ways to write the same relationship. You will also see that one equation can be used to check the other.

Finding Selling Price. The expanded formula for selling price, SP = C + OE + NP, can be used to solve for SP.

EXAMPLES:

Use the formula for SP to find the selling price. Use the formula for NP to check your work.

a. The cost is $15.65; the operating expense is $8.23; the net profit is $2.89.

   Solve: SP = C + OE + NP
   = $15.65 + $8.23 + $2.89
   = $26.77

   Check: NP = SP − (C + OE)
   = $26.77 − ($15.65 + $8.23)
   = $26.77 − $23.88
   = $2.89 (given NP)

b. The cost is $37.75; the operating expense is $31.08; the net profit is $10.85.

   Solve: SP = C + OE + NP
   = $37.75 + $31.08 + $10.85
   = $79.68

   Check: NP = SP − (C + OE)
   = $79.68 − ($37.75 + $31.08)
   = $79.68 − $68.83
   = $10.85 (given NP)

Finding Net Profit. There are two methods that can be used to calculate net profit (NP) when selling price (SP), cost (C), and operating expenses (OE) are given. Use the expanded equation for selling price, or use the equation for NP given at the beginning of Section 10.2. Both methods are illustrated in the example that follows.

EXAMPLE:

Find the net profit when the selling price is $41, the cost is $24, and the operating expense is $12.

Method 1: Use the expanded selling price equation.

$$SP = C + OE + NP$$
$$\$41 = \$24 + \$12 + NP$$
$$\$41 = \$36 + NP$$
$$\$5 = NP$$

In the third line, $36 was subtracted from $41 to calculate NP ($5).

Method 2: Use the equation for NP.

$$NP = SP - (C + OE)$$
$$= \$41 - (\$24 + \$12)$$
$$= \$41 - \$36$$
$$= \$5$$

With both methods, the same two steps of arithmetic are used. First, $12 and $24 are added together to get $36. Then, the $36 is subtracted from $41 to obtain $5. You should use whichever method is easier for you; both always result in the same answer. Also, you may prefer to write the equation for net profit without the parentheses, as follows:

$$NP = SP - C - OE$$
$$= \$41 - \$24 - \$12$$
$$= \$5$$

In this form, both C and OE are subtracted from SP. With or without parentheses, the two equations for NP are equivalent and always give the same answer. Use the one that your prefer.

Finding the Cost and Operating Expenses. If selling price (SP), operating expenses (OE), and net profit (NP) are all given, you can calculate the cost (C). If selling price, cost, and net profit are all given, you can calculate the operating expenses.

For each situation, there are two methods of solution. Substitute the given values directly into the expanded equation for selling price (SP) and then solve, or first rewrite the expanded equation and then substitute the given values into the new equation. The new equation should have the unknown amount—either cost (C) or operating expenses (OE)—on the left side and all the given values on the right side. Both methods are illustrated in the examples that follow.

EXAMPLE:

Find the cost when the selling price is $88, the operating expense is $25, and the net profit is $8.

Method 1: Use the expanded selling price equation.

$$SP = C + OE + NP$$
$$\$88 = C + \$25 + \$8$$
$$\$88 = C + \$33$$
$$\$55 = C$$

In the third line, $33 was subtracted from $88 to calculate C ($55).

Method 2: Rewrite the equation.

$$C = SP - (OE + NP)$$
$$= \$88 - (\$25 + \$8)$$
$$= \$88 - \$33$$
$$= \$55$$

In each method, $25 and $8 are added together to get $33. The $33 is then subtracted from $88 to obtain $55. With no parentheses in the equation for C, the solution is:

$$C = SP - OE - NP$$
$$= \$88 - \$25 - \$8$$
$$= \$55$$

EXAMPLE:

Find the operating expense when the selling price is $105, the cost is $68, and the net profit is $13.

Method 1: Use the expanded selling price equation.

$$SP = C + OE + NP$$
$$\$105 = \$68 + OE + \$13$$
$$\$105 = \$81 + OE$$
$$\$24 = OE$$

In the third line, $81 was subtracted from $105 to calculate OE ($24).

Method 2: Rewrite the equation, with OE alone on the left side.

$$OE = SP - (C + NP)$$
$$= \$105 - (\$68 + \$13)$$
$$= \$105 - \$81$$
$$= \$24$$

In each method, $68 and $13 are added together to get $81. The $81 is then subtracted from $105 to obtain $24. With no parentheses in the equation for OE, the solution is:

$$OE = SP - C - NP$$
$$= \$105 - \$68 - \$13$$
$$= \$24$$

PART B Net Loss

A business makes a net profit (NP) on an item when selling price (SP) is greater than the sum of cost and operating expenses (C + OE). When selling price is less than the sum of cost and operating expenses, there is a net loss (NL).

For example, when cost is $60, operating expense is $30, and selling price is only $80, there is a net loss of $10. Cost of the merchandise plus operating expense is $60 + $30, or $90. Therefore, $90 is spent, but only $80 is earned, thus creating the loss. To determine the size of the loss, subtract selling price (SP) from the sum of cost and operating expense (C + OE), which gives the following equation:

$$NL = (C + OE) - SP$$
$$= (\$60 + \$30) - \$80$$
$$= \$90 - \$80$$
$$= \$10$$

Without parentheses, the equation for net loss is written as follows:

$$NL = C + OE - SP$$

You can rewrite this equation in a form similar to the expanded equation for selling price that was given in Part A. It becomes:

$$SP = C + OE - NL$$

If you use this equation, you must remember to subtract the net loss. In the preceding example, the problem becomes:

$$SP = \$60 + \$30 - \$10$$
$$= \$80$$

Finding Selling Price. When the cost (C), operating expense (OE), and net loss (NL) are all given, you can find the selling price (SP) by using the new expanded equation for selling price: SP = C + OE − NL.

EXAMPLE:

Find the selling price when the cost is $72, the operating expense is $28, and the net loss is $9.

$$SP = C + OE - NL$$
$$= \$72 + \$28 - \$9$$
$$= \$91$$

Finding Net Loss. When the selling price (SP), cost (C), and operating expense (OE) are all given, you can calculate the net loss (NL) by two methods. Use the equation for NL, or use the new expanded equation for SP. Both methods are illustrated in the following example.

EXAMPLE:

Find the net loss when the selling price is $54, the cost is $39, and the operating expense is $19.

Method 1: Use the equation for NL.

$$NL = C + OE - SP$$
$$= \$39 + \$19 - \$54$$
$$= \$4$$

Method 2: Use the new expanded equation for selling price.

$$SP = C + OE - NL$$
$$\$54 = \$39 + \$19 - NL$$
$$\$54 = \$58 - NL$$

Thus, NL = $4. The left side of the equation is $54. To make the right side $54, NL must be $4. You calculate $4 by subtracting $54 from $58.

Finding the Cost and Operating Expenses. If selling price (SP), operating expense (OE), and net loss (NL) are all given, you can calculate the cost (C). If selling price, cost, and net loss are all given, you can calculate the operating expense. Use the new expanded equation for selling price:

$$SP = C + OE - NL$$

For both cost and operating expense, there are two methods of solution. Substitute the given values directly into the new expanded equation for selling price and then solve, or first rewrite the expanded equation and then substitute the given values into the new equation. The new equation should have the unknown (either C or OE) on the left side and all the given values on the right. Both methods are illustrated in the following examples.

EXAMPLE:

Find the cost when the selling price is $106, the operating expense is $43, and the net loss is $21.

Method 1: Use the new expanded equation for selling price.

 SP = C + OE − NL
$106 = C + $43 − $21
$106 = C + $22
 $84 = C

In the third line, $22 was subtracted from $106 to calculate C ($84).

Method 2: Rewrite the equation, with C alone on the left side.

C = SP − OE + NL
 = $106 − $43 + $21
 = $84

EXAMPLE:

Find the operating expense when the selling price is $150, the cost is $120, and the net loss is $20.

Method 1: Use the new expanded equation for selling price.

 SP = C + OE − NL
$150 = $120 + OE − $20
$150 = $100 + OE
 $50 = OE

In the third line, $100 was subtracted from $150 to calculate OE ($50).

Method 2: Rewrite the equation, with OE alone on the left side.

OE = SP − C + NL
 = $150 − $120 + $20
 = $50

Calculators and Net Loss. In Part A of this section, the equation for calculating net profit was given as NP = SP − (C + OE) or NP = SP − C − OE (see page 333). When you substitute the values of SP = $80, C = $60, and OE = $30 into this equation, NP is $80 − $60 − $30. If you work this problem on any calculator, the calculator will display an answer of −10. The −10 is called *negative 10* or *minus 10*.

In business practice, the −10 can be stated in two ways. There is either a net profit of −$10 or a net loss of $10. In this text, we will only use the second expression, because it is easier for most people to visualize a loss than a negative profit.

PART C Adjusting the Selling Price

Up to this point, the selling price has been determined as simply the cost plus the markup. Manufacturers and wholesale distributors normally quote this as the unit price for a product. Retailers, however, are more likely to adjust this unit price to the *even dollar* or to some popular price, such as one ending in 5, 8, or 9.

For example, if the cost plus markup on a product amounts to $26.32, a retailer may offer the product at an adjusted selling price of $26.00, $26.35, $26.49, $26.75, $26.98, etc. The selling price actually selected will depend on the store's policy and, to some extent, the manner in which competing businesses set their prices.

Table 10-3 illustrates a typical adjustment table that might be used by a retailer. To use the table, find the given cost plus markup in one of the price ranges on the left, and then adjust the selling price according to the ending given on the right.

EXAMPLES:

Adjust the selling price according to Table 10-3.

Cost Plus Markup	Adjusted Selling Price
a. 8¢	8¢
b. 23¢	25¢
c. 67¢	69¢
d. $2.44	$2.49
e. $5.07	$5.39
f. $6.73	$6.98
g. $12.36	$12.49
h. $23.64	$23.98
i. $32.17	$34.98
j. $46.90	$49.98
k. $80.06	$79.98
l. $271.82	$279.98

Cost Plus Markup	Adjusted Selling Price
15¢ or under	same as determined
16¢ to 99¢	ends in the next nearest 5¢ or 9¢
$1 to $4.99	ends in the next 9¢
$5 to $9.99	ends in the next nearest .39, .69, or .98
$10 to $19.99	ends in the next nearest .49 or .98
$20 to $29.99	ends in the next .98
$30 to $49.99	ends in the next nearest 4.98 or 9.98
$50 and up	ends in 9.98, either up or down

TABLE 10-3 A Selling Price Adjustment Table

Items under 50¢ often are priced in quantity on the basis of the adjusted price. For example, two items may be priced at twice the adjusted price less 1¢, such as 25¢ each or 2 for 49¢; three items may be priced at three times the adjusted price less 2¢, such as 19¢ each or 3 for 55¢; and four items may be priced at four times the adjusted price less 3¢, such as 13¢ each or 4 for 49¢.

EXAMPLES:

Adjust the given quantity prices according to the preceding policies.

	Cost Plus Markup	Adjusted Unit Selling Price	Adjusted Quantity Price
a.	8¢	8¢	2 for 15¢
b.	22¢	25¢	2 for 49¢
c.	18¢	19¢	3 for 55¢
d.	37¢	39¢	3 for $1.15
e.	13¢	13¢	4 for 49¢
f.	24¢	25¢	4 for 97¢

Occasionally, a retailer tempts a customer to buy in quantities that offer no real savings. For example, *25¢ each or 4 for $1.00* and *15¢ each or 10 for $1.50* are the same price, whether the items are bought singly or in quantity. In packaged or bottled goods that are sold by weight or fluid measure, larger sizes sometimes have a higher unit price than smaller sizes. For example, 16 ounces or 454 grams at 69¢ costs more per unit of measure than 12 ounces or 340 grams at 49¢. The alert consumer should be able to determine which prices actually offer savings.

PART D Using Price Tag Codes

Some merchandise (such as coats, dresses, jewelry, furniture, large appliances, etc.) may have the selling price on one tag, while another tag has the cost of the item in letters or in code. In the case of some items, such as expensive jewelry, both the cost and the selling price may be in code. Having the cost on the tag enables the retailer to quickly adjust the selling price, even at the time of purchase, without going below the cost. It is also helpful at inventory time.

A code may consist of a word or phrase which has a combination of 10 different letters, such as REDISCOUNT, REPUBLICAN, COME AND BUY, etc. The first letter of the word is used for the digit 1, the second letter is used for the digit 2, etc. The last letter is used for 0. An additional, different letter may be used as a *repeater* when the same digit occurs in succession in the cost or selling price. This also makes it more difficult for someone to break the code.

EXAMPLES:

Using REPUBLICAN as the code word and D for a repeater, write the given numbers in code.

R E P U B L I C A N
1 2 3 4 5 6 7 8 9 0

a. $5.95..BAB

b. $11.88..RDCD

c. $94.49..AUDA

In a numerical code, the cost figures are manipulated according to a scheme devised by the company. For example, a cost of $23.14 may be coded 282643, according to the following scheme: Double the cost, reverse the digits of the result, and then place these digits between any two meaningless digits which are never the same.

EXAMPLES:

Write the given numbers in a numerical code according to the preceding scheme.

a. $5.95..709112

b. $11.88..667324

c. $94.49...5898813

COMPLETE ASSIGNMENT 32

ASSIGNMENT 32
SECTION 10.2

Name

Date

Profit and Loss

	Perfect Score	Student's Score
PART A	24	
PART B	36	
PART C	20	
PART D	20	
TOTAL	100	

PART A Net Profit

DIRECTIONS: Of selling price, cost, operating expense, and net profit, three are given and one is unknown. Solve for the unknown. Show your work, and write your answers on the lines provided (3 points for each correct answer).

1. C = $8.25; OE = $3.56;
   NP = $1.13

   SP = _____

2. SP = $19.30; C = $12.51;
   OE = $4.28

   NP = _____

3. SP = $36.37; C = $23.92;
   NP = $3.90

   OE = _____

4. SP = $51.53; OE = $11.85;
   NP = $5.25

   C = _____

5. SP = $244.10; OE = $73.09
   NP = $22.18

   C = _____

6. SP = $196.94; C = $115.53;
   NP = $21.87

   OE = _____

7. SP = $140.80; C = $89.15;
   OE = $41.40

   NP = _____

8. C = $71.74; OE = $19.43;
   NP = $8.42

   SP = _____

STUDENT'S SCORE _____

PART B Net Loss

DIRECTIONS: Of selling price, cost, operating expense, and net loss, three are given and one is unknown. Solve for the unknown. Show your work, and write your answers on the lines provided (3 points for each correct answer).

9. C = $12.22; OE = $4.15;
   NL = $1.09

   SP = _____

10. SP = $21.49; C = $17.25;
    OE = $5.82

    NL = _____

11. SP = $41.79; C = $27.63;
    NL = $4.12

    OE = _____

12. SP = $62.30; OE = $17.48;
    NL = $7.97

    C = _____

13. SP = $259.93; OE = $103.75;
    NL = $17.18

    C = _____

14. SP = $201.49; C = $122.35;
    NL = $27.98

    OE = _____

15. SP = $162.45; C = $110.10;
    OE = $68.72

    NL = _____

16. C = $84.29; OE = $28.50
    NL = $7.23

    SP = _____

DIRECTIONS: The selling price, the cost, and the operating expense are given. Determine whether there is a net profit or a net loss, and then calculate it. Show your work, and write your answers, with the proper label (NP or NL), on the lines provided (3 points for each correct answer).

17. SP = $62.14; C = $37.83;
    OE = $26.19

    __ = _____

18. SP = $562.70; C = $328.84;
    OE = $185.68

    __ = _____

19. SP = $374.34; C = $227.58;
    OE = $115.14

    __ = _____

20. SP = $159.29; C = $98.50;
    OE = $77.67

    __ = _____

STUDENT'S SCORE _____

Chapter 10 Selling Goods   **341**

PART C Adjusting the Selling Price

DIRECTIONS: *Find the adjusted selling prices according to the schedule given. Write your answers on the lines provided (1 point for each correct answer).*

	Cost Plus Markup	Adjusted Selling Price
21.	$0.021	_____
22.	$0.067	_____
23.	$0.91	_____
24.	$0.27	_____
25.	$0.86	_____
26.	$0.35	_____
27.	$0.72	_____
28.	$0.48	_____
29.	$0.63	_____
30.	$2.27	_____
31.	$3.52	_____
32.	$6.69	_____
33.	$9.32	_____
34.	$11.75	_____
35.	$17.32	_____
36.	$82.18	_____
37.	$126.75	_____
38.	$0.17; 2 for	_____
39.	$0.11; 3 for	_____
40.	$0.25; 4 for	_____

Cost Plus Markup	Adjusted Selling Price
$ 0 to 0.15	nearest cent
0.16 to 0.19	19¢
0.20 to 0.25	25¢
0.26 to 0.29	29¢
0.30 to 0.35	35¢
0.36 to 0.39	39¢
0.40 to 0.45	45¢
0.46 to 0.49	49¢
0.50 to 0.55	55¢
0.56 to 0.59	59¢
0.60 to 0.65	65¢
0.66 to 0.69	69¢
0.70 to 0.75	75¢
0.76 to 0.79	79¢
0.80 to 0.85	85¢
0.86 to 0.89	89¢
0.90 to 0.95	95¢
0.96 to 0.99	99¢
$1 to $4.99	ending in next 9¢
$5 to $9.99	ending in next .49 or .99
$10 to $19.99	ending in next .99
$20 and up	ending in next 4.95 or 9.95

For quantities of 2: (2 × Selling Price) − 1¢
For quantities of 3: (3 × Selling Price) − 2¢
For quantities of 4: (4 × Selling Price) − 3¢

STUDENT'S SCORE _____

PART D Using Price Tag Codes

DIRECTIONS: Use a phrase code or a numerical code to determine the coded prices for the prices given. For the numerical code, double the given price, reverse the digits, and write the result between two meaningless digits. Show your work, and write the coded prices on the lines provided (2 points for each correct answer).

	Price	Type of Code	Coded Price
41.	$3.35	REDISCOUNT (repeater: B)	_____
42.	$15.99	EXCHAMPION (repeater: T)	_____
43.	$7.75	COME AND BUY (repeater: R)	_____
44.	$44.00	BOY AND GIRL (repeater: E)	_____
45.	$227.39	PURCHASING (repeater: V)	_____
46.	$3.35	Numerical code	_____
47.	$15.99	Numerical code	_____
48.	$7.75	Numerical code	_____
49.	$44.00	Numerical code	_____
50.	$227.39	Numerical code	_____

STUDENT'S SCORE _____

CHAPTER 11

Calculating Gross Pay for Payrolls

A **payroll** is a record of the earnings and deductions of a company's employees. Some deductions, such as income taxes and social security taxes, are required. Others, such as union dues, health insurance premiums, payments to credit unions, etc., are *employee authorized*.

Large companies that prepare their own payrolls use a computer. Today, there are specialized companies that sell a *computerized payroll service* to their client companies. The client companies pay to have their payrolls prepared for them. Many banks sell such payroll services. Certainly, there are many companies that still prefer to prepare their own payrolls, with or without a computer.

Gross pay is the total amount of money earned by an employee before any deductions are made. **Net pay** is the amount actually paid to the employee after all deductions from gross pay have been made. Deductions vary significantly from company to company and from state to state, so this chapter concentrates only on the common features of calculating gross pay.

Gross pay is usually based on time worked and/or actual productivity during a specified period, called the **pay period**. When time worked is the basis for pay, the unit of time may be an hour, day, week, month, etc. The term **wages** usually refers to payments for time worked on an hourly basis, and the term **salary** usually refers to payments for time worked on a weekly, monthly, or yearly basis.

Section 11.1 discusses various details of calculating gross pay based on time worked.

Measuring productivity for payment purposes depends on the job that an employee performs. For a salesperson, productivity is measured by the dollar amount of the items sold. When the salesperson receives a percent of sales as pay (or partial pay), the payment is called a **commission**. In manufacturing, productivity may be measured by the number of items, or pieces, that the employee actually produces. When the employee's pay is based on pieces produced, it is called **piecework**, and the payment may be called wages. Section 11.2 discusses payment based on commissions and piecework.

SECTION 11.1 Time-Basis Payment and Payroll Deductions

A **time card** or **time sheet** is used to keep a record of the total hours worked by an employee paid on a time basis. When a time clock is used, each company has its own time limits for punching in and out without penalty for the regular work period. **Overtime,** which is time worked beyond the regular hours, is punched in separate columns.

PART A Straight-Time (or Day-Rate) Pay

When the hours worked are all regular, or **straight time,** the gross pay equals the number of hours worked times the given hourly rate. This hourly rate is often called the **day rate,** especially in combination with a piecework payment plan (see Section 11.2).

EXAMPLE:

Jane Wright is paid $9.35 an hour for straight time. Find her gross pay for a week in which she worked 36 hours.

Gross pay = 36 × $9.35
= $336.60

PART B Overtime Pay on a Weekly Basis

A **standard workweek** is usually 40 hours, but it may be longer or shorter. **Regular pay** is the pay for the standard workweek hours at the given rate. When an overtime rate is paid for hours worked beyond the standard workweek, this is called **overtime on a weekly basis**. **Overtime pay** is the pay for the overtime hours at the overtime rate.

Most companies pay overtime at 1.5 times the regular rate (called *time-and-a-half*), and some pay two or three times the regular rate for work done on Saturdays, Sundays, or holidays.

In accordance with the Federal Wage and Hour Law, companies that engage in interstate commerce must pay time-and-a-half for all hours worked over 40 hours in one week. State laws and most union contracts also specify when overtime must be paid.

To find the overtime rate, multiply the hourly rate by 1.5 (or 2, or 3). To find the overtime pay, multiply the number of overtime hours by the overtime rate. The gross pay is the sum of the regular pay and the overtime pay.

EXAMPLES:

a. Carol Davis earns $11.50 per hour for a 40-hour standard workweek and time-and-a-half for overtime. Find her gross pay for a week in which she worked 45 hours.

Overtime rate = $11.50 × 1.5
= $17.25 per hour

Regular pay: 40 × $11.50 = $460.00
Overtime pay: 5 × $17.25 = + 86.25
Gross pay: $546.25

b. Ed Winston earns $10.80 per hour, with overtime paid on a weekly basis. According to the union contract, all hours over 40 worked on Monday through Friday are paid at time-and-a-half; all hours worked on Saturday are double time; and all hours worked on Sunday or a legal holiday are triple time. Find his gross pay for a week in which he worked these hours:

S	M	T	W	Th	F	Sa
4	7	9	10	8	8	3

Monday through Friday:	42 hours
Straight time:	40 hours
Overtime hours:	
Time-and-a-half:	2 hours
Double time:	3 hours
Triple time:	4 hours

Overtime rates:
Time-and-a-half: $10.80 × 1.5 = $16.20
Double time: $10.80 × 2 = $21.60
Triple time: $10.80 × 3 = $32.40

Regular pay: 40 × $10.80 = $432.00
Overtime pay: 2 × $16.20 =   32.40
3 × $21.60 =   64.80
4 × $32.40 = 129.60
Gross pay: = $658.80

PART C Overtime Pay on a Daily Basis

A company may pay for **overtime on a daily basis**, which means that all hours worked in one day beyond the regular workday hours (usually 8) are paid at the overtime rate. Although some companies may have a regular 40-hour week of four 10-hour days, a regular workday of 8 hours is used in the following examples.

By federal law, companies with federal contracts must pay time-and-a-half for any hours worked over 8 in a day or 40 in a week, *whichever is greater*. Many union contracts specify that all time worked over the regular workday hours in the standard workweek (usually Monday through Friday, but not including holidays) and all time worked on the remaining two days (usually Saturday and Sunday) and holidays must be paid at the overtime rate. Even though the total hours worked during the week may be less than 40, the overtime rate must be paid as specified for time worked beyond 8 hours a day.

Chapter 11 Calculating Gross Pay for Payrolls

EXAMPLES:

a. Janice Wells is paid $9.10 per hour, with overtime on a daily basis at time-and-a-half. Find her gross pay for the following time worked:

S	M	T	W	Th	F	Sa
0	7	10	8	11	6	0

	S	M	T	W	Th	F	Sa	
Straight time:	0	7	8	8	8	6	0	= 37
Overtime:	0	0	2	0	3	0	0	= 5

Overtime rate = $9.10 × 1.5
              = $13.65

Regular pay: 37 × $9.10 = $336.70
Overtime pay: 5 × $13.65 =   68.25
Gross pay: $404.95

b. Joe Lopez is paid $15.85 per hour, with overtime on a daily basis at time-and-a-half for Monday through Friday, double time for Saturday, and triple time for Sunday or a legal holiday. Find his gross pay for the following time worked:

S	M	T	W	Th	F	Sa
2	8	12	8	6	10	4

	S	M	T	W	Th	F	Sa	
Straight time:	0	8	8	8	6	8	0	= 38

Overtime hours:
  Time-and-a-half: 6
  Double time: 4
  Triple time: 2

Overtime rates:
  Time-and-a-half: $15.85 × 1.5 = $23.775
  Double time: $15.85 × 2 = $31.70
  Triple time: $15.85 × 3 = $47.55

Regular pay: 38 × $15.85 = $602.30
Overtime pay: 6 × $23.775 = 142.65
                4 × $31.70 = 126.80
                2 × $47.55 =   95.10
Gross pay: $966.85

If a rate has an exact half-cent ending, such as $23.775, all three decimal places are used in the calculation.

PART D Overtime Pay for Salaried Employees

Salaried employees above a certain management level are classified as **exempt** and do not receive overtime pay. However, there are many *nonexempt salaried employees,* who *are* eligible for overtime pay. To calculate their pay, first convert their weekly or monthly salary to an hourly rate. Then, determine the overtime hours on either a weekly or a daily basis, according to the company's policy.

To find the hourly rate for a *weekly* salary, divide the salary by the number of hours in the regular workweek.

EXAMPLE:

Paul Steinway is paid $490 for a regular workweek of 40 hours, with overtime paid at time-and-a-half. Find his gross pay for a week in which he worked 8 hours of overtime.

Hourly rate = $490 ÷ 40
            = $12.25
Overtime rate = $12.25 × 1.5
                = $18.375

Regular pay: $490.00
Overtime pay: 8 × $18.375 =   147.00
Gross pay: $637.00

For the purpose of calculating an hourly overtime rate for nonexempt salaried employees, a month is usually considered to have either $4\frac{1}{3}$ weeks or 22 workdays. Because $4\frac{1}{3}$ weeks is a better approximation, it will be used in this book. To determine an hourly rate for a monthly salary, divide the salary by the product of $4\frac{1}{3}$ and the number of hours in the regular workweek.

EXAMPLE:

Mary Ming earns $2,145 a month and is paid semimonthly. Her regular workweek is 36 hours, with overtime paid at time-and-a-half. Find her gross pay for a half-month in which she worked 6 hours of overtime.

Hourly rate = $2,145 ÷ (36 × $4\frac{1}{3}$)
            = $2,145 ÷ 156
            = $13.75
Overtime rate = $13.75 × 1.5
                = $20.625

Regular pay: $2,145 ÷ 2   = $1,072.50
Overtime pay: 6 × $20.625 =   123.75
Gross pay: $1,196.25

PART E Payroll Deductions

Only deductions required by law or union contracts or those authorized by the individual employee should be subtracted from the gross pay. Because of the many variations in payroll deductions and records, only those deductions required by law and a few general items on the subject are discussed here. For this reason, this material is not included in the assignments at the end of the chapter.

FICA Taxes. The Federal Insurance Contributions Act (FICA), commonly called Social Security, applies to most people who work in the United States. Individuals who are employees in the occupations covered by this act must have a Social Security number and must pay a certain percent of their gross pay toward Social Security and Medicare. The employer is required to deduct the FICA tax from the employee's earnings and to match it with an equal amount.

Self-employed people are also covered by FICA, but they must make 100% of the contribution themselves. Booklets explaining the FICA provisions are available without charge from the Social Security Administration (SSA).

The U.S. Congress has increased the FICA tax rate and maximum earnings base many times since Social Security went into effect in 1937. To make the correct calculations, you should get the current rates and maximum earnings base from either the SSA or the Internal Revenue Service (IRS).

Federal Income Taxes. The employer is also required to withhold, or deduct, federal income taxes for employees in occupations covered by income tax law. The amount deducted each pay period depends on the employee's gross earnings, the current income tax rate, and the number of exemptions claimed by the employee on a W-4 form. The W-4 form is filled out whenever an employee starts a new job.

State and City Income Taxes. Not all states and cities impose income taxes. Where such taxes are required, they are deducted from the employee's gross earnings. In addition, any employee contributions to accident or disability funds are also deducted.

Annual Tax Guides. The IRS prepares an annual *Employer's Tax Guide*, which shows the amounts that should be withheld for federal income taxes and FICA taxes. The revenue departments of those states and cities with an income tax also prepare annual tax guides. However, many companies with a computerized payroll use simplified tables or formulas, rather than the tax guides, to compute income tax and FICA tax deductions.

Employer's Tax Reports. Employers are required to deposit all amounts withheld for income taxes and Social Security, as well as their matching FICA contributions, in an authorized bank within a specific time. They must also file a quarterly report with the IRS which shows the wages paid to each employee during the quarter. As employers, they must pay federal and state taxes for unemployment and disability funds and file the necessary quarterly returns. The state rates for these taxes vary considerably, and the federal rate is subject to change when necessary.

Employee's Earnings Record. Because of the required government reports, most employers maintain an *employee's earnings record* for each employee, in addition to other payroll records. This record shows the employee's name, address, Social Security number, number of exemptions claimed, pay rate, gross earnings, deductions, net pay for each pay period, and the total to date of gross earnings and taxes withheld. Shortly after the end of the year, the employer is required by law to give each employee two or more copies of his or her W-2 form. The W-2 form shows the total amounts of gross pay and taxes withheld, and one copy of it must be attached to each of the employee's income tax returns.

COMPLETE ASSIGNMENT 33

_____
Name

_____
Date

ASSIGNMENT 33
SECTION 11.1

Time-Basis Payment and
Payroll Deductions

	Perfect Score	Student's Score
PARTS A & B	30	
PART C	25	
PART D	45	
TOTAL	100	

PARTS A AND B Straight-Time Pay and Overtime Pay on a Weekly Basis

DIRECTIONS: Bovio Supply Company has a regular workweek of 40 hours, with overtime paid on a weekly basis at time-and-a-half. Find the overtime rate, the overtime hours, the regular pay, the overtime pay, and the gross pay for each employee. Show your work, and write your answers on the lines provided (1 point for each correct answer).

Name and Time Worked	Hourly Rate	Overtime Rate	Overtime Hours	Regular Pay	Overtime Pay	Gross Pay
1. J. Olsen	$8.00	_____	_____	_____	_____	_____

S M T W Th F Sa
0 8 8 9 8   9 2

| **2.** H. White | $10.50 | _____ | _____ | _____ | _____ | _____ |

S M T   W Th F Sa
0 8 10 8 9   8 4

| **3.** M. Beard | $5.75 | _____ | _____ | _____ | _____ | _____ |

S M T   W Th F Sa
0 9 11 9 10 9 0

| **4.** C. Chou | $8.25 | _____ | _____ | _____ | _____ | _____ |

S M T W Th F Sa
4 8 8 8 8   8 4

| **5.** T. Smith | $12.10 | _____ | _____ | _____ | _____ | _____ |

S M T W Th F Sa
2 7 8 7 7   9 6

| **6.** R. Hurst | $9.90 | _____ | _____ | _____ | _____ | _____ |

S M T W Th F Sa
0 8 0 9 10 8 7

STUDENT'S SCORE _____

348 Unit 2 Applications

PART C Overtime Pay on a Daily Basis

DIRECTIONS: Swanson & Kaplan Demolition Co. pays overtime on a daily basis at time-and-a-half for all time worked over 8 hours per day, Monday through Friday, and double time for any hours worked on Saturdays, Sundays, or legal holidays. For each of the employees listed, find the regular pay, the overtime pay at time-and-a-half, the overtime pay at double time, the total overtime pay, and the gross pay. Show your work, and write your answers on the lines provided (1 point for each correct answer).

Name and Time Worked	Hourly Rate	Regular Pay	Overtime Pay @ 1.5×	Overtime Pay @ 2×	Total O.T. Pay	Gross Pay
7. L. Jones	$8.25					

S	M	T	W	Th	F	Sa
0	8	9	8	9	7	1

8. J. West	$10.70					

S	M	T	W	Th	F	Sa
0	9	11	8	8	8	2

9. C. Allen	$9.60					

S	M	T	W	Th	F	Sa
2	8	9	6	10	8	6

DIRECTIONS: Ortega Manufacturing Company pays overtime on a daily basis at time-and-a-half for all time worked over 8 hours per day, Monday through Friday, double time for any hours worked on Saturdays, and triple time for any hours worked on Sundays or legal holidays. For each of the following employees, find the regular pay, the overtime pay at time-and-a-half, the overtime pay at double time, the overtime pay at triple time, and the gross pay. Show your work, and write your answers on the lines provided (1 point for each correct answer).

Name and Time Worked	Hourly Rate	Regular Pay	Overtime Pay @ 1.5×	Overtime Pay @ 2×	Overtime Pay @ 3×	Gross Pay
10. A. Evans	$10.90					

S	M	T	W	Th	F	Sa
2	7	10	8	6	9	5

11. K. Ijichi	$9.40					

S	M	T	W	Th	F	Sa
3	8	8	6	10	9	4

STUDENT'S SCORE _____

Chapter 11   Calculating Gross Pay for Payrolls

PART D   Overtime Pay for Salaried Employees

12. David Blevins earns $400 for a regular workweek of 40 hours. Overtime is paid at time-and-a-half. Find his gross pay for a week in which he worked 6 hours of overtime (5 points).

Gross Pay _____

13. Lewis Devlin earns $490 for a regular workweek of 35 hours. Overtime is paid at time-and-a-half for Monday through Friday and double time for Saturdays and Sundays. Find his gross pay for a week in which he had 4 hours of overtime during the week and 6 hours of overtime on the weekend (6 points).

Gross Pay _____

14. Silvana Cometta earns $558 for a regular workweek of 36 hours. Overtime is paid at time-and-a-half for Monday through Friday and double time for Saturdays and Sundays. Find her gross pay for a week in which she had 3 hours of overtime during the week and 5 hours of overtime on the weekend (6 points).

Gross Pay _____

15. Mary Ann Fong earns $480 for a regular workweek of 40 hours. Overtime is paid at time-and-a-half for Monday through Friday, double time for Saturdays, and triple time for Sundays. Find her gross pay for a week in which she had 4 hours of overtime during the week, 4 hours of overtime on Saturday, and 3 hours of overtime on Sunday (8 points).

Gross Pay _____

16. Martha Ramirez earns $2,275 a month for a regular workweek of 35 hours (7 hours per day) and is paid semimonthly. Overtime is paid at time-and-a-half for work past 7 hours on Monday through Friday and double time for work on Saturdays, Sundays, and holidays. Calculate her gross pay for the half-month she worked, as shown below (8 points).

Date	1	2	3	4	5	6	7	8	9	10	11	12	13	14	15
Day	W	Th	F	Sa	S	M	T	W	Th	F	Sa	S	M	T	W
Hours	7	9	7	0	4	7	9	7	8	7	4	1	7	8	9

Gross Pay _____

17. Joseph O'Reilly earns $1,950 a month for a regular workweek of 36 hours (8 hours per day Monday through Thursday, and 4 hours on Friday). Overtime is paid at time-and-a-half for Monday through Friday, double time for Saturdays, and triple time for Sundays and legal holidays. Shown below is the calendar for July and the hours that Joseph worked each day. Calculate his gross pay for the entire month. Remember that July 4 is a legal holiday (12 points).

Day	Sun	Mon	Tue	Wed	Thu	Fri	Sat
Date / Hours		1 / 8	2 / 9	3 / 9	4 / 8	5 / 4	6 / 0
Date / Hours	7 / 0	8 / 8	9 / 10	10 / 8	11 / 8	12 / 4	13 / 6
Date / Hours	14 / 0	15 / 8	16 / 8	17 / 10	18 / 9	19 / 4	20 / 0
Date / Hours	21 / 4	22 / 8	23 / 8	24 / 9	25 / 9	26 / 4	27 / 2
Date / Hours	28 / 2	29 / 8	30 / 10	31 / 9			

Gross Pay _____

STUDENT'S SCORE _____

Chapter 11 Calculating Gross Pay for Payrolls

SECTION 11.2 Commissions and Piecework Methods of Payment

Section 11.2 discusses commission and piecework methods of calculating earnings. Both of these methods are similar, because the employee is *paid for performance*. When earnings are based on commission, you earn more when you sell more. When earnings are based on piecework, you earn more when you produce more. Part A discusses commission, and Part B discusses piecework.

PART A Various Plans of Commission Payment

A commission basis is frequently used to pay both wholesale and retail salespeople, as well as employees in some service industries. A commission is generally a percent of the individual's sales for a given period, but it may also be a fixed amount per item sold.

If traveling expenses for the salesperson are involved, the company may advance a **drawing account** (or simply a **draw**), which is a fixed amount allowed for payment of expenses as they occur. This gives the salesperson company money in advance to use for expenses in the coming month. Any unused money in the drawing account at the end of the month must, of course, be returned to the company. Whether or not there is a drawing account, an expense sheet must be submitted to the company by the salesperson, since travel expenses are paid by the company in addition to the commission.

Salespersons find their total sales for a given period from copies of the invoices or sales slips for orders placed with them. From these documents, they can calculate their commissions. To calculate gross amounts due, a salesperson adds the travel expenses to the commission and then subtracts the drawing account from that sum.

The reason for using the term *gross amount* rather than *gross pay* or *gross earnings* is that the company accounts for employee travel expenses differently than it accounts for the commission payment. In fact, there will often be two different checks from different bank accounts.

Methods of payment may be based on straight commission, commission and bonus, or salary and commission.

Straight Commission Plan In a straight commission plan, the amount due is calculated as a given percent of the total sales for the period.

EXAMPLES:

a. Rhonda Reaves receives a 6% commission on her total monthly sales. Find her commission for the month in which her sales were $58,000.

Commission: $58,000 × 6% = $3,480

b. Tom Burke earns a 5% commission on his total sales. Travel expenses are paid by the company, and he is given a drawing account of $1,000 a month. Find his gross amount due for a month in which his sales totaled $60,000 and his expense sheet totaled $945.25.

Commission: $60,000 × 5% =	$3,000.00
Travel Expenses:	+ 945.25
	$3,945.25
Drawing account:	− 1,000.00
Gross amount due:	$2,945.25

Commission and Bonus Plan. In a commission and bonus plan, a straight commission is paid on sales up to a given amount, and graded bonus commissions are paid for sales above that amount. The gross amount due is the sum of the commissions for each bracket.

EXAMPLES:

a. Barbara Nudd receives a 3% commission on the first $30,000 of monthly sales, 4% on the next $40,000, and 5% on all sales above $70,000. Find her gross amount due for a month in which her sales totaled $72,400.

$30,000 × 3% =	$ 900.00
$40,000 × 4% =	1,600.00
$2,400 × 5% =	120.00
Gross amount due	$2,620.00

b. Dave English receives a 4% commission on the first $20,000 of monthly sales, 4.5% on the next $15,000, and 5% on all sales above $35,000. He receives an advance drawing account of $800 a month. Find his gross amount due for a month in which his sales were $47,800 and his travel expenses were $1,024.82.

Commission:
$20,000 × 4% =	$ 800.00
$15,000 × 4.5% =	675.00
$12,800 × 5% =	640.00
Total commission earned:	$2,115.00
Travel expenses:	+ 1,024.82
	$3,139.82
Drawing account:	− 800.00
Gross amount due:	$2,339.82

Commission and Salary Plan. In a salary and commission plan, the gross amount due is the sum of the base salary and the commission earned on total sales.

EXAMPLE:

Donna Guinta receives a base salary of $375 a week, plus a 1.5% commission on her total weekly sales. Find her gross amount due for a week in which her sales totaled $8,650.

Commission: $8,650 × 1.5% =	$129.75
Base salary:	+ 375.00
Gross amount due:	$504.75

Some companies with a commission and salary plan establish a sales goal, or *quota*, for each salesperson. A salesperson who does not meet the quota receives only the base salary. The commission is paid only on the amount which exceeds the quota.

EXAMPLE:

Weston Marshal receives a base salary of $1,250 per month. He is paid a commission of 2% on sales that exceed his monthly quota of $25,000. Find his gross amount due for a month in which his sales totaled $64,200.

$64,200 − $25,000 = $39,200 above quota

Commission: $39,200 × 2% =	$ 784.00
Base salary:	+ 1,250.00
Gross amount due:	$2,034.00

PART B Various Plans of Piecework Payment

Piecework payment can be based on straight piecework, piecework with bonus, or piecework with an hourly base rate.

For each of these three plans, daily reports of the work done by each employee are necessary not only to compute the payroll but also to find total labor expenses (for pricing the finished products). The most common form of daily record is called a **time ticket** or **labor ticket**. This record contains all the information needed for payroll and cost purposes, such as the employee's name and clock number, the date, the pay rate, time worked, pieces produced, order number, job number, job code, etc. The employee or a production clerk records the time worked on each job, the number of pieces produced, and all other necessary information. The labor ticket must be signed by the employee's supervisor.

In the problems in Part B, the employee is assumed to work at only one pay rate. In some shops, however, the employee may work at several jobs, with different pay rates, in the course of one day. The labor ticket in Figure 11–1 is for a job in which 9 pieces were produced in 2.5 hours at a rate of $1.57 per piece.

Many companies now use a computerized production control card instead of separate labor tickets. The computer prints out all the information needed to process an order. There are enough spaces on the front and back of the card to record the production of all the employees who have worked on that order. Upon completion of the order, the information is fed into the computer for payroll and accounting purposes.

Straight Piecework. When payment is based on the total number of pieces produced in a regular workday, the gross daily wages equal the product of the number of pieces produced and the rate per piece.

Many companies, however, combine piecework with a day rate, so that the employee receives whichever amount is greater as that day's wages. This type of payment plan guarantees that the employee receives a reasonable income, even for days on which there is not much work to do. The

LABOR TICKET

Order No. _90-09-2-0015_   Quantity Required _15_   Drawing No. _B-7-117_   Part _pin_

Clock No. _34_

Name _D. Schmidt_   Date _9/2/--_

Job No.	Description of Work	Hours	Rate	Pieces Produced	IN	OUT
4	lathe and trim	2.5	1.57	9	1:30	4:00

Balance Pieces Due _6_   Supervisor _J. B. Gomes_

FIGURE 11-1 A Labor Ticket

---

hourly rate must be high enough to retain good employees but not so high that an employee can work slowly and not worry about the number of pieces produced.

EXAMPLES:

a. Carla Boles receives $1.60 for each piece assembled in a regular 8-hour day or a day rate of $7.50 per hour, whichever is greater. Find her gross pay for the week in which she produced the following numbers of pieces:

M T W Th F
61 68 34 65 59

Day rate: 8 × $7.50 = $60.00

Mon: 61 × $1.60 =	$ 97.60
Tue: 68 × $1.60 =	108.80
Wed: 34 × $1.60	
Use day rate:	60.00
Thu: 65 × $1.60 =	104.00
Fri: 59 × $1.60 =	94.40
Gross pay:	$464.80

b. Dan Slade receives $0.45 for each piece produced in a regular 8-hour day or a day rate of $6.25 per hour, whichever is greater. Find his gross pay for the week in which he produced the following numbers of pieces:

M T W Th F
175 110 202 196 112

Day rate: 8 × $6.25 = $50.00

Mon: 175 × $0.45 =	$ 78.75
Tue: 110 × $0.45	
Use day rate:	50.00
Wed: 202 × $0.45 =	90.90
Thu: 196 × $0.45 =	88.20
Fri: 112 × $0.45 =	50.40
Gross pay:	$358.25

Piecework with Bonus. A straight piecework plan may be combined with a bonus rate for all pieces produced beyond a standard minimum per day. This plan provides an incentive to employees to produce more pieces in a workday.

The daily wages earned under this plan equal the product of the total number of pieces produced and the straight piecework rate, plus the product of the number of pieces over the standard minimum and the bonus rate.

EXAMPLE:

Evan Norris receives $0.35 for each piece assembled in a regular 8-hour day, plus a $0.10 bonus for each piece assembled over the standard minimum of 200 pieces per day. Find his gross pay for the week in which he produced the following numbers of pieces:

M	T	W	Th	F
230	225	195	240	235

Mon: 230 × $0.35 = $80.50
      30 × $0.10 = + 3.00 $83.50

Tue: 225 × $0.35 = $78.75
      25 × $0.10 = + 2.50   81.25

Wed: 195 × $0.35 = $68.25
     No bonus   68.25

Thu: 240 × $0.35 = $84.00
      40 × $0.10 = + 4.00   88.00

Fri: 235 × $0.35 = $82.25
      35 × $0.10 = + 3.50   85.75

Gross pay: $406.75

Piecework with an Hourly Base Rate. A piecework plan of payment may include some time-basis payment at an hourly base rate. The hourly base rate should not be confused with the straight-time or day rate; it is generally much lower than the hourly day rate.

Under this plan, the daily gross earnings equal the product of the number of hours worked in the regular workday and the base rate, plus the product of the number of pieces produced and the piece rate. For overtime, both the base rate *and* the piece rate are multiplied by 1.5 (or 2, or 3) for payment of both the hours worked and the pieces produced during the overtime. This plan may also be combined with a day rate; that is, when an employee does not make enough money on piecework, he or she is paid the day rate for eight hours.

In the following example, a weekly summary card shows regular hours, overtime hours, and pieces for every day. Daily wages can easily be calculated with a table. You can check the weekly total by first summarizing the daily amounts in each pay category and then adding them. You can perform another check by first adding the regular hours and pieces and then calculating the weekly totals.

EXAMPLE:

William Felski is paid $1.30 for each piece produced during a regular 8-hour day and a base rate of $4.00 per hour. Overtime is paid at time-and-a-half. From the information given, calculate his daily wages and the weekly total. Check the calculations by another method.

	M	T	W	Th	F
Regular hours	8	6	8	8	8
Regular pieces	32	25	35	31	34
O.T. hours	2	0	1	0	3
O.T. pieces	7	0	5	0	13

Regular rates:
  Per hour: $4.00
  Per piece: $1.30
Overtime rates:
  Per hour: $4.00 × 1.5 = $6.00
  Per piece: $1.30 × 1.5 = $1.95

	Reg. Hourly	Reg. Piece	O.T. Hourly	O.T. Piece	Daily Total
Mon	8 × 4.00 32.00	32 × 1.30 41.60	2 × 6.00 12.00	7 × 1.95 13.65	$ 99.25
Tue	6 × 4.00 24.00	25 × 1.30 32.50			56.50
Wed	8 × 4.00 32.00	35 × 1.30 45.50	1 × 6.00 6.00	5 × 1.95 9.75	93.25
Thu	8 × 4.00 32.00	31 × 1.30 40.30			72.30
Fri	8 × 4.00 32.00	34 × 1.30 44.20	3 × 6.00 18.00	13 × 1.95 25.35	119.55
Weekly Total	$ 152.00	$ 204.10	$ 36.00	$ 48.75	$440.85

Total reg. hours: 38 × $4.00 = $152.00
Total reg. pcs.: 157 × $1.30 =   204.10
Total O.T. hours: 6 × $6.00 =   36.00
Total O.T. pcs.: 25 × $1.95 =   48.75
Weekly gross pay: $440.85

Notice that the weekly gross pay of $440.85 can be calculated in three different ways: by finding gross pay for each day, by finding the daily amounts in each category and adding them, and by first finding weekly totals for hours and pieces and then finding the weekly amounts in each category.

COMPLETE ASSIGNMENT 34

Name _____

Date _____

ASSIGNMENT 34
SECTION 11.2

Commissions and Piecework Methods of Payment

	PERFECT SCORE	STUDENT'S SCORE
PART A	40	
PART B	60	
TOTAL	100	

PART A Various Plans of Commission Payment

DIRECTIONS: Each salesperson for Bovio Supply Company has a different commission plan, depending on the employee's preference, experience, and time with the company. Find the gross amount due each employee (5 points for each correct answer).

	Employee	Sales	Commission Rate	Salary	Gross Amount Due
1.	B. Cash	$9,000	2% over $5,000	$1,800	_____
2.	F. Cox	$45,000	3% first $30,000 4% over $30,000	$1,250	_____
3.	D. Green	$88,400	3% first $20,000 4% next $40,000 5% over $60,000	$1,000	_____
4.	R. House	$42,000	3% first $25,000 4% over $25,000	$1,200	_____
5.	B. Mason	$109,500	4% first $25,000 5% next $25,000 6% over $50,000	$ 0	_____
6.	N. Miles	$31,000	2% over $10,000	$1,500	_____

	Employee	Sales	Commission Rate	Salary	Gross Amount Due
7.	W. Smart	$4,000	0.5% on all sales	$1,600	_____
8.	T. Young	$56,000	3% first $30,000 4% next $30,000 5% over $60,000	$ 900	_____

STUDENT'S SCORE _____

PART B Various Plans of Piecework Payment

DIRECTIONS: Garden Plastics Manufacturing pays its employees either hourly wages or piecework wages, whichever is greater for the 8-hour day. Piecework wages include a bonus for every piece exceeding a standard minimum. Find the daily earnings of each employee (5 points for each correct answer).

	Employee	Hourly Rate	Piece Rate	Bonus Rate	Standard Minimum	Pieces Produced	Daily Earnings
9.	G. Tower	$5.65	$3.75	$1.90	17	17	_____
10.	C. Chu	$4.75	$0.20	$0.07	250	317	_____
11.	D. Young	$5.30	$1.20	$0.40	45	34	_____
12.	B. Tong	$6.45	$0.85	$0.30	80	85	_____
13.	V. Bates	$4.20	$0.60	$0.20	74	57	_____
14.	J. Cruz	$5.15	$0.45	$0.15	120	138	_____

Chapter 11 Calculating Gross Pay for Payrolls

DIRECTIONS: Warner Implement Company pays its employees on a piecework basis combined with an hourly base rate for work produced in an 8-hour day. Overtime is paid at 1.5 times the piece rate and the hourly base rate. Find the daily earnings of each employee (5 points for each correct answer).

	Employee	Hourly Base	Hours Worked Regular	Hours Worked O.T.	Piece Rate	Pieces Produced Regular	Pieces Produced O.T.	Daily Earnings
15.	M. Lang	$5.35	6	0	$0.10	123	0	_____
16.	N. Romig	$6.05	7	0	$0.25	72	0	_____
17.	P. White	$6.45	8	0	$0.45	38	0	_____
18.	W. Beard	$5.25	8	0	$0.55	57	0	_____
19.	U. Hardy	$6.20	8	2	$0.70	24	6	_____
20.	L. True	$6.80	8	3	$1.20	12	4	_____

STUDENT'S SCORE _____

CHAPTER 12

Inventory Valuation, Cost of Goods Sold, and Depreciation

An important function of accounting is providing the information needed to prepare a company's financial statements. A **fiscal period** is that period of business at the conclusion of which the operations are summarized and presented in financial statements for analysis. A fiscal period can be a period of time such as a month, a quarter, a half-year, or a year.

In Chapter 10, we introduced the terms *gross profit* and *net profit*. Gross profit is the difference between sales revenue and the cost of the merchandise. In accounting, the term *cost* is often called the **cost of goods sold (CGS)**. In order to find the CGS, you must know the value of the inventory at the beginning of the fiscal period, the net cost of the goods purchased, and the value of the inventory at the end of the fiscal period. Finding the ending inventory value and the cost of goods sold is discussed in Section 12.1.

Net profit in a fiscal period is calculated by subtracting the operating expenses for that period from the gross profit. Most expenses, such as payroll, supplies, utilities, insurance, etc., are straightforward. A more complicated expense category is **depreciation**. Some assets owned by a business are used for several years and normally decrease in value as they age. Although such an asset is actually purchased in a single fiscal period, it is necessary to determine how much of that asset's cost is used up as a business expense during each period. The systematic procedure to allocate these periodic expenses is called *depreciation*. Section 12.2 discusses some common methods of calculating depreciation expenses.

SECTION 12.1 Inventory Valuation and Cost of Goods Sold

Merchandise inventory is the term for all the goods owned by a business and available for sale in the hope of making a profit. Most companies try to keep a running inventory, or **perpetual inventory**, of the stock on hand. In a large company, the inventory is recorded in a computer; in a small company, it may be recorded on stock record cards. As a check on the perpetual inventory, a *physical inventory* is taken at least once a year. During the physical inventory, each item in stock is counted, weighed, or measured by hand.

After the physical inventory, the value of the merchandise in stock is computed. The calculations may be based on either the *unit cost* of the goods or their *current market value*. In Parts A through D of this section, only a physical inventory is considered in finding the cost of ending inventory.

Although the quantities of items in stock can usually be easily determined, the actual cost of each unit is more difficult to determine. The cost of each unit may even be impossible to know when all of the items (such as screwdrivers, pliers, scissors, silverware, boxes of paper clips, etc.) look the same and do not have identifying serial marks. The inventory at the end of the year will probably contain some identical items that were purchased at different times during the year and *at different prices*. The problem, therefore, may not be finding the actual cost of each item but determining what unit cost to *assign* to each item.

There are four common methods of inventory valuation, with the choice of method depending on the kind of business. These methods are:

1. Specific identification
2. First-in, first-out (FIFO)
3. Last-in, first-out (LIFO)
4. Average cost

PART A Specific Identification

The **specific identification method** is used if the actual cost of each unit in the ending inventory is available. For example, each item may have an attached price tag, with the cost in code or a specific invoice reference. Finding the ending inventory cost is then only a matter of multiplying the quantities of each item by its unit cost.

EXAMPLE:

Find the cost of the ending inventory by specific identification.

10 units @ $12 =	$120
17 units @ $9 =	$153
26 units @ $14 =	$364
Cost of ending inventory:	$637

PART B First-In, First-Out (FIFO)

If the cost of individual units cannot be specifically identified, the only way to determine an inventory value is to make an assumption about which units have been sold during the fiscal period.

In the **first-in, first-out (FIFO)** method, it is assumed that merchandise is sold in the order in which it was received. That is, the oldest goods in stock at the beginning of the fiscal period (first-in) are assumed to be the first ones sold (first-out); then, the goods purchased during the fiscal period are assumed to be sold in the order that they were bought. Thus, the unsold goods at the end of the fiscal period, when the ending inventory is taken, are assumed to consist of the most recent purchases.

In order to show the different inventory values obtained with the three methods (first-in, first-out; last-in, first-out; and average cost), the following example will be used to illustrate each method. In this example, assume that an actual count reveals that there are 250 units in the ending inventory.

Item		Number of Units	Unit Cost
Beginning inventory	(Jan.)	150	$6.00
Purchase 1	(Mar.)	400	5.50
Purchase 2	(June)	200	7.00
Purchase 3	(Aug.)	600	6.50
Purchase 4	(Nov.)	200	7.75

EXAMPLE:

Find the cost of the ending inventory by using FIFO.

The most recent purchases are the units remaining in the ending inventory. Thus, the 250 units in the ending inventory consist of 200 units from Purchase 4 and 50 units from Purchase 3.

200 × $7.75 =	$1,550.00
50 × $6.50 =	325.00
Cost of ending inventory =	$1,875.00

PART C Last-In, First-Out (LIFO)

In the **last-in, first-out (LIFO)** method, it is assumed that the most recently purchased merchandise (last-in) has been sold (first-out) and that the ending inventory consists of the oldest stock. Since market prices usually rise during a fiscal period, the LIFO method will cause the cost of goods sold to more accurately reflect the actual cost of replacing those goods.

EXAMPLE:

Find the cost of the ending inventory by using LIFO.

The 250 units in the ending inventory consist of the 150 units in the beginning inventory and 100 units from Purchase 1.

150 × $6.00 =	$ 900.00
100 × $5.50 =	550.00
Cost of ending inventory =	$1,450.00

It is easy to confuse the procedures used in the FIFO and LIFO methods. It may be helpful to lightly cross out those units that are assumed to have been sold and thus removed from the physical inventory.

In the FIFO method, the *first* units to be removed from the inventory are the first ones in the inventory. Thus, cross out the units in the beginning inventory, because these units are the *first out*. Cross out the additional units that have been sold, until only those in the ending inventory remain. Cross out the units in the order that they were purchased.

EXAMPLE:

Find the cost of the ending inventory in the example from Part B by using FIFO.

Item	Number of Units	Unit Cost	
~~Beginning inventory~~	~~150~~	~~$6.00~~	
~~Purchase 1~~	~~400~~	~~5.50~~	
~~Purchase 2~~	~~200~~	~~7.00~~	
Purchase 3	600	6.50	 50
Purchase 4	200	7.75	200
			250

50 × $6.50 = $ 325.00
200 × $7.75 = 1,550.00
Cost of ending inventory = $1,875.00

In the LIFO method, the *most recent* purchase (last-in) will be the first one crossed out (first-out). Cross out the additional units that have been sold, until only those in the ending inventory remain. Cross out the units in the *reverse* order of their purchase.

EXAMPLE:

Find the cost of the ending inventory in the preceding example by using LIFO.

Beginning inventory	150	$6.00 150
Purchase 1	400	5.50 100
~~Purchase 2~~	~~200~~	~~7.00~~
~~Purchase 3~~	~~600~~	~~6.50~~
~~Purchase 4~~	~~200~~	~~7.75~~

100 × $5.50 = $ 550.00
150 × $6.00 =   900.00
Cost of ending inventory = $1,450.00

Note that the results are not the same. In this case, FIFO gives an ending inventory cost of $1,875, and LIFO gives $1,450.

PART D Average Cost

FIFO or LIFO may be used when the cost of inventory items cannot be specifically identified. However, if the costs of the items are indistinguishable, there is no guarantee that either the oldest *or* the newest items are sold first. Another method of inventory valuation is to find the average cost of each item that has been in the inventory during the fiscal period. This is called the **average cost** or **weighted average cost method**.

To calculate the average cost per unit, you must first find the total cost of all units and the total number of units. To do this, multiply each unit cost by the number of units that were purchased at that cost. The total cost is the sum of all these products. The total number of units is the beginning inventory plus all the units that were purchased during the fiscal period.

The average cost per unit is equal to the total cost divided by the total number of units:

$$\text{Avg. cost/unit} = \frac{\text{Total cost of all units}}{\text{Total number of units}}$$

After finding the average cost per unit, multiply it by the number of units in the ending inventory. This product is the cost of the ending inventory by the average cost method.

$$\begin{matrix}\text{Cost of ending}\\ \text{inventory}\end{matrix} = \begin{matrix}\text{Number of units in ending}\\ \text{inventory} \times \text{avg. cost/unit}\end{matrix}$$

EXAMPLE:

Find the cost of the ending inventory in the previous example by using the average cost method.

150 × $6.00 = $ 900
400 × $5.50 = 2,200
200 × $7.00 = 1,400
600 × $6.50 = 3,900
200 × $7.75 = 1,550
1,550 $9,950

$9,950 ÷ 1,550 = $6.4194 *or* $6.42
                (avg. cost/unit)

250 × $6.42 = $1,605.00 (cost of ending inventory)

PART E Finding the Cost of Goods Sold

To calculate gross profit, subtract the **cost of goods sold (CGS)** from sales revenue (see Chapter 10). The purpose of determining the ending inventory cost in Parts A through D was to enable us to calculate the CGS. The formula for finding CGS is:

CGS = Beginning inventory + Net purchases − Ending inventory

Net purchases is equal to total purchases minus adjustments to purchases.

Net purchases = Purchases − Adjustments

An example of an adjustment is *purchases returns*, which occur when an item is unsatisfactory and must be sent back to the manufacturer or vendor. Another adjustment to purchases is *purchases discounts*, which occur when the vendor offers a discount for prompt payment. *Ending inventory* may be found by any of the methods discussed in Parts A through D.

EXAMPLE:

From the information given for Bovio Supply Company, find the CGS for the period from January 1 to December 31.

Inventory, Jan. 1	$57,400
Purchases	79,200
Purchases returns	1,600
Inventory, Dec. 31	50,600

Net purchases = $79,200 − $1,600
            = $77,600

CGS = $57,400 + $77,600 − $50,600
    = $84,400

Figure 12-1 illustrates the placement of the cost of goods sold from this example in a partial income statement.

BOVIO SUPPLY COMPANY
Income Statement
For Year Ended December 31, 19—

Income:			
Sales			$189,750
Less: Sales returns and allowances		− 3,250	
Net sales			$186,500
Cost of goods sold:			
Inventory, January 1		$57,400	
Purchases	$79,200		
Less: Purchases returns	− 1,600		
Net purchases		77,600	
Goods available for sale		135,000	
Less: Inventory, December 31		− 50,600	
Cost of goods sold			84,400
Gross profit on sales			$102,100

FIGURE 12-1 A Partial Income Statement

Chapter 12 Inventory Valuation, Cost of Goods Sold, and Depreciation

EXAMPLE:

From the information given, find the CGS for the period from April 1 to June 30. Note that there are two adjustments to Purchases.

Inventory, April 1	$251,200
Purchases	175,000
Purchases returns	4,800
Purchases discounts	7,600
Inventory, June 30	239,400

Total adjustments = $4,800 + $7,600
 = $12,400

Net purchases = $175,000 − $12,400
 = $162,600

CGS = $251,200 + $162,600 − $239,400
 = $174,400

PART F Estimating Inventory Value at Cost

There are times in the operation of a business (other than the end of a fiscal period) when the value of the current inventory is needed. A physical inventory may be too time-consuming, too expensive, or even impossible, as in the case of fire. There are two commonly used methods of estimating the current inventory value at cost without taking a physical count: the gross profit method and the retail method.

Gross Profit (or Markup) Method. In accounting, the term *gross profit* is used instead of *markup*. The gross profit method is based on the following markup equation:

Cost = Sales − Markup

or

Cost = Sales − Gross Profit

To estimate the cost of the current inventory with the gross profit method, follow these four steps:

1. Find the approximate amount of gross profit to date.

   Multiply the net sales to date (this period) by the gross profit rate from the previous period. This product is the approximate amount of gross profit to date.

2. Find the CGS to date.

   Subtract the approximate amount of gross profit to date (found in Step 1) from net sales to date. This difference is the CGS to date.

3. Find the cost of goods available for sale to date.

   Add the net purchases to date to the cost of the beginning inventory. This sum is the cost of goods available for sale to date.

4. Find the cost of the current inventory.

   Subtract the CGS to date (found in Step 2) from the cost of goods available for sale to date (found in Step 3). This difference is the cost of the current inventory.

EXAMPLES:

a. Estimate the current inventory value at cost from the following information:

Beginning inventory	$ 900
Net purchases to date	9,050
Net sales to date	11,800
Gross profit rate	30%

Step 1.	Net sales to date	$11,800
	Gross profit rate	× 0.30
	Est. gross profit to date	$ 3,540

Step 2.	Net sales to date	$11,800
	Est. gross profit to date	− 3,540
	CGS to date	$ 8,260

Step 3.	Beginning inventory	$ 900
	Net purchases to date	+ 9,050
	Cost of goods available for sale to date	$9,950

Step 4.	Cost of goods available for sale to date	$9,950
	CGS to date	− 8,260
	Current inventory at cost (approximate)	$1,690

b. Estimate the current inventory value at cost from the following information:

Beginning inventory	$ 96,000
Net purchases to date	164,000
Net sales to date	240,000
Gross profit rate	25%

Step 1. Net sales to date $240,000
Gross profit rate × 0.25
Est. gross profit to date $60,000

Step 2. Net sales to date $240,000
Est. gross profit to date − 60,000
CGS to date $180,000

Step 3. Beginning inventory $ 96,000
Net purchases to date + 164,000
Cost of goods available
  for sale to date $260,000

Step 4. Cost of goods available
  for sale to date $260,000
CGS to date − 180,000
Current inventory at cost $ 80,000
  (approximate)

Retail Method. Most department stores, supermarkets, and other retail enterprises prepare monthly income statements in order to analyze sales. The retail method is used to estimate the current inventory value at cost.

As goods are received, they are marked with the expected retail prices. Records are kept to show both the net purchases and the expected sales on this merchandise.

To estimate the cost of the current inventory with the retail method, follow these five steps:

1. Find the goods available for sale at *cost* to date. This is the same as the *cost of goods available for sale.*

   Add the net purchases to date (at cost) to the beginning inventory (at cost). This sum is the goods available for sale at *cost* to date.

2. Find the goods available for sale at *retail* to date.

   Add the net purchases to date (at retail) to the beginning inventory (at retail). This sum is the goods available for sale at *retail* to date.

3. Find the cost/selling price ratio.

   Divide the goods available for sale at *cost* to date (found in Step 1) by the goods available for sale at *retail* to date (found in Step 2). This quotient is the cost/selling price ratio.

4. Find the current inventory at *retail*.

   Subtract the net sales to date from the goods available for sale at *retail* to date (found in Step 2). This difference is the current inventory at *retail*.

5. Find the current inventory at *cost*.

   Multiply the current inventory at *retail* (found in Step 4) by the cost/selling price ratio (found in Step 3). This product is the approximate current inventory at *cost*.

EXAMPLES:

a. Use the retail method to estimate the current inventory value at *cost* from the following information:

	Cost	Retail
Beginning inventory	$ 85,100	$133,200
Net purchases to date	160,600	244,800
Net sales to date		252,000

Step 1. Beg. inventory (cost) $ 85,100
Net purchases (cost) + 160,600
Goods available for sale $245,700
  at *cost* to date

Step 2. Beg. inventory (retail) $133,200
Net purchases (retail) + 244,800
Goods available for sale $378,000
  at *retail* to date

Step 3. Cost/selling price ratio:
$245,700 ÷ $378,000 = 0.65

Step 4. Goods available for sale
  at *retail* to date $378,000
Net sales to date − 252,000
Current inventory at *retail* $126,000

Step 5. Current inventory at *retail* $126,000
Cost/selling price ratio × 0.65
Current inventory at *cost* $ 81,900
  (approximate)

b. Use the retail method to estimate the current inventory value at *cost* from the following information:

	Cost	Retail
Beginning inventory	$ 32,000	$ 53,000
Net purchases to date	136,000	227,000
Net sales to date		230,000

Step 1. Beg. inventory (cost) $ 32,000
        Net purchases (cost)   + 136,000
        Goods available for sale $168,000
          at *cost* to date

Step 2. Beg. inventory (retail) $ 53,000
        Net purchases (retail)   + 227,000
        Goods available for sale $280,000
          at *retail* to date

Step 3. Cost/selling price ratio:
      $168,000 ÷ $280,000 = 0.60

Step 4. Good available for sale
         at *retail* to date $280,000
        Net sales to date   − 230,000
        Current inventory at *retail* $ 50,000

Step 5. Current inventory at *retail* $ 50,000
        Cost/selling price ratio   × 0.60
        Current inventory at *cost* $ 30,000
         (approximate)

Other factors, such as markups and markdowns, also affect the current inventory value at cost; however, they are not considered in this chapter.

COMPLETE ASSIGNMENT 35

ASSIGNMENT 35
SECTION 12.1

Name _____

Date _____

Inventory Valuation and Cost of Goods Sold

	Perfect Score	Student's Score
PART A	8	
PARTS B, C, D	48	
PART E	20	
PART F	24	
TOTAL	100	

PART A Specific Identification

DIRECTIONS: Find the cost of ending inventory with the specific identification method (4 points for each correct final answer).

1. 12 units @ $2.80 = _____
   50 units @ $3.17 = _____
   125 units @ $0.95 = _____
   40 units @ $1.43 = _____

   Ending inventory cost = _____

2. 316 units @ $0.62 = _____
   490 units @ $0.41 = _____
   625 units @ $0.29 = _____
   910 units @ $0.56 = _____

   Ending inventory cost = _____

STUDENT'S SCORE _____

PARTS B, C, AND D FIFO, LIFO, and Average Cost

DIRECTIONS: Find the cost of ending inventory by using the FIFO, LIFO, and average cost methods (4 points for each correct answer).

	Units	Cost/Unit
Inventory: Jan. 1, 19--	175	$6.20
Purchase 1 (Jan. 31)	375	6.30
Purchase 2 (Feb. 23)	250	6.45
Purchase 3 (March 18)	125	6.50

Inventory on March 31, 19--: 290 units

3. FIFO _____   4. LIFO _____   5. Avg. cost _____

	Units	Cost/Unit
Inventory: April 1, 19--	75	$3.20
Purchase 1 (April 15)	100	3.25
Purchase 2 (May 1)	75	3.30
Purchase 3 (May 25)	100	3.35
Purchase 4 (June 15)	50	3.30

Inventory on June 30, 19--: 200 units

6. FIFO _____ **7.** LIFO _____ **8.** Avg. Cost _____

	Units	Cost/Unit
Inventory: Oct. 1, 19--	300	$4.52
Purchase 1 (Oct. 20)	400	4.52
Purchase 2 (Nov. 5)	600	4.55
Purchase 3 (Nov. 20)	400	4.58
Purchase 4 (Dec. 20)	400	4.56

Inventory on Dec. 31, 19--: 900 units

9. FIFO _____ **10.** LIFO _____ **11.** Avg. Cost _____

	Units	Cost/Unit
Inventory: July 1, 19--	1,000	$9.00
Purchase 1 (Aug. 8)	600	9.20
Purchase 2 (Aug. 22)	1,200	8.80
Purchase 3 (Sept. 12)	400	9.30
Purchase 4 (Sept. 28)	800	9.40

Inventory on Sept. 30, 19--: 1,800 units

12. FIFO _____ **13.** LIFO _____ **14.** Avg. Cost _____

STUDENT'S SCORE _____

Chapter 12　Inventory Valuation, Cost of Goods Sold, and Depreciation

PART E Finding the Cost of Goods Sold

DIRECTIONS:　Find the CGS. Show your work in income-statement form in the space to the right (4 points for each correct answer).

15. Inventory, Oct. 1　　　　$190,000
　　Inventory, Dec. 31　　　 160,000
　　Net purchases　　　　　 580,000

　　CGS　　　　　　　　 _____

16. Inventory, July 1　　　　$173,800
　　Inventory, Sept. 30　　　 143,500
　　Purchases　　　　　　　561,000
　　Purchases returns　　　　24,300

　　CGS　　　　　　　　 _____

17. Inventory, April 1　　　　$ 85,600
　　Inventory, June 30　　　　74,400
　　Purchases　　　　　　　278,800
　　Purchases discounts　　　 3,200

　　CGS　　　　　　　　 _____

18. Inventory, Jan. 1　　　　$18,800
　　Inventory, June 30　　　　14,300
　　Purchases　　　　　　　99,800
　　Purchases returns　　　　3,700
　　Purchases discounts　　　 900

　　CGS　　　　　　　　 _____

19. Inventory, Jan. 1　　　　$ 54,000
　　Inventory, Dec. 31　　　　47,500
　　Purchases　　　　　　　219,000
　　Purchases returns　　　　8,500
　　Purchases discounts　　　1.5% of gross
　　　　　　　　　　　　　purchases

　　CGS　　　　　　　　 _____

STUDENT'S SCORE _____

PART F Estimating Inventory Value at Cost

DIRECTIONS: Find the estimated inventory values. Show your work in the space provided. In Problems 20–22, use the gross profit method; in Problems 23–25, use the retail method (4 points for each correct answer).

20. Net sales to date $120,000   Est. inventory value _____
      Inventory, Jan. 1   48,000
      Purchases   82,000
      Purchases returns   2,000
      Gross profit %, last period   30%

21. Net sales to date $112,500   Est. inventory value _____
      Inventory, July 1   21,400
      Purchases   79,500
      Purchases returns   3,200
      Purchases discounts   1,200
      Gross profit %, last period   35%

22. Net sales to date $640,000   Est. inventory value _____
      Inventory, Jan. 1   72,000
      Purchases   422,000
      Purchases returns   5,000
      Purchases discounts   2,000
      Gross profit %, last period   40%

Chapter 12 Inventory Valuation, Cost of Goods Sold, and Depreciation

23.

	Cost	Retail	Est. inventory value
Inventory, Jan. 1	$110,000	$170,000	_____
Net purchases to date	202,000	310,000	
Net sales to date		315,000	

24.

	Cost	Retail	Est. inventory value
Inventory, Apr. 1	$ 23,400	$ 46,500	_____
Net purchases to date	111,800	213,500	
Net sales to date		195,400	

25.

	Cost	Retail
Inventory, Oct. 1	$11,300	$ 22,400
Net purchases to date	93,200	167,600
Net sales to date		160,000

Est. inventory value _____

STUDENT'S SCORE _____

SECTION 12.2 Calculating Depreciation

Businesses and individuals often need to determine their **net worth**, which is essentially the difference between the value of their *assets* (what they own) and their *liabilities* (what they owe to others). Small businesses and individuals need to know their net worth whenever they go to a bank for a loan. Large businesses need to report their net worth to their stockholders.

To determine the net worth of a business, you must first determine the value of assets such as trucks, machinery, tools, office furniture, typewriters, calculators, and computers. Most of these assets will not be brand new. Furthermore, it is unlikely that they were all purchased at the same time or that they all have the same life expectancy. A truck will usually wear out before a trailer; a computer will be obsolete before it is physically worn out. Each of a business's assets is at a different stage in its own particular useful life span.

In attempting to make a reasonable estimate of the declining value of its assets, a business deducts part of the cost of its assets every year. This is called *depreciation*, and the business is said to be *depreciating* its assets. There are several methods of calculating depreciation; two important ones are the *straight-line (SL) method* and the *declining-balance (DB) method*. Another method, traditional but somewhat less important today because of new tax laws, is the *sum-of-the-years-digits (SOYD) method*. All of these methods are discussed in this section.

For businesses, another important reason to calculate depreciation is to accurately complete their income tax returns each year. The government requires a business to pay a percentage of its income as taxes. Before calculating the percentage, however, the business subtracts from its income all of its legitimate expenses during the year. A business that wants to maximize its after-tax profit tries to minimize its taxes.

Suppose that a business buys a pickup truck for $15,000 in 1990. The government will not permit that business to claim an expense of $15,000 in 1990. Because the truck will be used more than one year, only *part* of the $15,000 may be deducted in 1990. Another part of the $15,000 may be deducted in 1991, and so on. The part that is deducted each year is also called *depreciation*. The tax laws concerning depreciation are complex, as are all tax laws affecting businesses.

Prior to 1981, tax laws involved the depreciation methods mentioned previously. In 1981, Congress passed the Economic Recovery Tax Act, which introduced a new method of depreciation—the *accelerated cost recovery system (ACRS)*. Then, Congress passed the Tax Reform Act of 1986, which introduced another new method of depreciation—the *modified accelerated cost recovery system (MACRS)*. Both the straight-line (SL) and the declining-balance (DB) methods play important roles in MACRS. Although it is beyond the scope of this book to explain the new tax laws, Part E will illustrate how the SL and DB methods might be used in MACRS.

The objective of Section 12.2 is to introduce some common concepts involved in calculating depreciation for both financial statements and income tax purposes. Some companies use one method to calculate depreciation for their financial statements and another method (as specified by the government) for their income tax returns.

There are some terms that you must know to understand the fundamentals of depreciation. Some of them are:

1. The *original cost* includes not only the purchase price of the asset but also any sales taxes, shipping costs, installation charges, etc.
2. The *estimated life* of the asset is an estimate of the length of time that the asset will be useful. It is usually stated in years, but it may be stated in some other unit, such as miles driven, hours used, or pieces produced.
3. The estimated *salvage*, *scrap*, or *resale value* is an estimate of the asset's worth at the end of its useful life. However, scrap value is not taken into consideration under MACRS.
4. The *accumulated depreciation* is the total of all the annual depreciation amounts since the purchase of the asset.
5. The *book value* is the difference between the original cost of the asset and the accumulated depreciation. It is merely an estimate of the value of an asset at a certain point in time. The book value may or may not be close to the actual market value of the asset at that time.

PART A Straight-Line Method (SL)

The simplest way to calculate depreciation is to assume that the asset decreases in value by the same amount each year. This is called the **straight-line (SL) method**. When the original cost of the asset, its estimated life, and its scrap value are known, the annual depreciation can be calculated with the following formula:

$$\text{Annual Depreciation} = \frac{\text{Original cost} - \text{Scrap value}}{\text{Estimated life}}$$

EXAMPLES:

a. Use the SL method to determine the annual depreciation of an assset with an original cost of $9,000, an estimated useful life of 3 years, and an estimated scrap value of $600 at the end of three years.

$$\text{Annual Depreciation} = \frac{\$9,000 - \$600}{3}$$
$$= \frac{\$8,400}{3}$$
$$= \$2,800$$

b. For the asset in *Example a*, construct a table showing its annual depreciation, its accumulated depreciation, and its book value at the end of each year.

Year	Annual Deprec.	Accumulated Deprec.	Ending Book Value
1	$2,800	$2,800	$6,200
2	2,800	5,600	3,400
3	2,800	8,400	600

Notice that the book value of the asset at the end of three years is equal to its estimated scrap value. This is also true with each of the other traditional methods of calculating depreciation (the SOYD method and the DB method).

PART B Sum-of-the-Years-Digits Method (SOYD)

Depreciation is easy to calculate with the SL method, because it is assumed that the asset decreases in value at the same rate every year. However, it is not usually realistic to make this simple assumption. As anyone who has compared the prices of used and new cars knows, assets often decline in value more rapidly at the beginning of their lives. Methods of depreciating assets more rapidly in the beginning are called *accelerated* methods.

A traditional method of calculating accelerated depreciation is the **sum-of-the-years-digits (SOYD) method**, which is found in most accounting textbooks because it may be used for financial statements. Prior to 1981, SOYD was acceptable for income tax purposes. If an asset was purchased prior to 1981 and was depreciated with SOYD, then SOYD must still be used for the entire life of that asset.

When using SOYD, you must know the asset's original cost, estimated life, and estimated scrap value. To calculate the annual depreciation with SOYD, follow these steps:

1. Add up all the digits between 1 and the life of the asset, *inclusive*. For an asset with a five-year life, the digits are 1, 2, 3, 4, and 5. Adding these digits gives a sum of 15 (1 + 2 + 3 + 4 + 5).
2. Divide each of the digits given in Step 1 by the sum of those digits, obtained in Step 1. The resulting fractions are the annual depreciation rates, but they must be arranged in decreasing order. For an asset with a five-year life, this would be $\frac{5}{15}, \frac{4}{15}, \frac{3}{15}, \frac{2}{15},$ and $\frac{1}{15}$.
3. Find the difference between the original cost of the asset and its estimated scrap value. Then, multiply this difference by the appropriate depreciation rate to find the amount of annual depreciation. If the cost of an asset with a five-year life is $6,500 and its estimated scrap value is $500, the difference would be $6,000 ($6,500 − $500). The depreciation for the first year would be $2,000 ($6,000 × $\frac{5}{15}$).

EXAMPLES:

a. Use the SOYD method to determine the annual depreciation for an asset with a cost of $9,000, an estimated useful life of 3 years, and an estimated scrap value of $600 at the end of three years.

Step 1. For a three-year life, add the digits 1, 2, and 3 to get 6.

Step 2. Divide each digit by 6 to get the annual depreciation rates. These rates, in decreasing order, are $\frac{3}{6}, \frac{2}{6},$ and $\frac{1}{6}$.

Step 3. The difference between the cost and the estimated scrap value is $8,400 ($9,000 − $600). When you multiply $8,400 by each rate, the annual depreciation for each year is as follows:

Year	Deprec. Rate		Cost Minus Scrap Value		Deprec.
1	$\frac{3}{6}$	×	$8,400	=	$4,200
2	$\frac{2}{6}$	×	$8,400	=	2,800
3	$\frac{1}{6}$	×	$8,400	=	1,400

Total depreciation = $8,400

b. For the asset in *Example a*, construct a table showing its annual depreciation, its accumulated depreciation, and its book value at the end of each year.

Year	Annual Deprec.	Accumulated Deprec.	Ending Book Value
1	$4,200	$4,200	$4,800
2	2,800	7,000	2,000
3	1,400	8,400	600

PART C Declining-Balance Method (DB)

The **declining-balance (DB) method** is perhaps the most widely used accelerated method, and it has several variations. Not only is it used for financial statements, but it is also very important in MACRS. In the DB method, *the depreciation each year is always the same percentage of the beginning book value that year.* Furthermore, the percentage rate can vary for different assets, in order to increase or decrease the annual depreciation and more accurately reflect the decreasing value of each asset.

The procedure for calculating depreciation with the DB method is outlined in the following material. The most important item in the DB method is the depreciation rate itself. In the SOYD method, the rate decreases every year; in the DB method, the rate is constant.

1. *Calculating the depreciation rate.* There are two steps involved in calculating the actual depreciation rate. To get an idea about the rate, first recall *Example a* on page 374 in Part A. In that example, $8,400 was divided by 3 years. Equivalently, $8,400 could have been multiplied by $\frac{1}{3}$. For this reason, $\frac{1}{3}$ is sometimes called the *straight-line rate for 3 years*.

The first step is to determine the SL rate. Then, to *accelerate* the depreciation, multiply the SL rate by a number greater than 1, such as 2 (200%) or 1.5 (150%). Other numbers, such as 1.25, are possible, but not common; numbers greater than 2 are not used.

If you use the factor 2 with the SL rate of $\frac{1}{3}$, the DB depreciation rate is $\frac{2}{3}$ ($\frac{1}{3} \times 2$). If you use the factor 1.5 with the SL rate of $\frac{1}{3}$, the DB depreciation rate is $\frac{1}{2}$ ($\frac{1}{3} \times 1.5$ or $\frac{1}{3} \times \frac{3}{2}$).

When the SL rate is multiplied by the factor 2, this depreciation method is often called the *double-declining-balance method*. Essentially, the SL rate is being *doubled*. Likewise, when the SL rate is multiplied by the factor 1.5, this method may be called the *one-and-one-half-times-declining-balance method*. The federal income tax guides refer to these two methods as the *200%-declining-balance (200%-DB)* and the *150%-declining-balance (150%-DB)* methods, respectively.

Even though these methods are named with the term *percent*, taxpayers are instructed to multiply the SL rate (called the *basic rate*) by 2 or 1.5, respectively. You may find references to 125%-declining-balance calculations, because the government required them on certain properties a few years ago. The SL rate was simply multiplied by 1.25.

EXAMPLES:

For assets with the given life expectancies, find the depreciation rates for both the 200%-DB and the 150%-DB methods. $\frac{3}{2}$ will be used in the 150%-DB method, because $150\% = 1.5 = \frac{3}{2}$.

Expected Life	SL Rate	200%-DB Rate	150%-DB Rate
a. 3 years	$\frac{1}{3}$	$\frac{1}{3} \times 2 = \frac{2}{3}$	$\frac{1}{3} \times \frac{3}{2} = \frac{1}{2}$
b. 5 years	$\frac{1}{5}$	$\frac{1}{5} \times 2 = \frac{2}{5}$	$\frac{1}{5} \times \frac{3}{2} = \frac{3}{10}$
c. 10 years	$\frac{1}{10}$	$\frac{1}{10} \times 2 = \frac{1}{5}$	$\frac{1}{10} \times \frac{3}{2} = \frac{3}{20}$
d. 15 years	$\frac{1}{15}$	$\frac{1}{15} \times 2 = \frac{2}{15}$	$\frac{1}{15} \times \frac{3}{2} = \frac{1}{10}$
e. 20 years	$\frac{1}{20}$	$\frac{1}{20} \times 2 = \frac{1}{10}$	$\frac{1}{20} \times \frac{3}{2} = \frac{3}{40}$

IMPORTANT! The depreciation rate is calculated at the beginning, when the asset is new. That *same* rate is used to calculate the annual depreciation for that asset every year.

2. *Calculating the annual depreciation.* To find the depreciation of an asset in a particular year with the DB method, first find the desired rate, as previously described. Then, multiply the rate by the book value of the asset at the *beginning* of that year.

3. *Calculating the new book value.* After finding the annual depreciation for a particular year, subtract it from the book value at the *beginning* of that year. The difference is the asset's new book value at the *end* of that year. Of course, this is also the asset's book value for the beginning of the next year.

4. *Adjust the depreciation if necessary.* If there is a scrap value, it must be equal to the ending book value in the final year. Thus, the amount of depreciation in the final year will usually have to be adjusted slightly. This is illustrated in the following examples. As mentioned earlier, MACRS does not take scrap value into consideration. A different adjustment for MACRS will be explained in Part E.

EXAMPLES:

An asset has a cost of $9,000, an estimated life of 3 years, and a scrap value of $600. Construct a table to show the annual depreciation with the 200%-DB and the 150%-DB methods.

a. For a 3-year asset, the depreciation rate with the 200%-DB method is $\frac{2}{3}$ ($\frac{1}{3} \times 2$).

Year	Beginning Book Value	Deprec. Rate	Annual Deprec.	Ending Book Value
1	$9,000	$\frac{2}{3}$	$6,000	$3,000
2	3,000	$\frac{2}{3}$	2,000	1,000
3	1,000	$\frac{2}{3}$	400*	600

*In Year 3, $\frac{2}{3}$ times $1,000 equals $667. However, that is too much depreciation, because the ending book value must equal the scrap value of $600. Therefore, the annual depreciation is adjusted to $400 (or $1,000 − $600).

b. For a 3-year asset, the depreciation rate with the 150%-DB method is $\frac{1}{2}$ or 0.5 ($\frac{1}{3} \times 1.5$ or $\frac{1}{3} \times \frac{3}{2}$).

Year	Beginning Book Value	Deprec. Rate	Annual Deprec.	Ending Book Value
1	$9,000	0.5	$4,500	$4,500
2	4,500	0.5	2,250	2,250
3	2,250	0.5	1,650*	600

*In Year 3, 0.5 times $2,250 equals $1,125. However, $1,125 is not enough depreciation, because the ending book value must equal the scrap value of $600. Therefore, the annual depreciation is adjusted to $1,650 (or $2,250 − $600).

Constructing a table to summarize depreciation and accumulated depreciation is the same with the DB method, the SL method, and the SOYD method.

EXAMPLES:

For the two preceding examples, construct tables showing the annual depreciation, the accumulated depreciation, and the book value at the end of each year.

Original cost: $9,000
Estimated life: 3 years
Estimated scrap value: $600

a. Use the 200%-DB method.

Year	Annual Deprec.	Accumulated Deprec.	Ending Book Value
1	$6,000	$6,000	$3,000
2	2,000	8,000	1,000
3	400	8,400	600

b. Use the 150%-DB method.

Year	Annual Deprec.	Accumulated Deprec.	Ending Book Value
1	$4,500	$4,500	$4,500
2	2,250	6,750	2,250
3	1,650	8,400	600

PART D Comparison of Depreciation Methods

Parts A through C in this section illustrated four methods of depreciation: SL, SOYD, 200%-DB, and 150%-DB. The differences in these four methods are most apparent when the annual depreciation is displayed in tabular form, as follows:

Original cost: $9,000
Estimated life: 3 years
Estimated scrap value: $600

Year	SL	SOYD	200%-DB	150%-DB
1	$2,800	$4,200	$6,000	$4,500
2	2,800	2,800	2,000	2,250
3	2,800	1,400	400	1,650
Total	$8,400	$8,400	$8,400	$8,400

If a business wants to accurately evaluate its assets, it will use the method that gives the most realistic results for those assets. Some assets decrease in value more rapidly than others and should therefore be depreciated more rapidly.

If all of these methods were permissible for tax purposes (and they are *not*), the method selected would depend on when the business wants to take the largest tax deduction—in the beginning or later in the life of the asset.

EXAMPLE:

An asset has an estimated life of 5 years, an original cost of $50,000, and an estimated scrap value of $5,000. Compare the annual depreciation amounts from each of the four different methods: SL, SOYD, 200%-DB, and 150%-DB.

a. Find the annual depreciation with the SL method.

$$\text{Annual Depreciation} = \frac{\$50,000 - \$5,000}{5}$$
$$= \frac{\$45,000}{5}$$
$$= \$9,000$$

b. Find the annual depreciation with the SOYD method.

Step 1. $1 + 2 + 3 + 4 + 5 = 15$
Step 2. The annual rates, in decreasing order, are $\frac{5}{15}, \frac{4}{15}, \frac{3}{15}, \frac{2}{15}$, and $\frac{1}{15}$.

Year	Deprec. Rate		Cost Minus Scrap Value		Annual Deprec.
1	$\frac{5}{15}$	×	$45,000	=	$15,000
2	$\frac{4}{15}$	×	45,000	=	12,000
3	$\frac{3}{15}$	×	45,000	=	9,000
4	$\frac{2}{15}$	×	45,000	=	6,000
5	$\frac{1}{15}$	×	45,000	=	3,000

c. Find the annual depreciation with the 200%-DB method.

SL rate: $\frac{1}{5}$ or 0.2
2 × SL rate: 0.4 (2 × 0.2)

Year	Beginning Book Value	Deprec. Rate	Annual Deprec.	Ending Book Value
1	$50,000	0.4	$20,000	$30,000
2	30,000	0.4	12,000	18,000
3	18,000	0.4	7,200	10,800
4	10,800	0.4	4,320	6,480
5	6,480	0.4	1,480*	5,000

*In Year 5, 0.4 times $6,480 equals $2,592. However, $2,592 is too much depreciation, because the ending book value must equal the scrap value of $5,000. Therefore, the annual depreciation is adjusted to $1,480 (or $6,480 − $5,000).

d. Find the annual depreciation with the 150%-DB method.

SL rate: $\frac{1}{5}$ or 0.2

1.5 × SL rate: 0.3 (1.5 × 0.2)

Year	Beginning Book Value	Deprec. Rate	Annual Deprec.	Ending Book Value
1	$50,000	0.3	$15,000	$35,000
2	35,000	0.3	10,500	24,500
3	24,500	0.3	7,350	17,150
4	17,150	0.3	5,145	12,005
5	12,005	0.3	7,005*	5,000

*In Year 5, 0.3 times $12,005 equals $3,601.50. However, $3,601.50 is not enough depreciation, because the ending book value must equal the scrap value of $5,000. Therefore, the annual depreciation must be adjusted to $7,005 (or $12,005 − $5,000). Notice that there is more depreciation in the fifth year than there is in the fourth year. Usually, a change will be made to avoid this situation; that is, there will be conversion in the fourth year to the SL method.

e. In a table, compare the annual depreciation amounts from each of the four methods.

Original cost: $50,000
Estimated life: 5 years
Estimated scrap value: $5,000

Year	SL	SOYD	200%-DB	150%-DB
1	$ 9,000	$15,000	$20,000	$15,000
2	9,000	12,000	12,000	10,500
3	9,000	9,000	7,200	7,350
4	9,000	6,000	4,320	5,145
5	9,000	3,000	1,480	7,005
Total	$45,000	$45,000	$45,000	$45,000

PART E Modified Accelerated Cost Recovery System (MACRS)

The **Modified Accelerated Cost Recovery System (MACRS)** is that part of the Tax Reform Act of 1986 which relates to depreciation of assets. The word *recovery* is used because a company is *recovering* a part of the original cost of an asset when it claims the annual depreciation of that asset for a year. The word *accelerated* is used because MACRS provides for larger depreciation amounts in the earlier years. The word *modified* is used because the 1986 law changed the previous depreciation system, the **Accelerated Cost Recovery System (ACRS)**.

MACRS is a set of rules telling the taxpayer the kind of assets that can be depreciated for *income tax purposes*, the assumed life expectancy of a particular type of asset, and the depreciation methods that can be used. Under ACRS, there were tables of the percentages used to multiply the book value of the asset. Basically, MACRS simply tells the taxpayer which depreciation method is applicable to a specific category of assets.

There are also alternate depreciation methods that may be used and guidelines for using them. It is acceptable to use the SL method instead of a DB method. For some assets, such as those used entirely outside the United States, the SL method is required.

Scrap Value. Under MACRS, the taxpayer assumes that an asset will *not* have any scrap value at the end of its life. If the taxpayer keeps the asset all of its life and is still able to sell it later, the amount of the sale must be reported as income in the year of the sale. This removes guesswork from the tax computation process.

Mid-Year Convention. No business buys all of its assets on the first day of the year. Assets that are bought early in the year are used longer and should qualify for more depreciation than assets bought later in the year.

For most assets, MACRS states that *no matter when you started using an asset, you must assume that it occurred at the midpoint of the first year*. Essentially, this means that you may only deduct half of the first-year depreciation on the first tax return. This is another rule that makes all tax returns more standard and easier for the Internal Revenue Service (IRS) to check.

There are also *mid-quarter* and even *mid-month* conventions, which are similar to the mid-year convention. The MACRS guidelines specify when these conventions must be used instead of the mid-year convention.

Chapter 12   Inventory Valuation, Cost of Goods Sold, and Depreciation

Summary of the MACRS Asset Classes. Under MACRS, assets are defined as belonging to a certain class. The classification of an asset determines its life expectancy and the depreciation method that can be used. Complete information is available from the IRS in *Publication 534: Depreciation*. Remember that this system applies to assets put into use after 1986. For assets put into use in 1986 or earlier, other methods are used.

Every asset falls into one of these classes: 3-year, 5-year, 7-year, 10-year, 15-year, 20-year, 27.5-year, or 31.5-year. The following list is a summary, with some examples. For every class, there are important rules that need to be studied carefully before depreciation for a tax return is calculated.

3-year class. This is a relatively restricted class. Basically, it only includes tractors that will operate on roads and racehorses that do not begin racing until they are more than 2 years old.

5-year class. This class includes trucks, automobiles, computers, office machines, and any property that is used in research and experiments.

7-year class. Office furniture and fixtures are in this class. Also in this class is any asset that is not designated as part of some other class.

10-year class. The major items in this class are water transportation vehicles, such as barges and tugboats.

15-year class. Included in this class are designated municipal water-treatment plants.

20-year class. This class includes farm buildings.

27.5-year class. This is the class of residential rental property. It includes houses and apartments but not hotels or motels, unless they are occupied on a relatively permanent basis.

31.5-year class. This is the class of all property with a life expectancy of more than 27.5 years. This class specifically excludes residential rental property. Office buildings and warehouses are in this class.

Depreciation Methods. The MACRS rules specify which depreciation method can be used for the assets in each class, as follows:

Class	Method
3-year	200%-DB (and then SL)
5-year	200%-DB (and then SL)
7-year	200%-DB (and then SL)
10-year	200%-DB (and then SL)
15-year	150%-DB (and then SL)
20-year	150%-DB (and then SL)
27.5-year	SL
31.5-year	SL

Under MACRS, you must calculate each year's depreciation with *both* the DB and the SL methods. Eventually, the SL method will give a larger annual depreciation than the DB method. In that year, you change to the SL method and use it for the remaining life of the asset.

Calculating Depreciation. The actual calculations of depreciation under MACRS are straightforward. They follow the procedures explained in Parts A and C of this section. With the last two classes (where the SL method is used), the mid-month convention must be used. With the other classes, the mid-year or the mid-quarter convention must be used.

In the following two examples, the mid-year convention will be used. Calculate the first-year depreciation with the 200%-DB method, and then multiply it by $\frac{1}{2}$, or 0.5. Therefore, there will be an extra half-year of depreciation at the end of the life of the asset.

EXAMPLE:

A 5-year asset costs $50,000. Construct a table showing the annual depreciation. There is no scrap value.

SL rate: $\frac{1}{5}$ or 0.2

2 × SL rate: 0.4 (2 × 0.2)

First-year rate: 0.2 (0.4 × 0.5)

Tax Year	Remain. Life Yrs.	Beg. Book Value	Deprec. Rate	Annual Deprec.	Ending Book Value
1	5.0	$50,000	0.2	$10,000	$40,000
2	4.5	40,000	0.4	16,000	24,000
3	3.5	24,000	0.4	9,600	14,400
4	2.5	14,400	SL/0.4*	5,760	8,640
5	1.5	8,640	SL	5,760	2,880
6	0.5	2,880	SL	2,880	0

*In Year 4, the asset has a book value of $14,400 and 2.5 years of life remaining. Therefore, the SL

depreciation is $5,760 per year ($14,400 ÷ 2.5) for *each of the remaining 2.5 years*. By coincidence, $5,760 happens to be exactly the same as the depreciation calculated with the 200%-DB method for Year 4.

In Year 5, the SL amount is still $5,760, as calculated in Year 4. The actual amount of MACRS depreciation is $5,760, because the 200%-DB amount is $3,456 ($8,640 × 0.4), and $3,456 is smaller than $5,760.

In Year 6, there is only a half-year of life remaining. Half of $5,760 (the SL depreciation for one year) is $2,880. The book value for Year 6 is also $2,880. Therefore, the depreciation amount for the final year is $2,880.

The following example also illustrates the changeover from the DB method to the SL method. For the 15-year class, the 150%-DB method must be used.

EXAMPLE:

A 15-year asset costs $50,000. Construct a table showing the annual depreciation. There is no scrap value.

SL rate: $\frac{1}{15}$

1.5 × SL rate: 0.1 (1.5 × $\frac{1}{15}$)

First-year rate: 0.05 (0.1 × 0.5)

Tax Year	Remain. Life Years	Beg. Book Value	Deprec. Rate	Annual Deprec.	Ending Book Value
1	15	$50,000	0.05	$2,500	$47,500
2	14.5	47,500	0.1	4,750	42,750
3	13.5	42,750	0.1	4,275	38,475
4	12.5	38,475	0.1	3,848	34,628
5	11.5	34,628	0.1	3,463	31,165
6	10.5	31,165	0.1	3,117	28,048
7	9.5	28,048	SL	2,952	25,096
8	8.5	25,096	SL	2,952	22,144
9	7.5	22,144	SL	2,952	19,192
10	6.5	19,192	SL	2,952	16,240
11	5.5	16,240	SL	2,952	13,288
12	4.5	13,288	SL	2,952	10,366
13	3.5	10,366	SL	2,952	7,384
14	2.5	7,384	SL	2,952	4,432
15	1.5	4,432	SL	2,952	1,480
16	0.5	1,480	SL	1,480	0

To decide *when* to convert to the SL method, calculate the SL depreciation each tax year. In Year 6, the amount of 150%-DB depreciation is $3,116.50 ($31,165 × 0.1); the amount of SL depreciation is $2,968 ($31,165 ÷ 10.5). The 150%-DB depreciation of $3,116.50 is larger than the SL depreciation of $2,968; therefore, the 150%-DB method is still used.

In Year 7, the amount of 150%-DB depreciation is $2,804.80 ($28,048 × 0.1); the amount of SL depreciation is $2,952.42 ($28,048 ÷ 9.5). Since the SL depreciation of $2,952.42 is larger than the 150%-DB depreciation of $2,804.80, you should convert to the SL method in this year.

The depreciation in the final tax year ($1,480) is not exactly half of $2,952 because of rounding. The rounded depreciation amount of $2,952 is used instead of the more exact $2,952.42.

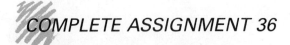
COMPLETE ASSIGNMENT 36

Name _____

Date _____

ASSIGNMENT 36
SECTION 12.2

Calculating Depreciation

	Perfect Score	Student's Score
PARTS A & B	24	
PART C	28	
PART D	20	
PART E	28	
TOTAL	100	

PARTS A AND B Straight-Line and Sum-of-the-Years-Digits Methods

DIRECTIONS: Find the annual depreciation, the accumulated depreciation, and the ending book value for the given year. Use the SL method for Problems 1–5 and the SOYD method for Problems 6–8 (1 point for each correct answer).

	Original Cost	Estimated Life (years)	Estimated Scrap Value	Year of Use	Annual Deprec.	Accumulated Deprec.	Ending Book Value
1.	$9,600	10	$ 600	1			
2.	$8,100	7	$ 225	2			
3.	$3,800	5	$ 300	3			
4.	$60,000	30	$ 0	4			
5.	$154,000	28	$ 0	5			
6.	$9,900	3	$ 900	2			
7.	$750	4	$ 50	2			
8.	$19,250	5	$1,250	3			

STUDENT'S SCORE _____

PART C Declining-Balance Method

DIRECTIONS: Complete the following tables. Use the 200%-DB method for Problems 9 and 10; use the 150%-DB method for Problems 11 and 12. These calculations are NOT for income tax purposes. Do not adjust the first-year depreciation (1 point for each correct answer).

Problem number:	9.	10.	11.	12.
Original cost of asset:	$60,000	$100,000	$175,000	$240,000
Estimated life (years):	5	10	15	20
Estimated scrap value:	$5,000	$0	$10,000	$0

	Year	Beginning Book Value	Annual Depreciation	Accumulated Depreciation	Ending Book Value
9.	1	$ 60,000	_____	_____	_____
	2	_____	_____	_____	_____
10.	1	$100,000	_____	_____	_____
	2	_____	_____	_____	_____
11.	1	$175,000	_____	_____	_____
	2	_____	_____	_____	_____
12.	1	$240,000	_____	_____	_____
	2	_____	_____	_____	_____

STUDENT'S SCORE _____

Chapter 12   Inventory Valuation, Cost of Goods Sold, and Depreciation

PART D Comparison of Depreciation Methods

DIRECTIONS: An asset has an original cost of $120,000, an estimated useful life of 5 years, and an estimated scrap value of $9,600. Find the annual depreciation for each of the 5 years. Use the SL method, the 200%-DB method, the 150%-DB method, and the SOYD method. This is NOT for income tax purposes. Do not make any adjustments to the depreciation amounts (1 point for each correct answer).

	Year	SL	200%-DB	150%-DB	SOYD
13.	1				
14.	2				
15.	3				
16.	4				
17.	5				

STUDENT'S SCORE _____

PART E Modified Accelerated Cost Recovery System (MACRS)

DIRECTIONS: Complete the following tables by using the MACRS guidelines. Be sure to use the mid-year convention in all four problems. Scrap value is not a consideration. Use the proper DB method for the asset's class (1 point for each correct answer).

Problem number:	**18.**	**19.**	**20.**	**21.**
Original cost of asset:	$330,000	$88,000	$640,000	$158,000
Asset Classification:	15-year	5-year	20-year	10-year

	Tax Year	Beginning Book Value	Annual Depreciation	Accumulated Depreciation	Ending Book Value
18.	1	$330,000	_____	_____	_____
	2	_____	_____	_____	_____
19.	1	$88,000	_____	_____	_____
	2	_____	_____	_____	_____
20.	1	$640,000	_____	_____	_____
	2	_____	_____	_____	_____
21.	1	$158,000	_____	_____	_____
	2	_____	_____	_____	_____

STUDENT'S SCORE _____

ANSWERS TO ODD-NUMBERED PROBLEMS

CHAPTER 1 Working with Decimals

Assignment 1

PART A

1. eight hundred fifty-seven
3. ten thousand forty-seven
5. seven million seventy-seven

PART B

7. 89,100
9. 3,000,070

PART C

11. one hundred twenty-two and eight hundred forty-five thousandths
13. four thousand and four ten-thousandths
15. nine hundred seven and forty-seven thousandths

PART D

17. 400,045.0404
19. 200,600,002.04

PART E

21. 0.078   0.78
23. 0.32   0.325
25. 0.01   0.0091
27. 0.5   0.500
29. 0.330   0.033
31. 0.2   0.1634
33. 0.802   0.80100
35. 0.751   0.7501
37. 0.675   0.675001
39. 300.125   301.025
41. 80.06   80.60
43. 6.177   61.77
45. 43.67   43.76
47. 712,590   712,509
49. 5.41825   5.4183
51. 444.
    400.40
    344.
    44.4
    4.440
    4.4
    3.44
    0.44

53. 980.3
980.29
520.5
520.15
520.05
520.
173.08
173.008
0.4164
0.4146

55. 772.
771.99
504.721
504.712
483.81
483.801
483.000
72.77
0.9255
0.92505

PART F

57. 1 2 0 0 0
59. 9 2 4 5 . 3
61. · 0 0 9 2
63. · 6 7 5
65. 8 3 9 9

Assignment 2

PART A

1. 2.22
3. 2.83422
5. 446.0806

PART B

7. a. $ 1,162.70
   b. $ 1,986.57
   c. $ 3,149.27
9. a. $ 793.82
   b. $10,718.53
   c. $ 2,551.37
   d. $14,063.72

PART C

11. 2,258.41
13. 2,674.86
15. 1,045.59
17. 1,911.07
19. 3,010.36
21. 12,506.06

PART D

23. $130.30
25. 48.8892

Assignment 3

PART A

1. 24.85
3. 4,178.29
5. 13,579.24
7. 91.03
9. 7,934.93

PART B

11. 3.25
13. 6.17
15. 380.2704

PART C

17. 28.72
19. 586.84

PART D

21. 0.683
23. 75,518.15
25. 6.9346

Answers to Odd-Numbered Problems

PART E

27.	3,281.40	37.	✔
29.	12,131.96	39.	X
31.	X	41.	X
33.	✔	43.	✔
35.	X		

PART F

45. $733.41 (decrease)
47. $1,997.19 (increase)
49. −$32.81
51. +$1,670.23
53. +$5,889.46

Assignment 4

PART A

1. 2.4
3. 1.584
5. 4.05536
7. 105.08
9. 30,344.49

PART B

11. 0.049
13. 0.010011
15. 0.0068612

PART C

17. 16.9
19. 100.0
21. 8.67
23. 0.03
25. 0.01
27. 54.614
29. 0.149
31. 11.9859
33. 101.0000
35. 29.2929

PART D

37. 1,001.2
39. 3,100
41. 19,900
43. 2
45. 140,000
47. 9,635.1
49. 110.083
51. 643,250
53. 370,400
55. 4,750

57. $3,000 \times 2 = 6,000$
59. $1,010 \times 6 = 6,060$
61. $50.6 \times 125 = 6,325$
63. $4.2 \times 72 = 302.4$
65. $6,875 \times 49 = 336,875$

PART E

67. 83,420.025
69. 18.954
71. 16.22
73. 77,216.64
75. 0.02

PART F

77. $15.3375
79. $130.625
81. $186.375
83. $118.15
85. $1,005.54

Assignment 5

PART A

1. 218
3. 92

PART B

5. 0.84
7. 0.166
9. 0.018

PART C

11. 700
13. 128

PART D

15. 0.217
17. 0.02
19. 1.159

PART E

21. 5.5714
23. 0.44

PART F

25. 62.7
27. 0.284
29. 0.00844
31. 0.00291
33. 0.0101
35. 16.522
37. 0.0405
39. 0.0218

PART G

41. 19.375
43. 3.7222222
45. 0.3157894
47. 19.38
49. 3.72
51. 0.32

PART H

53. 2.95 lb

CHAPTER 2 Working with Fractions

Assignment 6

PART A

1. $3/9$
3. $18/27$
5. $80/140$
7. $14/49$
9. $42/54$
11. $90/110$
13. $48/52$
15. $143/165$
17. $117/180$
19. $48/81$

PART B

21. $2/5$, G.C.D. = 3
23. $1/2$, G.C.D. = 8
25. $4/7$, G.C.D. = 4
27. $2/7$, G.C.D. = 5
29. $3/7$, G.C.D. = 7
31. $1/2$, G.C.D. = 85
33. $13/23$, G.C.D. = 5
35. $14/23$, G.C.D. = 7
37. $28/125$, G.C.D. = 4
39. $4/125$, G.C.D. = 16

PART C

41. $14/3$
43. $102/5$
45. $115/7$
47. $29/8$
49. $173/16$

51. $9\tfrac{1}{2}$
53. 8
55. $1\tfrac{16}{19}$
57. $27\tfrac{1}{9}$
59. $17\tfrac{1}{3}$

Assignment 7

PART A

1. $5\frac{2}{3}$
3. $15\frac{1}{20}$
5. $16\frac{3}{8}$

PART B

7. $1\frac{1}{5}$
9. 19

PART C

11. 30
13. 84
15. 756

PART D

17. $\frac{2}{3}$
19. $1\frac{23}{48}$

PART E

21. $13\frac{5}{24}$
23. $78\frac{59}{60}$
25. $92\frac{13}{30}$

PART F

27. $12\frac{1}{4}$ hrs

Assignment 8

PART A

1. $\frac{2}{3}$
3. $3\frac{3}{5}$
5. $11\frac{7}{16}$
7. $\frac{1}{12}$
9. $2\frac{13}{24}$

PART B

11. $\frac{1}{3}$
13. $8\frac{1}{6}$
15. $2\frac{11}{20}$
17. $5\frac{5}{12}$
19. $28\frac{16}{45}$

PART C

21. $60\frac{7}{12}$
23. $87\frac{1}{6}$
25. 32

Assignment 9

PART A

1. $\frac{1}{6}$
3. $\frac{3}{80}$
5. $\frac{1}{2}$
7. $\frac{1}{6}$
9. $\frac{1}{2}$

PART B

11. 1
13. $4\frac{4}{5}$
15. $7\frac{1}{5}$
17. $11\frac{1}{4}$
19. $19\frac{1}{2}$

PART C

21. $\frac{7}{9}$
23. 9
25. $55\frac{4}{5}$
27. $2\frac{9}{20}$
29. 21

PART D

31. 462
33. 665
35. $918\frac{3}{4}$

PART E

37. $130.50
39. $23.25

Assignment 10

PART A

1. $1\frac{1}{5}$
3. $\frac{5}{6}$
5. $2\frac{1}{3}$
7. $17\frac{1}{2}$
9. 10
11. 9
13. 80
15. 28

PART B

17. $\frac{1}{24}$
19. $\frac{7}{30}$

PART C

21. $\frac{3}{4}$
23. $\frac{5}{14}$
25. $\frac{8}{15}$
27. $1\frac{2}{3}$
29. $5\frac{5}{8}$
31. $\frac{2}{3}$
33. $1\frac{5}{14}$
35. $\frac{7}{16}$
37. 2
39. $19\frac{1}{3}$

PART D

41. $\frac{3}{4}$
43. $\frac{33}{50}$
45. $\frac{7}{30}$

PART E

47. $37.50

Assignment 11

PART A

1. 0.9167
3. 0.5417
5. 0.7188
7. 0.7273
9. 0.1389

PART B

11. $\frac{3}{50}$
13. $\frac{1}{8}$
15. $4\frac{17}{400}$

PART C

17. $27\frac{1}{4}$
19. $74\frac{7}{40}$
21. 13.25
23. 7.875
25. 12.225

PART D

27. $37\frac{1}{2}$
29. 9.3502

PART E

31. 96
33. 10,875
35. 0.1272

PART F

37. $0.4375
39. 8.135

CHAPTER 3 Working with Percents

Assignment 12

PART A
1. 0.5
3. 0.04
5. 0.2
7. 0.475
9. 0.0375
11. 0.4167
13. 0.815
15. 0.0067
17. 0.0002
19. 0.009
21. 1.01
23. 5
25. 1.1133
27. 2.12
29. 3.03

PART B
31. ¼
33. ³⁄₅₀₀
35. ⅓
37. ¹⁄₂₄
39. ⅛
41. ¹⁄₃₂
43. 10
45. ¹⁄₃₀₀

PART C
47. 25%
49. 1%
51. 100%
53. 67.5%
55. 56.67% or 56⅔%
57. 20%
59. 0.275%
61. 60%
63. 101%
65. 150%

PART D
67. 350%
69. 0.6%
71. 46.67% or 46⅔%
73. 540%
75. 150%

Assignment 13

PART A
1. 6.1
3. $12.75
5. $0.22
7. $0.80
9. $122.55

PART B
11. 37
13. $6.03
15. $400
17. 0.28
19. 0.07

PART C
21. 0.37
23. 16
25. $6
27. $30
29. $0.01

PART D
31. $11.25
33. $67.48
35. 3.24
37. $10,080
39. 8,800

PART E
41. Women's Apparel: $2,868,000
    Men's Apparel: 2,424,000
    Children's Apparel: 1,320,000
    Cosmetics: 1,605,600
    Shoes: 2,282,400
    Linens: 1,500,000

43.
	Percent	Amount
Electrical work	15.0	$10,770.00
Insurance	0.5	359.00
Labor	32.0	22,976.00
Materials and supplies	30.0	21,540.00
Plumbing	12.5	8,975.00
Miscellaneous expenses	2.0	1,436.00
Profit	8.0	5,744.00

Assignment 14

PART A
1. 12%
3. 5.3%
5. 7.3%
7. 64.3%
9. 66.7%

PART B
11. 0.625% or ⅝%
13. 0.27%
15. 0.8% or ⅘%
17. 0.25% or ¼%
19. 0.5% or ½%

PART C
21. 200%
23. 600%
25. 175%
27. 402.6%
29. 227.3%

PART D
	Amount of Change	Percent of Change
31.	$3,000	+25.0%
33.	−$4,000	−16.7%
35.	−$3,500	−12.5%

Answers to Odd-Numbered Problems

Assignment 15

PART A
1. 966.67
3. 165
5. $96.80
7. $32.19
9. $33
11. 2
13. 1.4
15. 0.2

PART B
17. $300
19. $15,000

PART C
21. 4
23. 1,800
25. $0.38
27. $0.46
29. $1,600

PART D
31. $742.40
33. $84.60
35. $59.98

CHAPTER 4 Working with Weights and Measures

Assignment 16

PART A
1. 99 t
3. 52 oz
5. 15 min
7. 1,000 lb
9. 3 ft 9 in.
11. 1 1/4 lb *or* 1.25 lb
13. 11/12 h *or* 0.92 h
15. 3 1/3 cu yd *or* 3.33 cu yd
17. 0.382 g
19. 750 mL

PART B
21. 3.66 m
23. 9.08 qt
25. 4.54 kg
27. 21.14 qt
29. 0.7075 m^3

PART C
31. 1 h 30 min
33. 8 sq yd 5 sq ft 95 sq in.
35. 990 cm

PART D
37. 8 cu yd 18 cu ft 1,168 cu in.
39. 483.7 mm

PART E
41. 26 gal 3 qt 1 pt
43. 5 sq yd 8 sq ft 140 sq in.
45. 2 cu yd 21 cu ft
47. 9.375 kg
49. 1.6 m^2

PART F
51. 4 lb 3 oz
53. 1.274 L
55. 0.7 kg

Assignment 17

PART A
1. 110 ft
3. 220 m
5. 4.5 gal
7. $2.40 per sq ft
9. $63.33

PART B
11. $24
13. $17.60
15. $80
17. $0.75
19. $3.15
21. $54
23. $35
25. $500

PART C
27. $2.10
29. $35
31. $30
33. $2.72
35. $83.75
37. $37.50
39. $241.92

PART D
41. $25.50
43. $1.59
45. $19.08
47. $7,550
49. $9.11

CHAPTER 5 Estimations, Graphs, and Shortcuts

Assignment 18

PART A
1. 8
3. 9
5. 501
7. $98
9. $4,000

PART B
11. 30
13. 50
15. 9,830
17. $10
19. $110
21. 800
23. 0
25. 1,000
27. $100
29. $5,000
31. 1,000
33. 6,000
35. 10,000
37. $5,000
39. $15,000

PART C

41.	1,900	47.	320
43.	7,000	49.	4,620
45.	111,000		

PART D

51. 80 × 40 = 3,200
53. 3,000 × 1 = 3,000
55. 32,000 × 0.002 = 64
57. 0.007 × 60 = 0.42
59. 30,000 × 0.08 = 2,400

PART E

61. 600 ÷ 300 = 2 (2.13)
63. 90 ÷ 30 = 3 (3.11)
65. 60 ÷ 0.02 = 3,000 (3,449.44)
67. 0.02 ÷ 0.2 = 0.1 (0.09)
69. 8 ÷ 200 = 0.04 (0.04)

PART F

71.	3,380 X	77.	900 ✓
73.	4,400 ✓	79.	1,500 X
75.	1,800 X		

Assignment 19

PART A

1.	$30,000	5.	$20,000
3.	1986		

PART B

7.	$20,000	15.	$1,500,000
9.	1988	17.	$22,500
11.	$35,000	19.	Bill
13.	$300,000	21.	$10,000

23.

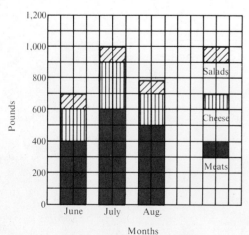

PART C

25. 15%
27. 100%
29. No, because CGS and OE are unknown

Assignment 20

PART A

1. 77
3. 69
5. 858
7. 102
9. 97
11. 94.7
13. 71.98
15. 110.93

PART B

17. 13
19. 511
21. 56
23. 209
25. 2.90
27. 2.26
29. 5.25

PART C

31. 5/6
33. 11/30
35. 7/10
37. 13/72
39. 17/60
41. 15/56
43. 1/6
45. 3/20

PART D

47. 1/6
49. 1/42
51. 2/9
53. 1/60
55. 1/6
57. 1/15
59. 1/16

PART E

61. 0.701
63. 2.77
65. 0.3206
67. 3,000
69. 80,200

PART F

71. 2,400
73. 4
75. 13,300
77. 11.64
79. 3.04

PART G

81. 275
83. 1,001
85. 517
87. 96.8
89. 0.913

PART H

91. 83
93. 51
95. 484
97. 868
99. 528

Answers to Odd-Numbered Problems

CHAPTER 6 Keeping a Checking Account

Assignment 21

1.

	Dollars	Cents
Balance Forward	3,690	33
Deposit	834	47
TOTAL	4,524	80
This Check	209	50
Balance	4,315	30

No. 1896 $209.50
Date Sept. 29 19--
To Ascot & Sons
For Plastic pipe

Check No. 1896, Sept. 29, 19--
Pay to the order of: Ascot & Sons — $209.50
Two hundred nine and 50/100 — DOLLARS
For: Plastic pipe
Signed: John Bovio

3.

	Dollars	Cents
Balance Forward	2,167	52
Deposit	361	48
TOTAL	2,529	00
This Check	150	00
Balance	2,379	00

No. 1906 $150.00
Date Oct. 7 19--
To Clean Corp.
For Janitorial services

Check No. 1906, Oct. 7, 19--
Pay to the order of: Clean Corp. — $150.00
One hundred fifty and 00/100 — DOLLARS
For: Janitorial services
Signed: John Bovio

5.

DEPOSIT TICKET
DATE October 10, 19--

BOVIO SUPPLY COMPANY
934 Rose Street
Menlo Park, CA 94025-4420

Western National Bank
San Francisco, CA 94131-2130

	DOLLARS	CENTS
CURRENCY	127	00
COIN	18	00
CHECKS	97	51
	211	43
	427	50
	198	25
TOTAL FROM OTHER SIDE		
TOTAL	1,079	69

Assignment 22

1.

Bovio Supply Company
Reconciliation Statement

Checkbook balance, Jan. 14		$2,928.31
Add:		
Unrecorded deposit, Dec. 17	$75.40	
Error on check stub, Jan. 6	18.00	
Total		+ 93.40
Adjusted checkbook balance		$3,021.71
Bank balance, Jan. 14		$3,881.37
Add:		
Unrecorded deposit, Jan. 14		+ 238.00
		$4,119.37
Subtract:		
#1524	$ 88.25	
#1541	147.90	
#1556	72.54	
#1557	387.25	
#1558	401.72	
Total		−1,097.66
Adjusted bank balance		$3,021.71

3.

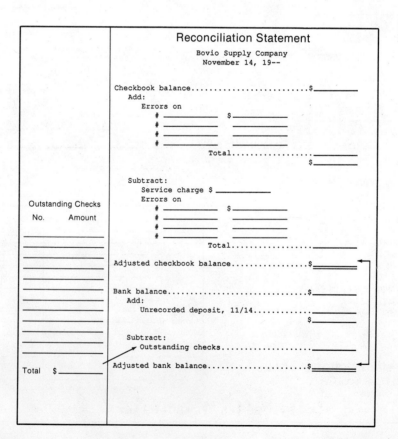

Answers to Odd-Numbered Problems

CHAPTER 7 Calculating Interest

Assignment 23

PART A

1. 62 days
3. 90 days
5. 76 days
7. 71 days
9. 120 days

PART B

	Due Date	Exact Time
11.	Jan. 23	15 days
13.	July 18	45 days
15.	July 5	91 days
17.	Jan. 2	151 days

PART C

	Due Date	Exact Time
19.	Apr. 15	31 days
21.	Apr. 20, 1994	90 days
23.	Apr. 10, 1996	152 days

Assignment 24

PART A

1. $112
3. $180
5. $11.25
7. $3,130

	Rate × Time	Interest
9.	0.01	$17.50
11.	0.03	$60
13.	0.02	$18
15.	0.01	$12.50
17.	0.01	$22.75

PART B

19. $48
21. $27.06
23. $30
25. $0.56

PART C

27. $72
29. $31.20
31. $252
33. $225

Assignment 25

PART A

1. Amount = $1,225.04
   Interest = $225.04

PART B

3. Amount = $10,778.85
   Interest = $9,528.85
5. Amount = $2,709.17
   Interest = $1,209.17
7. Amount = $8,354.50
   Interest = $6,354.50
9. Amount = $2,841.35
   Interest = $1,241.35
11. 1.276282
13. 1.125509
15. 1.061364

CHAPTER 8 Calculating Time-Payment Plans and Short-Term Loans

Assignment 26

PART A

	Total Time-Payment Price	Finance Charge
1.	$1,179.00	$179.00
3.	$1,746.00	$246.00
5.	$4,115.00	$465.00

	Total Amount Repaid	Finance Charge
7.	$2,812.50	$312.50
9.	$942.00	$142.00
11.	$1,920.00	$270.00

PART B

	1st Month's Interest	Finance Charge	Monthly Payment
13.	$24.00	$84.00	$280.67*
15.	$31.50	$252.00	$136.80
17.	$22.00	$110.00	$256.67*

PART C

	Total Time-Payment Price	Finance Charge	APR
19.	$2,580.00	$180.00	12.6%
21.	$3,895.00	$395.00	21.2%
23.	$2,040.00	$240.00	24.6%

Answers to Odd-Numbered Problems

Assignment 27

PART A

	Due Date	Amount of Interest	Maturity Value
1.	Aug. 16	$20.00	$1,020.00
3.	Feb. 14	$ 8.33	$1,258.33
5.	Nov. 15	$48.00	$1,648.00
7.	Feb. 25	$57.81	$1,907.81
9.	Oct. 8	$57.50	$2,557.50

PART B

	Bank Discount	Proceeds
11.	$64.17	$2,135.83
13.	$31.50	$1,368.50
15.	$66.67	$2,433.33
17.	$29.25	$1,770.75
19.	$35.00	$1,465.00
21.	$56.22	$1,943.78

23. Due date: Sept. 22
    Discount period: 83 days
    Bank discount: $47.96
    Proceeds: $2,552.04
25. Due date: Aug. 4
    Discount period: 50 days
    Bank discount: $25.00
    Proceeds: $1,775.00
27. Due date: July 30
    Discount period: 45 days
    Bank discount: $22.75
    Proceeds: $1,377.25

PART C

29. Due date: Aug. 30
    Interest amount: $38.00
    Maturity value: $1,938.00
    Discount period: 30 days
    Bank discount: $29.07
    Proceeds: $1,908.93
31. Due date: Sept. 25
    Interest amount: $73.60
    Maturity value: $3,273.60
    Discount period: 31 days
    Bank discount: $36.65
    Proceeds: $3,236.95

CHAPTER 9 Purchase Orders and Invoices, Cash Discounts, and Trade Discounts

Assignment 28

PART A

1.

PURCHASE ORDER				NO. 59747
Coastal Garden Stores				SHOW PURCHASE ORDER NUMBER ON PACKAGES, INVOICES, AND CORRESPONDENCE
1814 West Holmby, Suite 200				
Los Angeles, CA 90024-2418				

REQUISITION NO.	DATE	SHIP VIA	SHIP BY	REFER INQUIRES TO BUYER
89032	5/23/--	IME	6/15/--	Jan Mason

Bovio Supply Company
934 Rose Street
Menlo Park, CA 94025-4420

SHIP TO:
1729 Anderson Avenue
Palo Alto, CA 94303-3870

PLEASE SUPPLY THE FOLLOWING

QUANTITY	MODEL NO.	DESCRIPTION	PRICE
300	EJ-046-B	Elbow joint 5/8" (brass)	$76.00/C
20 doz	S-1321-P	Pop-up sprinkler (plastic)	8.50/doz
40	T-512-BD	Water-Miser timer	12.75 ea
5 doz	N-2004-B	Water nozzle (brass)	3.17 ea
		Per telephone quote 5/20/--	

PLEASE ACKNOWLEDGE RECEIPT OF THIS ORDER IMMEDIATELY. MAIL INVOICES IN DUPLICATE. IN ACCEPTING THIS ORDER YOU AGREE TO ALL TERMS AND CONDITIONS STATED ON THE REVERSE SIDE.

PART B

3. $ 456.50
       120.00
        30.24
        71.00
        64.80
       427.50
       455.00
       111.94
        34.00
    $1,770.98

Assignment 29

PART A

	Cash Discount	Net	Past Due
1.	✔		
3.			✔
5.		✔	
7.		✔	
9.	✔		
11.		✔	
13.		✔	
15.	✔		
17.	✔		
19.			✔

Answers to Odd-Numbered Problems

PART B

	Date of Remittance	Cash Discount	Amount of Remittance
21.	Sept. 7	$12.50	$1,237.50
23.	Jan. 8	$14.11	$ 456.22
25.	Sept. 6	$ 7.44	$ 736.46
27.	Dec. 11	$10.47	$2,207.53

	Amount Credited	Date of Remittance	Cash Discount
29.	$1,010.10	Apr. 20	$10.10

		Amount	Net
31.	5 doz PVC cap ½"	$ 53.40	
	3 doz PVC plug ½"	24.84	
	10 doz PVC threaded coupling ½"	54.00	
	Subtotal	$132.24	
	Less 25%, 10%, 10%.............		$ 80.34
	12 PVC pipe cutter	$155.40	
	8 Pipe wrench 10"	214.00	
	Subtotal	$369.40	
	Less 30%, 10%		232.72
	NET INVOICE DUE................		$313.06

CHAPTER 10 Selling Goods

Assignment 31

PART A

	Amount of Markup	Selling Price
1.	$ 60.00	$210.00
3.	$183.79	$367.58
5.	$341.10	$492.70

	Amount of Markup	MR
7.	$ 67.77	60%
9.	$110.50	125%

11. $8.88   17. $171.19
13. $58.75   19. $159.99
15. $115.50

Assignment 30

PART A

	Amount of Discount	Net Price
1.	$95.00	$380.00
3.	$33.90	$ 50.85
5.	$75.60	$ 92.40

7. $50.85
9. $412.50

PART B

	Amount of Trade Discount	Net Price
11.	$187.00	$238.00
13.	$ 68.84	$ 69.96
15.	$141.50	$ 98.50

17. $63.00
19. $33.47

PART C

	Net Price Factor	Net Price
21.	0.72	$108.00
23.	0.81	$ 89.10
25.	0.8075	$ 19.86
27.	0.432	$216.00
29.	0.336	$215.04

PART B

	Amount of Markup	Cost
21.	$ 1.63	$ 4.89
23.	$ 24.94	$24.94
25.	$224.22	$74.74

	Amount of Markup	MR
27.	$ 9.81	30%
29.	$190.28	80%

31. $6.35   37. $124.25
33. $17.50   39. $200.00
35. $64.75

PART C

41. 33⅓%   47. 28.57%
43. 100%   49. 40%
45. 300%

Assignment 32

PART A

1. SP = $12.94
3. OE = $8.55
5. C = $148.83
7. NP = $10.25

PART B

9. SP = $15.28
11. OE = $18.28
13. C = $173.36
15. NL = $16.37
17. NL = $1.88
19. NP = $31.62

PART C

21. $0.02 *or* 2¢
23. $0.95 *or* 95¢
25. $0.89 *or* 89¢
27. $0.75 *or* 75¢
29. $0.65 *or* 65¢
31. $3.59
33. $9.49
35. $17.99
37. $129.95
39. $0.31 *or* 31¢

PART D

41. DBS
43. DRA
45. UVSRN
47. 789136
49. 300887

CHAPTER 11 Calculating Gross Pay for Payrolls

Assignment 33

PARTS A AND B

	O.T. Rate	O.T. Hours	Regular Pay	O.T. Pay	Gross Pay
1.	$12.00	4	$320.00	$ 48.00	$368.00
3.	$8.625	8	$230.00	$ 69.00	$299.00
5.	$18.15	6	$484.00	$108.90	$592.90

PART C

	Regular Pay	O.T. Pay @ 1.5×	O.T. Pay @ 2×	Total O.T Pay	Gross Pay
7.	$321.75	$24.75	$ 16.50	$ 41.25	$363.00
9.	$364.80	$43.20	$153.60	$196.80	$561.60

	Regular Pay	O.T. Pay @ 1.5×	O.T. Pay @ 2×	O.T. Pay @ 3×	Gross Pay
11.	$357.20	$42.30	$75.20	$84.60	$559.30

PART D

13. Gross pay = $742.00
15. Gross pay = $756.00
17. Gross pay = $2,900.00

Answers to Odd-Numbered Problems

Assignment 34

PART A

1. $1,880
3. $4,620
5. $5,820
7. $1,620

PART B

9. $63.75
11. $42.40
13. $34.20
15. $44.40
17. $68.70
19. $91.30

CHAPTER 12 Inventory Valuation, Cost of Goods Sold, and Depreciation

Assignment 35

PART A

1. 12 units @ $2.80 = $ 33.60
    50 units @ $3.17 =   158.50
    125 units @ $0.95 =   118.75
    40 units @ $1.43 =   57.20
    Ending inventory cost = $368.05

PARTS B, C, AND D

3. FIFO = $1,876.75
5. Average cost = $1,841.50
7. LIFO = $647.50
9. FIFO = $4,111
11. Average cost = $4,095
13. LIFO = $16,280

PART E

15. $610,000
17. $286,800
19. $213,715

PART F

21. $23,375
23. $107,250
25. $16,500

Assignment 36

PARTS A AND B

	Annual Deprec.	Accumulated Deprec.	Ending Book Value
1.	$ 900	$ 900	$ 8,700
3.	$ 700	$ 2,100	$ 1,700
5.	$5,500	$27,500	$126,500
7.	$ 210	$ 490	$ 260

PART C

	Year	Beginning Book Value	Annual Deprec.	Accumulated Deprec.	Ending Book Value
9.	1		$24,000	$24,000	$ 36,000
	2	$ 36,000	$14,400	$38,400	$ 21,600
11.	1		$17,500	$17,500	$157,500
	2	$157,500	$15,750	$33,250	$141,750

PART D

	SL	200%–DB	150%–DB	SOYD
13.	$22,080	$48,000	$36,000	$36,800
15.	$22,080	$17,280	$17,640	$22,080
17.	$22,080	$ 5,952	$19,212	$ 7,360

PART E

	Tax Year	Beginning Book Value	Annual Deprec.	Accumulated Deprec.	Ending Book Value
19.	1		$17,600	$17,600	$ 70,400
	2	$ 70,400	$28,160	$45,760	$ 42,240
21.	1		$15,800	$15,800	$142,200
	2	$142,200	$28,440	$44,240	$113,760

GLOSSARY

A

ABA NUMBER The official number of a particular bank, which is usually printed in the upper right-hand corner of the checks that the bank makes available to its customers

ACCELERATED COST RECOVERY SYSTEM (ACRS) A feature of the Economic Recovery Tax Act of 1981 which made changes in the method of calculating depreciation for tax purposes. With this method, a few simple tables are used to find the appropriate depreciation expense.

ADDENDS The numbers being added

ADDITION The arithmetic operation that finds the total of two or more numbers

ALIGNMENT Writing decimals in a column so that the decimal points are vertically lined up

ANNUAL PERCENTAGE RATE (APR) An annual interest rate based on the total finance charge

AREA A surface within a set of lines that is expressed in square units

AUTOMATED CLEARINGHOUSE A system using EFT in which information regarding payment is put on computer storage devices and the translated data are run through the bank's computer to complete the desired transfer of funds

AUTOMATED TELLER MACHINES (ATM) Computer terminals that communicate with a bank or a savings and loan association by means of plastic cards with magnetic strips that identify the individual cardholder

AVERAGE A reasonable estimate of the size of each number in a given group of numbers

AVERAGE COST METHOD A method of inventory valuation which uses the average cost of each item that has been in the inventory during the fiscal period

B

BANK DISCOUNT The amount of interest from the face value of a note, which the bank deducts at the time the loan is made

BANKER'S INTEREST The loan interest calculated by assuming a year to have 360 days

BAR GRAPH A graph that uses horizontal or vertical bars to display data

BASE The number for which a percentage is determined

BILL A statement for goods that have been purchased or services that have been performed

BILL OF LADING A packing slip or a shipping ticket which describes a shipment of goods without any prices

BLANK ENDORSEMENT A form of endorsing checks where only the name of the payee is signed

BOARD FOOT The unit by which lumber is sold in the United States. A board foot is 1 foot long, 1 foot wide, and 1 inch thick.

C

C Per hundred

CANCELED CHECK A check that has been cleared and stamped

CASH DISCOUNTS Price reductions granted by sellers to encourage buyers to pay promptly for merchandise that has been delivered

CATALOG PRICE (OR LIST PRICE) The price of products as shown in a catalog

CELSIUS A scale for measuring temperatures where water boils at 100 degrees and freezes at 0 degrees

CHAIN DISCOUNTS A series of trade discounts

CHECK A written order to a bank to pay a stated amount of money to the individual or business named

CHECKING ACCOUNTS Bank accounts that represent money deposited and against which checks may be drawn

CIRCLE GRAPH A pie chart illustrating percentages or parts that comprise a total or whole

CIRCUMFERENCE The distance around a circle, which is found by multiplying the diameter by pi (3.1416)

CLEARINGHOUSE The place where a check is cleared when its amount is charged to the bank against which it was drawn

COD (CASH ON DELIVERY) Terms under which payment must be made upon delivery of the goods

COLLATERAL Something of value which is evidence of the ability of a borrower to pay back a loan and which the lender has the right to sell if the loan is not repaid

COMMERCIAL LOANS Loans made to businesses

COMMISSION Pay received by a salesperson as a percent of his or her sales

COMMON DENOMINATOR A number which is the denominator of two or more fractions

COMMON FRACTION A proper fraction or an improper fraction

COMMON MULTIPLE METHOD A method of simplifying a complex fraction by finding the common multiple of the two denominators

COMPLEMENT The difference between a trade discount rate and 100%

COMPLEX FRACTION A fraction whose numerator or denominator contains a fraction or a mixed number

COMPOUND AMOUNT The sum of the interest charged for a period and the principal for that period

COMPOUND AMOUNT FACTOR In compound-interest tables, the number where the Period row and the Rate column intersect; also called the accumulation factor; multiplied by the principal to find the compound amount

COMPOUND INTEREST Interest charged for one period and added to the principal for that period to create a new principal that becomes the basis for computing interest for the next period

CONSUMER LOANS Personal, family, or household loans that are often repaid with time-payment plans

CONVERSION A change of a number, a measurement, or an amount to an equivalent form

COST OF GOODS SOLD (CGS) The amount subtracted from sales revenue to calculate gross profit

CUBIC MEASURE The product of three linear measures; a measure of volume

CUSTOMARY SYSTEM (OR ENGLISH SYSTEM) System of weights and measures used largely in the United States

CWT Per hundredweight, or per hundred pounds

D

DATA Information such as numbers, percentages, or statistics

DATE OF ORIGIN The date when a borrower signs a promissory note and receives the money

DAY RATE Hourly rate based on straight time

DECEM The Latin word for ten

DECIMAL A number which contains a decimal point

DECIMAL EQUIVALENT The decimal form of a common fraction

DECLINING-BALANCE (DB) METHOD The most widely used accelerated method of calculating depreciation, which assumes that the depreciation each year is always the same percentage of the beginning book value for that year

DENOMINATION The name of a unit of measurement

DENOMINATOR The bottom number of a fraction

DEPRECIATION The systematic procedure that allocates part of the cost of an asset as a periodic business expense

DESCENDING ORDER The arrangement of numbers in a column from largest to smallest, with the largest number first and the smallest number last

DIAMETER The distance across a circle through the center

DIFFERENCE The result of subtraction

DIGIT A single numerical figure, such as 0, 1, 2, 3, 4, 5, 6, 7, 8, or 9

DISCLOSURE STATEMENT Detailed information in writing, which must be given to customers before a sale or a loan is completed

DISCOUNT PERIOD The time used in calculating a simple bank discount

DISCOUNT RATE The percentage rate charged by a bank when it discounts notes payable

DISCOUNTED A characteristic of a note when the interest on it is paid in advance to the bank that discounts the note

DISPLAY The window where numbers appear on a calculator

DIVIDEND The number which is being divided

DIVISION The arithmetic operation that finds how many times one number is contained in another

DIVISION METHOD A method of simplifying a complex fraction by considering the numerator to be a dividend and the denominator to be a divisor

DIVISOR The number by which another number is divided

DOZEN 12 units

DRAWING ACCOUNT A fixed amount advanced to a salesperson by his or her employer for payment of expenses as they occur

DUE DATE The date when a loan must be paid off

E

ELECTRONIC FUNDS TRANSFER (EFT) A checkless method of paying for goods or services, making deposits, withdrawing money, etc.

END-OF-MONTH (EOM) DATING Terms under which the discount period starts immediately after the end of the month in which the invoice has been dated

ENDORSE Sign a check on the back, across the short side at the top left

EQUIVALENT FRACTION A fraction changed to higher or lower terms from a given fraction

EQUIVALENT SINGLE DISCOUNT One discount equal to a series of discounts

ESTIMATE An approximate answer

EXACT INTEREST The loan interest calculated by assuming a year to have its actual 365 days (or 366 days during leap years)

EXACT TIME The actual number of days in a loan period

EXEMPT The classification of salaried employees above a certain management level

EXPONENT Power

EXTENSION (OR EXTENDING) The product of the number of items and the unit cost

EXTRA DATING Terms under which time is added to the period allowed for taking a cash discount when goods are sold before they are in season

F

FACE VALUE (OR FACE) The amount of a loan, as written on a note

FACTOR A multiplier or multiplicand; one of two numbers used to find a product

FAHRENHEIT A scale for measuring temperatures where water boils at 212 degrees and freezes at 32 degrees

FINANCE CHARGE A charge made by sellers to cover interest for the loan of money, bookkeeping expenses for recording payments, losses on goods if payment is not made, and insurance

FIRST-IN, FIRST-OUT (FIFO) A method of inventory valuation which assumes that the merchandise is sold in the order in which it was received

Glossary

FISCAL PERIOD A period of doing business (such as a month, a quarter, a half-year, or a year), at the conclusion of which operations are summarized and presented in financial statements for analysis

FRACTION One or more equal parts of a whole

G

GRAND TOTAL The sum of all subtotals

GRAPH A diagram showing the relationships between different data

GREATEST COMMON DIVISOR (G.C.D.) The largest number that will divide evenly into two or more numbers

GROSS 144 units, or 12 dozen

GROSS PAY The total amount of money earned by an employee before any deductions are made

GROSS PROFIT See markup

H

HIGHER TERMS The condition of a fraction when the numerator and denominator are both multiplied by the same number greater than 1

HORIZONTAL AXIS The horizontal line of a graph, usually representing increments of time

I

IMPROPER FRACTION A fraction whose numerator is equal to or greater than its denominator

INSTALLMENTS Regular monthly or weekly payments made under a time-payment plan

INTEREST The charge for the use of money

INTEREST AFTER DATE See interest due at maturity

INTEREST-BEARING The characteristic of a note which indicates a rate of interest that has to be paid either at maturity or as specified

INTEREST DUE AT MATURITY The interest on short-term loans, which is payable on the due date

INTEREST PERIOD The time (in days, months, years, or combinations of these) for which interest must be paid

INTEREST TO FOLLOW See interest due at maturity

INTERNATIONAL METRIC SYSTEM A system of weights and measures

INVERSE The opposite or reverse; for example, subtraction is the inverse operation of addition, and division is the inverse operation of multiplication

INVOICE A seller's document for goods that have been sold

K

KEY-ENTERING Putting information into a calculator by pressing the calculator keys

L

LAST-IN, FIRST-OUT (LIFO) A method of inventory valuation which assumes that the most recently purchased merchandise has been sold and that the ending inventory consists of the oldest stock

LEAST COMMON DENOMINATOR The smallest possible denominator common to two or more fractions

LIKE TERMS Fractions that have a common denominator

LINE GRAPH A graph which displays time series data by showing lines that connect plotted points

LINEAR MEASURES Measures of length or distance

LOAN PERIOD The time between the date of origin and the due date, or a specific amount of time after the date of origin

LONG-TERM LOANS Loans made for one year or longer

LOWEST TERMS The condition of a fraction when no whole number except 1 will divide evenly into both the numerator and the denominator

M

M Per thousand

MAKER The borrower or signer of a note

MARGIN ON SALES See markup

MARKON See markup

MARKUP The difference between the cost of an item and its selling price

MARKUP TABLES Tables that show what the selling price should be for an item with a given unit cost and a given markup rate based on the selling price

MARKUP WHEELS See markup tables

MATURITY VALUE The amount to be paid to a lender on the due date of a loan

MERCHANDISE INVENTORY All the goods owned by a business and available for sale in the hope of making a profit

METRIC SYSTEM An international system of weights and measures

MINUEND The number from which another number is subtracted

MIXED NUMBER A whole number followed by a proper fraction

MODIFIED ACCELERATED COST RECOVERY SYSTEM (MACRS) A set of rules that tells the taxpayer the kind of assets that can be depreciated for income tax purposes, the life expectancies of types of assets, and the depreciation methods that can be used

MONTHLY BANK STATEMENT A document indicating a depositor's bank balance as of a given date

MONTHLY STATEMENT A vendor's summarization of a purchaser's invoices or bills due and cash payments made

MULTIPLE A number that contains another number as a factor; for example, 100 is a multiple of 10, since 10 is a factor of 100.

MULTIPLICAND The first number in a multiplication problem; one of two factors

MULTIPLICATION The arithmetic operation that adds a number to itself a specified number of times

MULTIPLIER The number by which a multiplicand is multiplied; the second number in a multiplication problem; one of two factors

N

NEGOTIABLE Transferable to another person

NET LOSS The result when total operating expenses exceed gross profit

NET PAY The amount actually paid to an employee after all deductions have been subtracted from gross pay

NET PRICE The price after all discounts have been deducted from the list price

NET PRICE FACTOR The product of the complements of a series of discount rates, changed to its decimal form

NET PROFIT The difference between total operating expenses and gross profit

NET WORTH The difference between the value of a person's assets and the value of that person's liabilities

NON-INTEREST-BEARING The characteristic of a note which indicates that its maturity value is equal to its face value

NOTE PAYABLE A note signed by its maker, with a bank as the payee

NOTE RECEIVABLE A note which a company has received from a customer in settlement of an account

NSF (NOT SUFFICIENT FUNDS) A stamp placed on a check to indicate that the account against which the check was drawn has insufficient funds to cover the check

NUMERATOR The top number of a fraction

O

OPERATING EXPENSES Items (such as salaries, advertising expenses, office supplies, selling expenses, utilities, etc.) which are deducted from markup to determine net profit

ORDER FORM A vendor's internal document

ORDINARY DATING The period for taking cash discounts where the rate and number of days are followed by the total number of days allowed for paying the invoice without penalty

ORDINARY SIMPLE INTEREST The loan interest calculated by assuming a year to have only 360 days

OUTSTANDING CHECKS Checks that have not yet been cleared

OVERDRAWING Drawing on an account that has insufficient funds

OVERTIME Time worked beyond the regular work hours

OVERTIME ON A DAILY BASIS Overtime paid at the overtime rate for all hours worked in a day beyond the regular workday hours

OVERTIME ON A WEEKLY BASIS Overtime paid at the overtime rate for all hours worked in a week beyond the regular workweek hours

OVERTIME PAY Pay received for overtime hours worked at the overtime rate

P

PAST DUE The status of an account when it is not paid within the payment period

PAY PERIOD The specified time period on which gross pay is based

PAYEE The business or individual to whom a check or loan is payable

PAYROLL A company's record of earnings and deductions for each employee

PERCENT Per hundred, or part of a hundred

PERCENTAGE A part or fraction of a whole

PERIMETER The distance around a surface

PERPETUAL INVENTORY A running inventory recorded on a computer or stock record cards

PHYSICAL INVENTORY A counting and recording of all items on hand

PIECEWORK An employee's pay based on the number of pieces produced

PRIME NUMBER A number greater than 1 whose only whole-number factors are 1 and itself

PRIME RATE The interest rate charged by leading banks to their best, low-risk customers

PRINCIPAL The amount of money borrowed

PROCEEDS The amount of money actually received by a borrower after a bank discounts the borrower's note

PRODUCT The result of multiplication

PROMISSORY NOTE A written promise to pay the amount borrowed, plus interest, within a specified period

PROPER FRACTION A fraction whose numerator is less than its denominator

PROX. Another term for end-of-month (EOM) dating

PURCHASE ORDER A buyer's document that shows the particular goods requested and specifies quantities and unit prices

PURCHASE REQUISITION A signed statement requesting the purchase of needed materials

Q

QUOTIENT The result of division

R

RADIUS The distance from the center of a circle to its edge

Glossary

RATE A given percent, or the percent charged for the use of a principal for one year

RECAPITULATED (OR RECAPPED) Describes tabulated numbers whose totals are shown both horizontally and vertically

RECEIPT-OF-GOODS (ROG) DATING Terms under which the date of receipt of the goods, rather than the invoice date, is the first day of the discount period

RECEIVING RECORD A record of the goods delivered to a receiving department

RECIPROCAL In division, the inverted fraction of a given fraction

RECONCILIATION STATEMENT A document in which a depositor's current checkbook balance and the balance on the monthly bank statement are compared and adjusted to find the true cash balance in the account as of the date of the reconciliation

RECTANGLE A four-sided figure with four right angles

REGULAR PAY Pay for the standard workweek hours at a given rate

REMAINDER The final difference of a division which is not exact

REMITTANCE Payment for purchased goods

REPEATED DIVISION METHOD A method of finding the least common denominator of two or more fractions by dividing the denominators by prime numbers

REPEATING DECIMAL The decimal number that results when a number appears a second time as a remainder in long division

REQUISITION A request for stock to be taken out or issued

RESTRICTIVE ENDORSEMENT A check endorsement where the name of the person or business allowed to cash the check is written below the words *Pay to the order of*

ROUNDED OFF The characteristic of a decimal that is expressed to a certain number of decimal places

S

SALARY Payment for time worked on a weekly, monthly, or annual basis

SERVICE CHARGE An extra charge on an installment purchase

SHORT DIVISION A method of division used when the divisor is a single-digit number

SHORT-TERM LOANS Loans made for less than one year

SI *See* metric system

SIMPLE INTEREST Interest charged on only the original principal

SPECIFIC IDENTIFICATION METHOD A method of inventory valuation which uses the actual cost of each unit in the ending inventory

SQUARE MEASURE The product of two linear measures; a measure of area

STANDARD WORKWEEK Usually, a workweek of 40 hours

STOCK RECORD A card or sheet on which a record is kept of all the goods that are placed in a stockroom or warehouse after an order has been delivered

STRAIGHT-LINE (SL) METHOD A method of calculating depreciation which assumes that an asset decreases in value by the same amount each year

STRAIGHT TIME Regular work hours

SUBTOTALS Totals of smaller groups of numbers arranged in long columns for addition

SUBTRACTION The arithmetic operation that finds the difference between two numbers

SUBTRAHEND The number that is subtracted from another number

SUGGESTED RETAIL PRICE *See* catalog price

SUM (OR TOTAL) The result of addition

SUM-OF-THE-YEARS-DIGITS (SOYD) METHOD A traditional method of calculating accelerated depreciation, which is used for financial statements and in which the asset's original cost, estimated life, and estimated scrap value must be known

T

TERMS OF PAYMENT The conditions for a business transaction, which state the time within which the purchaser must pay the total amount of the invoice to avoid a penalty

TIME *See* interest period

TIME CARD (OR TIME SHEET) A record of the total hours worked by an employee who is paid on a time basis

TIME TICKET (OR LABOR TICKET) A record containing all the information needed for payroll and cost purposes, such as the employee's name, clock number, date, rate, time worked, pieces produced, order number, job number, job code, etc.

TRADE DISCOUNTS Reductions from list prices that enable sellers to give different prices to different customers without having to print several different catalogs

U

UNLIKE TERMS Fractions with different denominators

USURY An interest charge above the legal limit

USURY LAWS Laws which establish the maximum annual interest rate that can be charged to a consumer

V

VENDOR Seller

VERTICAL AXIS The vertical line at the left of a graph

VOLUME The inside space or capacity of a container, which is always expressed in cubic units

W

WAGES Payment for time worked on an hourly basis

WEIGHTED AVERAGE COST METHOD *See* average cost method

INDEX

A

ABA number, defined, 231
Accelerated, defined, 378
Accelerated Cost Recovery System (ACRS), 378; defined, 373
Accelerated methods, of depreciation, defined, 374
Accumulated depreciation, defined, 373
Addend, defined, 13
Addition: by 9, 217–218; defined, 13; of columns of decimals, 14; of decimals, 13–15; of decimals vertically, 13, 14; of decimals with a calculator, 14–15; of digits from left to right, 213; of fractions, 63–66; of fractions or mixed numbers to whole numbers, 63; of fractions whose numerators are 1, 215
Adjacent numbers, 79
Alignment, 6
American Banker's Association (ABA), 231
Annual depreciation, calculation of, 376
Annual percentage rate (APR), 282–283; defined, 279–280
Approximated value, 108
Area, 175; defined, 176
Assets, defined, 373
Automated clearinghouse: defined, 232; uses of the, 232
Automatic teller machines (ATM), defined, 232
Average, defined, 50
Average cost method, of inventory valuation, 361

B

Balance forward, 229
Bank discount, defined, 288
Banker's interest, defined, 287
Bank statement: monthly, 237; reconciliation statement and, 237–241
Bar graph: comparative, 202–203; composite, 201–202
Base, defined, 125
Base unit, 162
Basic rate, defined, 375
Bill-to-address, 300
Blank endorsement, defined, 230
Board foot, defined, 168
Book value, defined, 373
Borrowing, subtracting fractions by, 72–73

C

C, 179
Calculating depreciation, 373
Calculators: estimating answers, 190–191; multiplying with, 104–105; use of decimals by, 7–8
Canceled check, defined, 231
Cancellation: defined, 56; reducing fractions and, 256; shortcuts, 257
Carry-over, 13
Cash discount, 303–307; calculating, 305–307; defined, 297; discount period for, 303–305
Cash on delivery (COD), defined, 303

Catalog price, defined, 311
Cent, 116
Centi, 158, 162
Centimeter, 158
Chain discount, defined, 313
Charge, service, 279
Charge account, revolving, 279
Checking account: defined, 228; keeping records of a, 228–229
Checks: canceled, 231; clearing of, 231; deposits and, 228–232; endorsing, 230–231; outstanding, 239; writing, 229–230
Circle graph: approximating percents for a, 205; defined, 204
Circumference, defined, 176
City income taxes, defined, 346
Clearinghouse: automated, 232; defined, 231
Collateral, defined, 287
Commercial loan: defined, 279; short-term, 287–292
Commission, and bonus plan, 351; defined, 343; method of payment, 351; payment, plans of, 351; and salary plan, 352; straight, 351
Common denominator, 63–65; adding a fraction or mixed number with a, 63; defined, 63
Common divisor, greatest, 56
Common fraction, defined, 55
Common multiple method, defined, 91
Comparative bar graph, 202–203
Complement, defined, 312
Complex fractions, defined, 91
Composite bar graph, 201–202
Compound amount, defined, 267
Compound amount factors: calculating, 272–273; defined, 271
Compound interest, 255, 280; calculating, 267–273; defined, 247; total, 267
Compound interest tables, using, 268–272
Computerized payroll service, defined, 343
Consumer loan, defined, 279
Conversion: of a decimal between 0.01 and 1.00 to its percent equivalent, 119; of a decimal greater than 1 to its percent equivalent, 119; of a decimal less than 0.01 to its percent equivalent, 119; of a decimal to its percent equivalent, 119; of a fraction or mixed number to its percent equivalent, 119–120; of a percent between 1% and 100% to a decimal or whole number, 115–116; of a percent between 1% and 100% to a fraction, 117; of a percent greater than 100% to a decimal or whole number, 116; of a percent greater than 100% to a mixed or whole number, 117; of a percent less than 1% to a decimal, 116; of a percent less than 1% to a fraction, 117; of a percent to a fraction or mixed number, 116–119; percent, 115–123
Conversion rule, defined, 118
Cost: as the base for markup, 322–324; defined, 359; finding operating expenses and, 334–335; finding the, 323, 324
Cost of goods available for sale, defined, 364
Cost of goods sold (CGS), 359; calculation of, 362
Cost per unit of weight, 108
Credit plans, 279

Index

Cubic measure, defined, 167
Current market value, 359
Customary system, 155; weights and measures, 156
Customary system, working with the metric and, 155–168
Cwt, 179

D

Data, defined, 7
Date of origin, defined, 247
Dating: end-of-month, 304; extra, 304; ordinary, 303; receipt-of-goods, 304
Day rate, defined, 343
Day-rate pay, defined, 343
Decem, defined, 2
Deci, 158, 162
Decimal number system, 2–12
Decimals: adding and subtracting fractions and, 103; changing fractions to, 101–102; changing to fractions in lowest terms, 102–103; defined, 2; division by whole numbers, 46, 47; division of, 45; division of decimals by, 47–49; division of fractions and, 105–108; division with the calculator, 49–50; multiplying fractions and, 103–105; reading and writing, 4–5; repeating, 101; writing from words, 5–7
Declining-balance (DB) method, of depreciation, 373, 374
Deduction, employee-authorized, 343; payroll, 343
Deductions, payroll, 346
Deka, 158
Denomination, 155
Denominator, 72; adding a fraction or mixed number with a common, 63; adding a mixed number with an unlike, 65–66; adding fractions with an unlike, 65; common, 63; defined, 55; finding the least common, 63–65; least common, 64
Deposits: checks and, 228–232; making, 231
Depreciation, calculation of, 373; calculation of, declining balance (DB) method, 373, 374; calculation of, MACRS, 379; calculation of, straight-line (SL) method, 373, 374; calculation of, sum-of-the-years-digits (SOYD) method, 373, 374; defined, 359, 373
Depreciation methods, comparison of, 377–378; MACRS, 379
Depreciation rate, calculation of, 375
Descending order, defined, 6
Diameter, defined, 176
Difference: defined, 21; estimating sum and, 187–188
Dimension, 177
Disclosure statement, defined, 279
Discount: bank, 288; calculating, 305–307; cash, 297, 303–307; chain, 313; discount period for cash, 303–305; equivalent single, 315; trade, 297, 311–315
Discounted, defined, 288
Discount period, defined, 288
Discount rate: defined, 288; U.S. interest rate, 289
Display, defined, 7
Dividend, 89
Divisibility rules, 57
Division: by 50 and 25, 216–217; by 0.1, 0.01, 0.001, etc., 215–216; by ten and its multiples, 49; of decimals by whole numbers, 46; of decimals with the calculator, 49–50; of fractions, 89–92; of fractions and decimals, 105–108; with whole numbers, 45–46
Division method, defined, 91
Divisor, greatest common, 56
Double-declining-balance method, of depreciation, 375
Dozen, 178–179; defined, 178
Draw, defined, 351
Drawing account, defined, 351
Due date: and nonbusiness days, 250; defined, 247; finding the, 249–250, 287–288; finding when interest period is in days, 249–250; finding when interest period is in months, 250–251; in February, 251; nonbusiness, 256

E

Earnings record, employee, 346
Electronic funds transfer (EFT), 231–232; defined, 232
Employee authorized, defined, 343
Employee earnings record, defined, 346
Employer's Tax Guide, defined, 346
Employer's tax reports, defined, 346
Ending inventory, 362
End-of-month dating, defined, 304
Endorse, defined, 230
Endorsement: blank, 230; restrictive, 230
English system, 155
Equivalent forms, 101–108
Equivalent fraction, defined, 55
Equivalent single discount, defined, 315
Equivalents of metric and customary measures, 163
Estimated life, defined, 373
Estimating inventory value at cost, 363; gross markup method, 363; gross profit method, 363; retail method, 364
Estimation, 185–191; checking calculator answers by, 190–191; of products, 188–189; of quotients, 189–190; of sums and differences, 187–188
Exact interest, 255, 258–259
Exact time, 255; defined, 248; finding between two dates, 248–249; finding when interest period is in days, 249–250; finding when interest period is in months, 250–251
Exempt, defined, 345
Exponent, defined, 273
Exponential notation, 177
Exponentiation, 177
Extension, defined, 38
Extra dating, defined, 304

F

Face value, defined, 287
Factor, defined, 33
Federal Banking Act of 1987, 230
Federal income taxes, defined, 346
Federal Reserve System, 230
FICA taxes, defined, 346
15-year class (MACRS), defined, 379
Finance charge (FC): computing the total, 281; defined, 279; finding the amount of the, 280–281
First-in, first-out (FIFO) method of inventory valuation, 360
Fiscal period, defined, 359
5-year class (MACRS), defined, 379
For deposit only, 230

Fractional equivalent, of percent, 126–127
Fraction bar, 55
Fractions: adding, 63–66; adding and subtracting decimals and, 103; adding to a mixed or whole number, 63; adding to a whole number, 63; adding with unlike denominators, 65; changing mixed numbers into improper, 58–59; changing to decimals, 101–102; changing to higher terms, 55–56; common, 55; complex, 91; converting to its percent equivalent, 119–120; defined, 55; dividing, 89–92; dividing by, 89; dividing decimals and, 105–108; equivalent, 55; improper, 55; multiplying, 79–82; multiplying by two or more, 79; multiplying decimals and, 103–105; multiplying whole numbers by, 80; proper, 55; reducing to lower terms, 56–58; simplifying complex, 91–92; subtracting, 71–74; subtracting by borrowing, 72–73; subtracting without borrowing, 71–72; types of, 55–59
Functions, defined, 7

G

Grand total, defined, 14
Graph, 199–205; bar, 200–203; circle, 204–205; comparative bar, 202–203; line, 199–200
Greatest common divisor (G.C.D.), defined, 56
Gross, 178–179; defined, 178
Gross amount, 351
Gross earnings, 351
Gross pay, 351; defined, 343
Gross profit, defined, 321, 359
Gross profit method, 363

H

Hecto, 158
Higher terms, defined, 55
Horizontal axis, defined, 199

I

Improper fractions: defined, 55; changing mixed numbers into, 58–59
Income statement, illustrated, 362
Inspection, 64
Installment: defined, 279; finding the amount of the monthly, 281–282
Interest, 255; banker's, 287; calculating compound, 267–273; calculating simple, 255–259; compound, 247, 255, 280; compounded annually, 267; compounded daily, 268; compounded monthly, 268; compounded quarterly, 268; compounded semiannually, 267–268; compounding at different periods, 267–268; defined, 247; exact, 248, 255, 258, 259; ordinary, 255–257; ordinary simple, 248; simple, 247, 280; total compound, 267
Interest after date, defined, 287
Interest-bearing note: defined, 287; discounting, 290–292
Interest charge, finding the, 287–288
Interest due at maturity, defined, 287
Interest period, defined, 247
Interest rate: monthly, 259; discount rates, 289
Interest to follow, defined, 287

International Metric System (SI), defined, 155
Inventory, extending the, 38
Inventory valuation, average cost method, 360; and cost of goods sold, 359; first-in, first-out (FIFO) method, 360; last-in, first-out (LIFO) method, 360; specific identification method, 360
Inventory value at cost: estimation of, 363; estimation of, gross markup method, 363; estimation of, gross profit method, 363; estimation of, retail method, 364
Invoices: preparing and checking, 298–300; purchase orders and, 297–300

K

Key-entering, defined, 7
Kilo, 158
Kilometer, 158

L

Labor ticket, defined, 352; illustrated, 353
Last-in first-out (LIFO) method, of inventory valuation, 360
Least common denominator (L.C.D.): defined, 64; finding the, 63–65
Liabilities, defined, 373
Like terms, defined, 63
Line graph, defined, 199
Linear measure, defined, 167
List price, defined, 311
Loan: commercial, 279; consumer, 279; long-term, 247; short-term, 247; short-term commercial, 287–292
Loan period, defined, 247
Long-term loan, defined, 247
Loss: finding the net, 335; net, 321, 335–336; profit and, 333–338
Lowest terms, defined, 56

M

M, 179
MACRS, asset classes, summary of, 379
Maker, defined, 287
Margin on sales, defined, 321
Markon, defined, 321
Markup (M), defined, 321
Markup method, 363
Markup rate: converting from one base to another, 326–328; finding the, 322–323, 325
Markup table, defined, 325
Markup wheel, defined, 325
Maturity value: defined, 256; finding the, 256, 287–288
Measure: cubic, 167; dividing, 168; linear, 167; square, 167
Merchandise inventory, defined, 359
Metric system: of weights and measures, 157; working with the customary and, 155–168
Mid-month convention, defined, 378
Mid-quarter convention, defined, 378
Mid-year convention, defined, 378
Milli, 158, 162
Minuend, 22; defined, 21

Index

Mixed numbers: adding to whole numbers, 63; adding with unlike denominators, 65–66; an alternate method of multiplying, 81; changing improper fractions into, 58–59; converting to percent equivalent, 119–120; defined, 55; dividing by, 90–91; multiplying, 80–81
Modified Accelerated Cost Recovery System (MACRS), 378–380; defined, 373
Monthly bank statement, defined, 237
Monthly statement, defined, 299
Monthly installment, finding the amount of the, 281–282
Multiplicand, defined, 33
Multiplication: by 11, 217; by 50 and 25, 216–217; by 0.1, 0.01, 0.001, etc., 215–216; by percents between 1% and 100%, 125–126; by percents less than 1%, 127–128; by ten and its multiples, 35–37; defined, 33; of decimals, 33–38; of decimals by decimals or whole numbers, 34; of fractions, 79–82; of fractions by whole numbers, 80; of measures, 166–168; of two-digit numbers by other two-digit numbers, 218–219; of two or more fractions, 79; of whole numbers or decimals with a calculator, 37–38; with a calculator, 104–105
Multiplier, defined, 33

N

Negotiable, defined, 287
Net cost, 311
Net cost factor, 314
Net loss, 335–336; calculators and, 336; defined, 321; finding the, 335
Net pay, defined, 343
Net profit, defined, 359
Net price: defined, 311; finding with a chain of discounts, 313–314; finding with a single trade discount, 311–313
Net price factor: defined, 314; using for chain discount, 314–315
Net price factor table, using a, 315
Net profit, 333–335; finding the, 333–334
Net purchases, defined, 362
Net worth, defined, 373
New book value, calculation of, 376
Nonexempt salaried employee, defined, 345
Non-interest-bearing note, 288–290; defined, 287
Note: interest-bearing, 287; non-interest-bearing, 288–290; promissory, 287
Notes payable: defined, 288; finding the proceeds of discounted, 288–289
Notes receivable: defined, 288; finding the proceeds of discounted, 289–290
Not sufficient funds (NSF), defined, 237
Numbers: adjacent, 79; an alternate method of multiplying mixed, 81; dividing by mixed, 90–91; dividing by whole, 90; multiplying mixed, 80–81; multiplying whole, 80–81; prime, 57
Numerator, 72; defined, 55

O

One-and-one-half-times declining-balance method, of depreciation, 375
150%-declining-balance method, of depreciation, 375
Operating expenses (OE): defined, 321; cost and, 334–335
Operations, defined, 7
Order form, defined, 299

Ordinary dating, defined, 303
Ordinary interest, 255–257
Ordinary simple interest, defined, 248
Original cost, defined, 373
Ounces, 178
Outstanding checks, defined, 239
Overdrawing, defined, 237
Overtime, on a daily basis, 344; defined, 343; on a weekly basis, 344
Overtime pay, daily basis, 344; defined, 344; for salaried employees, 345; weekly basis, 344

P

Partial dividend, 45
Partial product, 33
Past due, defined, 303
Pay, day-rate, defined, 343; gross, defined, 343; net, defined, 343; straight-line, defined, 343
Pay period, defined, 343
Payee, defined, 229, 287
Payment: 90-day period for, 279; terms of, 303
Payroll, defined, 343
Payroll deduction, 343, 346
Pay to the order of, 230
Percent, 116; conversion, 115–123; converting to a decimal or whole number, 115–116; defined, 115; finding between 1% and 100%, 137–139; finding greater than 100%, 139–140; finding less than 1%, 139; multiplying between 1% and 100%, 125–126; multiplying by, 127–128; using fractional equivalents, 126–127
Percentage: defined, 125; finding the, 125; finding the base of a, 147–150
Percentage rate, finding the annual, 282–283
Percent equivalent: converting a decimal between 0.01 and 1.00 to its, 119; converting a decimal to its, 119
Perimeter, defined, 175
Perpetual inventory, defined, 359
Physical inventory, defined, 38, 359
Pi, defined, 176
Piecework, with bonus, 353; defined, 343; with an hourly base rate, 354; method of payment, 351; method of payment, plans of, 352; straight, 352
Point, used instead of and, 5
Pounds, 178
Prefix, 155–156
Price: catalog, 311; finding the selling, 322; list, 311; net, 311; retail, 311; selling, 321; wholesale, 311
Price tag code, using, 337–338
Prime number, defined, 57
Prime rate, defined, 287
Principal, 255; defined, 247
Proceeds, defined, 288
Products: defined, 33; estimating, 188–189
Profit: gross, 321; loss and, 333–338; net, 321, 333–335
Promissory note, 287; defined, 247
Proper fraction, defined, 55
Prox., defined, 304
Purchase orders: defined, 297; invoices and, 297–300; preparing and checking, 297–298
Purchase requisition, defined, 297
Purchases discounts, defined, 362
Purchases returns, defined, 362

Q

Quota, defined, 352
Quotients, estimating, 189–190

R

Radius, defined, 176
Rate, 255; annual percentage, 279–280, 282–283; defined, 125, 247; discount, 288; finding the, 137–141; finding the markup, 322–323; monthly interest, 259; prime, 287
Receipt-of-goods (ROG) dating, defined, 304
Received, 73
Reciprocal, defined, 89
Reconciliation statement: bank statement and, 237–241; defined, 237; preparing the, 239–241
Recovery, defined, 378
Rectangle: defined, 175; perimeter of, 175
Regular pay, defined, 344
Remittance, defined, 303
Repeating decimal, defined, 101
Requisition, defined, 74
Resale value, defined, 373
Restrictive endorsement, defined, 230
Retail price, defined, 311
Revolving charge account, 279
Rounding off, 186–187; defined, 35
Running meter, 176
Running yard, 176

S

Salary, defined, 343
Salvage value, defined, 373
Scrap value, 378; defined, 373
Selling price (SP), 321; adjusting the, 336–337; as base for markup, 324–326; finding the, 322, 325–326, 333, 335
Service charge, defined, 279
7-year class (MACRS), defined, 379
Ship-to-address, 300
Shortcuts, 213–219; adding and subtracting by 9, 217–218; adding digits from left to right, 213; adding two fractions whose numerators are 1, 215; multiplying and dividing by 0.1, 0.01, 0.001, etc., 215; multiplying a two-digit number by another two-digit number, 218–219; subtracting digits from left to right, 214–215; subtracting two fractions whose numerators are 1, 215
Short division, defined, 46
Short-term commercial loans, 287–292
Short-term loan, defined, 247
Simple interest, 280; calculating, 255–259; defined, 247
Specific identification method, of inventory valuation, 360
Square measure, defined, 167
Standard workweek, defined, 344
State income taxes, defined, 346
Statement: disclosure, 279; monthly, 299
Stock record, defined, 73
Straight commission plan, defined, 351
Straight-line method, of depreciation, 373, 374
Straight line rate for 3 years, defined, 375
Straight piecework, defined, 352
Straight time, defined, 343
Straight time pay, defined, 343
Subtotal, defined, 14
Subtraction: by 9, 217–218; defined, 21; of a decimal from a whole number, 21–22; of a whole number from a decimal, 22–23; of decimals, 21–25; of decimals with a calculator, 22–23; of digits from left to right, 214–215; of fractions, 71–74; of fractions without borrowing, 71–73; of measures, 166; of one decimal from another, 21; of two fractions whose numerators are 1, 215
Subtrahend, 22; defined, 21
Sum: defined, 13; estimating difference and, 187–188
Sum-of-the-years-digits (SOYD) method, of depreciation, 373, 374
Symbols, 155

T

Tax: city income, 346; federal income, 346; FICA, 346; guide, employer's, 346; guides, annual, 346; reports, employer's, 346; state income, 346
10-year class (MACRS), defined, 379
Terms: like, 63; unlike, 63
Terms of payment, defined, 303
31.5-year class (MACRS), defined, 379
3-year class (MACRS), defined, 379
Time, 255
Time-and-a-half, defined, 344
Time-basis payment, defined, 343
Time card, defined, 343
Time-payment plan, 279–283
Time sheet, defined, 343
Time ticket, defined, 352
Tons, 178
Total: defined, 13; grand, 14
Trade discount, 311–315; defined, 297
20-year class (MACRS), defined, 379
27.5-year class (MACRS), defined, 379
200%-declining-balance method, of depreciation, 375

U

Unit cost, 359
Units, 178–179
Unlike terms, defined, 63
Usury laws, defined, 280

V

Vendor, defined, 297
Vertical axis, defined, 199
Volume, 175; defined, 177

W

Wages, defined, 343
Weighted average cost method, of inventory valuation, 361
Weights and measures, applications using, 175–180
Whole number: adding a fraction or mixed number to a, 63; dividing a decimal by a, 46; dividing by a, 90; dividing by a decimal, 47; multiplying a, 80–81; multiplying a fraction by a, 80; reading and writing, 2–3
Wholesale price, 311

Z

Zeros, writing to the left of a product, 34